ELECTROCHEMICAL SUPERCAPACITORS FOR ENERGY STORAGE AND DELIVERY

FUNDAMENTALS AND APPLICATIONS

ELECTROCHEMICAL ENERGY STORAGE AND CONVERSION

Series Editor: Jiujun Zhang
National Research Council Institute for Fuel Cell Innovation
Vancouver, British Columbia, Canada

Published Titles

Electrochemical Supercapacitors for Energy Storage and Delivery: Fundamentals and Applications
Aiping Yu, Victor Chabot, and Jiujun Zhang

Forthcoming Titles

Electrochemical Polymer Electrolyte Membranes
Yan-Jie Wang, David P. Wilkinson, and Jiujun Zhang

Lithium-Ion Batteries: Fundamentals and Applications
Yuping Wu

Proton Exchange Membrane Fuel Cells
Zhigang Qi

Solid Oxide Fuel Cells: From Fundamental Principles to Complete Systems
Radenka Maric

ELECTROCHEMICAL ENERGY STORAGE AND CONVERSION

ELECTROCHEMICAL SUPERCAPACITORS FOR ENERGY STORAGE AND DELIVERY

FUNDAMENTALS AND APPLICATIONS

Aiping Yu, Victor Chabot,
and Jiujun Zhang

CRC Press
Taylor & Francis Group
6000 Broken Sound Parkway NW, Suite 300
Boca Raton, FL 33487-2742

First issued in paperback 2017

Version Date: 20121030

ISBN 13: 978-1-138-07711-9 (pbk)
ISBN 13: 978-1-4398-6989-5 (hbk)

Library of Congress Cataloging-in-Publication Data

Electrochemical supercapacitors for energy storage and delivery : fundamentals and applications / Aiping Yu ... [et al.].
p. cm. -- (Green chemistry and chemical engineering)
Includes bibliographical references and index.
ISBN 978-1-4398-6989-5 (hardback)
1. Supercapacitors. 2. Storage batteries. 3. Supercapacitors. I. Yu, Aiping.

TK7872.C65E45 2013
621.31'2424--dc23 2012032953

Visit the Taylor & Francis Web site at
http://www.taylorandfrancis.com

and the CRC Press Web site at
http://www.crcpress.com

Contents

Series Preface

Electrochemical Energy Storage and Conversion

Clean energy technologies, which include energy storage and conversion, will play the most important role in overcoming fossil fuel exhaustion and global pollution for the sustainable development of human society. Among clean energy technologies, electrochemical energy storage and conversion are considered the most feasible, environmentally friendly, and sustainable. Electrochemical energy technologies such as batteries, fuel cells, supercapacitors, hydrogen generation and storage as well as solar energy conversion have been or will be used in important application areas including transportation and stationary and portable/micro power. To meet the increasing demand in both the energy and power densities of these electrochemical energy devices in various new application areas, further research and development are essential to overcome major obstacles such as cost and durability, which are considered to be hindering their applications and commercialization. To facilitate this new exploration, we believe that a book series that covers and gives an overall picture of all the important areas of electrochemical energy storage and conversion technologies is highly desired.

This book series will provide a comprehensive description of electrochemical energy conversion and storage in terms of fundamentals, technologies, applications, and the latest developments including secondary (or rechargeable) batteries, fuel cells, supercapacitors, CO_2 electroreduction to produce low-carbon fuels, electrolysis for hydrogen generation/storage, and photoelectrochemical for water splitting to produce hydrogen. Each book in this series will be self-contained and written by scientists and engineers with excellent academic records and strong industrial expertise, who are at the top of their fields on the cutting-edge of technology. With a broader view of various electrochemical energy conversion and storage devices, this book series will be unique and an essential read for university students including undergraduates and graduates, scientists and engineers working in related fields. We believe that reading this book series will enable readers to easily locate the latest information concerning electrochemical technology, fundamentals, and applications.

Jiujun Zhang, PhD

Preface

In today's energy-dependent world, electrochemical devices for energy storage and conversion such as batteries, fuel cells, and electrochemical supercapacitors (ESs) have been recognized as the most important inventions among all energy storage and conversion technologies. The electrochemical supercapacitor, also known as a supercapacitor, ultracapacitor, or electrochemical double-layer capacitor, is a special type of capacitor that can store relatively high energy density compared to storage capabilities of conventional capacitors. ES devices possess several high-impact characteristics such as fast charging capabilities, long charge–discharge cycles, and broad operating temperature ranges. As a result, their use in hybrid and electrical vehicles, electronics, aircraft, and smart grids is widespread. Although ES systems still face some challenges, such as relatively low energy density and high cost, further development will allow ESs to work as power devices in tandem with batteries and fuel cells and also function as stand-alone high energy storage devices.

To facilitate research and development, we believe a book discussing both fundamentals and applications of ES technology is definitely needed. The best known book in the field is B. E. Conway's *Electrochemical Supercapacitors: Scientific Fundamentals and Technological Applications* published in 1999. Conway's book presented the first comprehensive coverage of the development of ESs in the 20th century.

Our book will focus on the introduction to electrochemical supercapacitors from more technical and practical aspects and the crystallization of their development in the past decade. It delivers the basic electrochemical theory and calculations and discusses components and characterization techniques. Further discussion is focused on the structure and options for device packing. Choices for electrode, electrolyte, current collector, and sealant materials are evaluated based on comparisons of available data. The book will help illuminate the practical aspects of understanding and applying the technology in industry and provide sufficient technical detail about new materials developed by experts in the field that may surface in the future. Furthermore, the technical challenges are also discussed in this book to give readers an understanding of the practical limitations and their associated parameters in ES technology.

We hope this book will provide readers with a comprehensive understanding of the components, designs, and characterizations of electrochemical supercapacitors. We anticipate industrial and academic scientists and engineers along with undergraduate and graduate students in the field will use it as a reference in their work and also to foster ideas for new devices that will help further the technology as it fills a larger role in mainstream energy storage.

We would like to acknowledge and express our sincere thanks for the contributions of all those involved in preparing and developing this book. In particular, we want to thank Aaron Davies, Drew Higgins, and Fathy Hassan for supplying some of the chapter materials and images in the text. We would like to express our appreciation to CRC Press for inviting us to lead this book project, and we thank Allison Shatkin and Jessica Vakili for their guidance and support in smoothing the book preparation process.

If there are any technical errors in the book, we would deeply appreciate constructive comments for further improvement.

Authors

Aiping Yu is an assistant professor at the University of Waterloo in Canada. She earned her PhD from the University of California–Riverside. Her research interests are materials and modeling development for energy storage and conversion, photocatalysts, and nanocomposites.

Dr. Yu has published over 35 papers in peer-reviewed journals such as *Science* and one book chapter relating to supercapacitors. She currently is the editorial member of the *Nature: Scientific Reports*. Her work has been featured by major media such as *Nature: Nanotechnology, Photonics.com,* and *Azonano. com*. Her patent for graphene nanomaterials has been licensed to a company in San Jose.

Victor Chabot received his bachelor's degree in nanoengineering from the University of Waterloo and currently is pursuing his graduate degree in chemical engineering at the University of Waterloo. His research focuses on nanomaterial development for high energy density supercapacitors.

Jiujun Zhang is a principal research officer and technical leader at the National Research Council of Canada's Institute for Fuel Cell Innovation (NRC-IFCI), now the council's Energy, Mining, and Environment (NRC-EME) portfolio. Dr. Zhang earned a BS and MSc in electrochemistry from Peking University in 1982 and 1985, respectively, and a PhD in electrochemistry from Wuhan University in 1988. After completing his doctorate, he took a position as an associate professor at the Huazhong Normal University for 2 years. Starting in 1990, he carried out three terms of postdoctoral research at the California Institute of Technology, York University, and the University of British Columbia.

Dr. Zhang has over 28 years of research and development experience in theoretical and applied electrochemistry, 14 of which were spent working on fuel cells at Ballard Power Systems and at NRC-IFCI. He also spent 3 years researching electrochemical sensors. Dr. Zhang holds adjunct professorships at the University of Waterloo, the University of British Columbia, and at Peking University.

To date, Dr. Zhang has co-authored or edited more than 300 publications including 190 refereed journal papers with approximately 4,700 citations, books, conference proceeding papers, book chapters, and 50 conference and invited oral presentations. He also holds over 10 patents worldwide along with 9 U.S. patent publications and has produced more than 80 industrial technical reports. Dr. Zhang serves as an editor or editorial board member for several international journals and is also the editor for CRC's Electrochemical Energy Storage and Conversion series of books. Dr. Zhang is an active member of the Electrochemical Society, the International Society of Electrochemistry, and the American Chemical Society.

1

Fundamentals of Electric Capacitors

1.1 Introduction

An electric capacitor has a sandwich structure containing two conductive plates (normally made of metal) surrounding a dielectric or insulator as shown in Figure 1.1a. Common dielectrics include air, oiled paper, mica, glass, porcelain, or titanate. An external voltage difference is applied across the two plates, creating a charging process. During charging, the positive charges gradually accumulate on one plate (positive electrode) while the negative charges accumulate on the other plate (negative electrode). When the external voltage difference is removed, both the positive and negative charges remain at their corresponding electrodes. In this way, the capacitor plays a role in separating electrical charges. The voltage difference between the two electrodes is called the cell voltage of the capacitor. If these electrodes are connected using a conductive wire with or without a load, a discharging process occurs—the positive and negative charges will gradually combine through the wire. In this way, the capacitor plays a role for charge storage and delivery. Before we start a deeper discussion about capacitors, explanations of their history and some fundamental concepts may be useful.

1.1.1 History

Thales of Miletus, a philosopher, discovered electric charges when he rubbed amber with a cloth and observed magnetic particle attraction. Since then, the act of rubbing two non-conducting materials together to induce a charge has been treated as a demonstration of the triboelectric effect. In 1745, a better understanding of electrostatics and electrochemistry led to the invention of a condenser, as shown in Figure 1.1b. In the first condenser, two important factors determined charge separation and charge storage: the dielectric thickness and the surface area of the conductive materials. With technological advancements in materials and manufacturing, condensers evolved into modern capacitors that are now used in electrical systems.

Early capacitors were created from a Leyden jar as shown in Figure 1.1b. A capacitor consisted of a glass vessel whose interior and exterior were coated

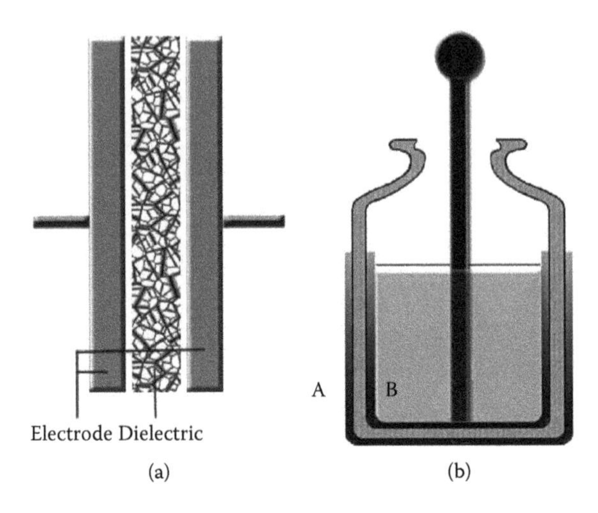

FIGURE 1.1
(See color insert.) (a) Simplified schematic of capacitor design. (b) Cross-sectional schematic of Leyden jar (water-filled glass jar containing metal foil electrodes on its inner and outer surfaces, (denoted A and B).

with metal foil. The foils acted as the electrodes and the jar acted as the dielectric. The foil coverings stopped before the jar's mouth to prevent arc discharge. When capacitors became more prevalent in the twentieth century, their structures were designed to be more practical and economical in storing electrostatic charges as shown in Figure 1.1a. This structural change was very important after both world wars when demands for electronic parts increased. Capacitors were used in complex electronic systems, resulting in greater production and standardization programs to ensure the reliability and quality of the capacitors. Significant effort to meet quality and reliability requirements contributed to the successful improvements of modern electronics. Smaller and lighter capacitors possess greater capabilities and stability in adverse conditions and over wide temperature ranges.

This chapter reviews the fundamentals of capacitors and emphasizes the critical parameters of dielectric materials and the construction of capacitors, as well as their operations in a variety of applications [1].

1.2 Electric Charge, Electric Field, and Electric Potential and Their Implications for Capacitor Cell Voltage

1.2.1 Electric Charge

The roles of a capacitor are to separate, store, and deliver electric charges and the concepts and properties of charges must be understood. In general, both positive and negative electric charges exist in all physical objects in the universe whether they are animate or inanimate. A physical object will display a neutral charge when there are equal numbers of positive charges (protons) and negative charges (electrons). However, a *net* electric charge can occur through unbalancing the charge equilibrium in some areas of an object. Thus some areas will have more negative than positive charges and vice versa.

In general, two positive charges reciprocally repel each other as do two negative charges; while opposite charges feel mutual attractive forces. Figure 1.2 shows positive and negative charges. Coulomb's law states that the electrostatic force between two charges can be described by

$$F = \frac{q_+(-q_-)}{4\pi\varepsilon_0 r^2} = -\frac{q_+ q_-}{4\pi\varepsilon_0 r^2} \tag{1.1}$$

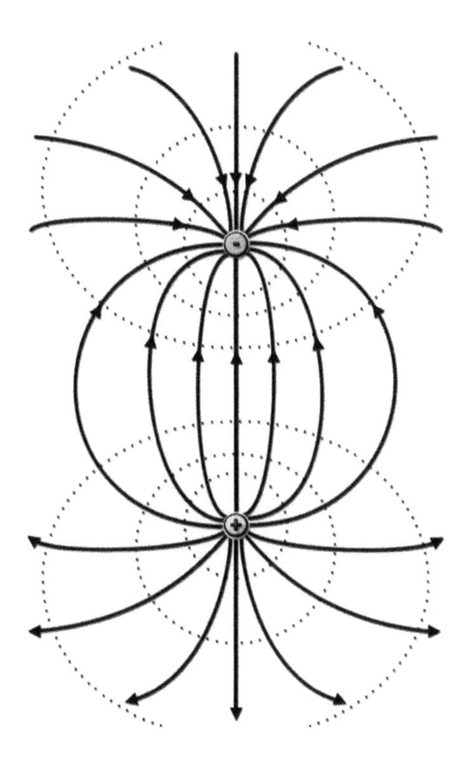

FIGURE 1.2
Field lines of force radiating outward from a positive charge or inward toward a negative charge.

where F represents the magnitude of electrostatic force present between the two charges; q_+ and q_- are the magnitudes of the positive and negative charges; r is the distance between these two charges; ε_0 is the dielectric constant of the an empty vacuum; and the negative sign represents the force as attractive rather than repelling. The dielectric constant is an important parameter for a capacitor and will be discussed in detail in Section 1.3.1. Note that the charge magnitude in Equation (1.1) can be quantized by n values of the elementary charge of an electron, allowing charge q to be written as

$$q = ne \tag{1.2}$$

where $n = \pm 1, \pm 2, \pm 3$, etc., and e is the elementary charge constant equal to 1.602×10^{-19} C.

1.2.2 Electric Field and Potential

A charge, for example, the positive charge (q_+) in Figure 1.2, can emit an electric flux on the surrounding area to form an electric field. Figure 1.2 is helpful for visualizing the electric field as lines of force that radiate outward (positive charge) or inward (negative charge). The strength of this field can be felt by the negative charge, and is expressed as

$$E = \frac{F}{q_-} \tag{1.3}$$

where E is the electric field strength of the positive charge q_+. Combining Equations (1.1) and (1.3), the electric field strength can be expressed alternatively as

$$E = -\frac{q_+ q_-}{4\pi\varepsilon_0 r^2}\frac{1}{(-q_-)} = \frac{q_+}{4\pi\varepsilon_0 r^2} \tag{1.4}$$

The corresponding electric potential (V_{q+}) at the position of the negative charge can be treated as the work done by moving the negative charge from its position through a distance r toward the positive charge and is expressed as

$$V_{q+} = Er \tag{1.5}$$

By combining Equations (1.4) and (1.5), the electric potential induced by the positive charge can be expressed as

$$V_{q+} = \frac{q_+}{4\pi\varepsilon_0 r} \tag{1.6}$$

Equation (1.6) indicates that the longer the distance from the charge, the smaller the electric potential, and the shorter the distance toward the charge, the larger the electric potential. In extreme cases, r becomes infinite and the electric voltage becomes zero; if the distance becomes zero, the electric potential will become infinite regardless of how large or small the charge magnitude is.

1.2.3 Implication of Electric Potential in Capacitor Cell Voltage

Figure 1.3 displays a capacitor with positive charge of Q_+ uniformly distributed on a positive planar electrode and negative charge of Q_- on the negative electrode. Here $|Q_+| = |Q_-|$. The planar surface is A and the vacuum space between the two planar electrodes has a dielectric constant of ε_0 and a distance of d [2]. According to Gauss' law, the electric field outside of the planar conductive surface can be expressed as

$$E = \frac{Q_+}{\varepsilon_0 A} \tag{1.7}$$

The potential difference between the electrodes can be treated as the work done to move the charge from the positive electrode to the negative electrode. Similar to Equation (1.5), the work or electric difference between two electrodes (ΔV) can be expressed as

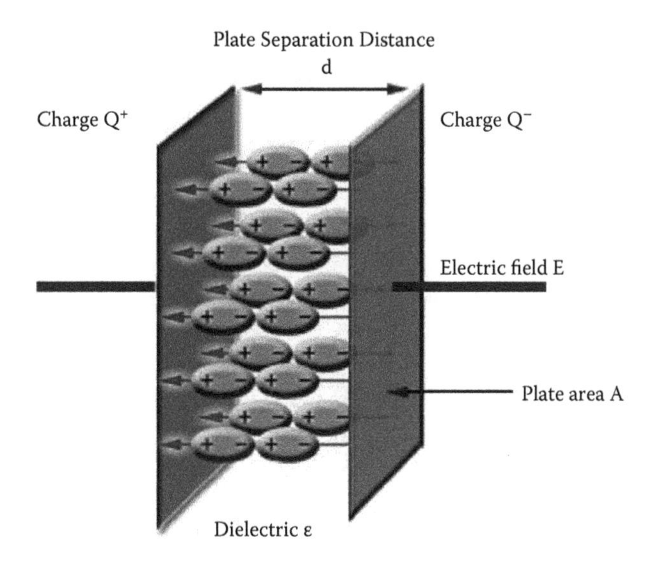

FIGURE 1.3
(See color insert.) Charged capacitor device.

$$\Delta V = Ed \tag{1.8}$$

Combining Equations (1.7) and (1.8), the following expression can be obtained:

$$\Delta V = \frac{Q_+ d}{\varepsilon_0 A} \tag{1.9}$$

The electric difference between the two electrodes of a capacitor can be alternatively expressed as V rather than ΔV:

$$V = \frac{Q_+ d}{\varepsilon_0 A} \tag{1.10}$$

1.3 Capacitance Definition and Calculation

A typical capacitor consists of two conducting parallel plates separated by a vacuum or dielectric material, as shown in Figures 1.1a and 1.3. When charging a capacitor, the plates will have equal and opposite magnitudes of charge. This indicates a zero net charge on the capacitor. Both plates are electrically conductive materials, so the uniform distributions of charge along their surfaces are reached easily, giving a potential difference V. Charge Q ($Q = |Q_+| + |Q_-|$) and potential V are related to each other through proportionality constant C:

$$Q = CV \tag{1.11}$$

A capacitor's proportionality constant refers to its capacitance, and provides a measurement of the amount of charge necessary to induce a specified potential between the plates. The magnitude of capacitance is independent of the charge accumulated across the plates in a linear capacitor. However, this capacitance is dependent upon the geometry of the charged plates. Capacitance is measured in farads (F) where 1 F = 1 coulomb/volt (C/V). Combining Equations (1.10) and (1.11), capacitance can be expressed as

$$C = \frac{Q}{V} = \frac{\varepsilon_0 A}{d} \tag{1.12}$$

Equation (1.12) demonstrates that capacitance is dependent on the dielectric constant of the dielectric material, the electrode surface area, and the distance between the two planar electrodes. However, Equation (1.12) is applicable only to capacitors having two parallel planar electrodes. For cylindrical, spherical, and isolate sphere capacitors, the capacitance expressions are different. For a long cylindrical coaxial capacitor of length L, the inner conducting cylinder has a radius of a and the outer cylinder has a radius of b. Its capacitance can be expressed as

$$C = 2\pi\varepsilon_0 \frac{L}{\ln(b/a)} \tag{1.13}$$

For a spherical capacitor with an inner sphere radius of a and an outer sphere radius of b, its capacitance is

$$C = 4\pi\varepsilon_0 \frac{ab}{b-a} \tag{1.14}$$

For an isolated sphere capacitor that only has a single isolated spherical conductor with a radius of R, its capacitance can be expressed as

$$C = 4\pi\varepsilon_0 R \tag{1.15}$$

Equations (1.12) through (1.15) demonstrate that capacitance is strongly dependent on the dielectric constant ε_0 if the configuration of the capacitor is fixed. This ε_0 is the dielectric constant of an empty vacuum. However, for a non-vacuum dielectric, the material's relative permittivity or relative dielectric constant is defined as the relative dielectric constant ($\varepsilon_r = \varepsilon/\varepsilon_0$) where ε is the dielectric constant of the material. Note that this equation will be discussed further in Section 1.3.1. Every dielectric material has a different dielectric constant resulting in a different capacitance. Table 1.1 lists the dielectric constants of common materials used in capacitors. All of them have larger dielectric constants compared to those of air or a vacuum. In this case, Equation (1.12) can be rewritten as

$$C = \frac{Q}{V} = \frac{\varepsilon A}{d} = \frac{\varepsilon_r \varepsilon_0 A}{d} \tag{1.12a}$$

Larger dielectric constants can effectively reduce the magnitude of the electric field of a charged object, as seen in Equations (1.4) and (1.7). However, larger dielectric constants have more energy storage because the energy stored in a capacitor is proportional to its capacitance.

The intrinsic dielectric strength describes the maximum electric field the capacitor can tolerate prior to the breakdown voltage, and is an important consideration when selecting dielectric materials. For example, when an applied voltage exceeds the dielectric strength of a material, the insulation of

TABLE 1.1

Dielectric Constants of Capacitor Materials

Dielectric	Dielectric Constant κ
Air	1.00054
Paper	2.0 to 6.0
Plastics	2.1 to 6.0
LDPE/HDPE	2.3
Mineral oil	2.2 to 2.3
Silicone oil	2.7 to 2.8
Quartz	3.8 to 4.4
Glass	4.8 to 8.0
Mica	7
Silicone	12
Germanium	16
Tantalum pentoxide	26
Ethanol	25
Ethylene glycol	3.7
Dipropyl ketone	12.6
Glycol	37
Glycerol	42.5
Water (20°C)	80.4
Acids	
Acetic	6.2
Butyric	3
Lactic	22
Cresol	5 to 11.5
Propanoic	3.1
Ceramics	12 to 400,000
Aluminum oxide	4.5 to 11.5
Titanium dioxide	14 to 110
Porcelain	5.1 to 8.0
Porcelain (zircon)	7.1 to 10.5
Titanates (Ba, Sr, Ca, Mg, Pb)	15 to 12,000

Note: See References 3 and 4.

the electrodes may break, resulting in a conductive path for electrons to flow through the capacitor. Dielectric strength can be affected by several factors such as the thickness of the dielectric material, temperature, humidity, and frequency. In applications, an extrinsic property is generally responsible for breakdown of the dielectric, and intrinsic properties serve more as guidelines for upper values under ideal conditions. A solid material that undergoes a breakdown generally experiences partial or complete impairment of its insulating properties.

1.3.1 Dielectric Materials and Constants

As mentioned above, a dielectric is the middle layer between the two conductive plates of a capacitor and is generally composed of an electric insulation material such as a vacuum, non-ionized gas, solid such as a ceramic or polymer, or a liquid (aqueous or non-aqueous electrolyte). Two electrodes located at both sides of the dielectric collect the current. Since a dielectric is the primary material responsible for the storage of charge in capacitor devices, it exhibits extremely low conduction currents from free electrons and ion impurities. Although solids and liquids are predominantly used as insulators, vacuums and non-ionized gases are better insulators.

There are two parameters used in characterizing dielectrics: (1) leakage conductivity σ_l and (2) relative dielectric constant ε_r. The leakage conductivity is determined by the resistance of the material R (Ω); thickness d (cm); and dielectric surface area A (cm^2). Equation (1.16) expresses the leakage conductivity.

$$\sigma_l = \frac{d}{RA} \tag{1.16}$$

The dielectric constant, also known as relative permittivity, can be defined. As described in Equation (1.1), if two charges q_1 and q_2 are separated from each other by a small distance r in a vacuum, the electrostatic force in vacuum (F_0) can be expressed as

$$F_0 = \frac{|q_1||q_2|}{4\pi\varepsilon_0 r^2} \tag{1.17}$$

In the above equation, ε_0 is the electrical permittivity or dielectric constant of a vacuum. If the separation medium between the charges is replaced by another material, Equation (1.17) will become

$$F_m = \frac{|q_1||q_2|}{4\pi\varepsilon\, r^2} \tag{1.18}$$

The relative dielectric constant of that material (ε_r) is found by dividing Equation (1.17) by Equation (1.18).

$$\varepsilon_r = \frac{F_0}{F_m} = \frac{\varepsilon}{\varepsilon_0} \tag{1.19}$$

Note that this ε_r is a dielectric constant relative to that of a vacuum. The relative dielectric constant of a vacuum is 1.00000. The physical meaning of the dielectric constant can be described as how effectively the dielectric allows a capacitor to store more charge. Every material used in a capacitor has its own dielectric constant, depending on the dielectric material. In practice, this dielectric constant can be measured. For example, if a capacitor with a vacuum as the dielectric is fully charged, the voltage difference across the two electrodes can be measured as V_0; if a material is used as the dielectric and is fully charged, the voltage difference between the two electrodes will be measured as V. The dielectric constant of this material can be calculated as a ratio of the cell voltage to the material in and out of the vacuum:

$$\varepsilon_r = \frac{V_0}{V} \tag{1.20}$$

Since V is always less than or equal to V_0, the dielectric constant is greater than or equal to 1. Equation (1.20) means that the dielectric constant of a

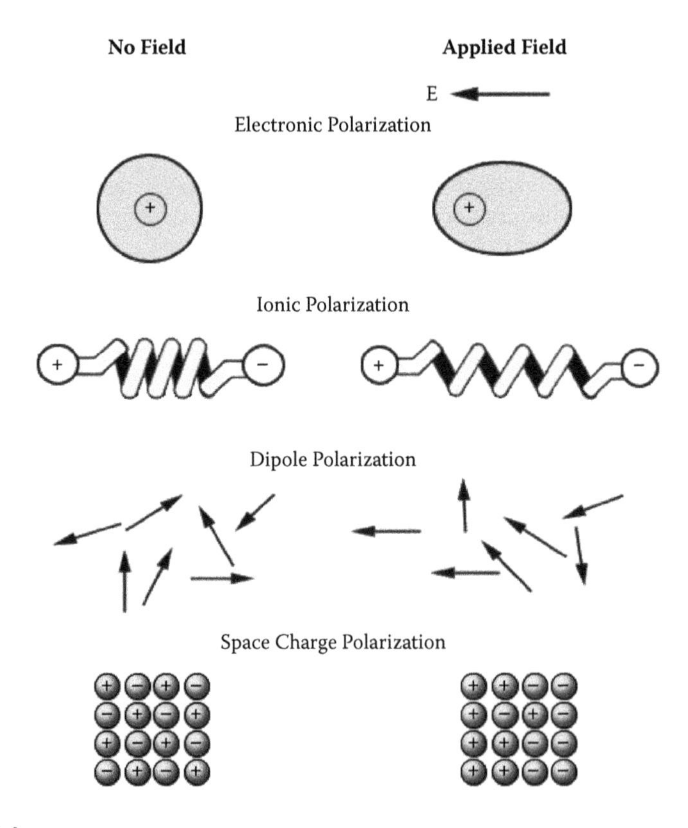

FIGURE 1.4
Representation of different mechanisms of polarization induced by an electric field (E). (*Source:* Von Hippel, A. R. 1954. Dielectrics and Waves. Cambridge, MA: MIT Press. pp. 90–100. With permission.)

vacuum is 1, represented as 1.00000, and other materials' dielectric constants are relative to this value. For example, if a metal was used for the dielectric instead of an insulator, the cell voltage would be zero according to Equation (1.20), and the corresponding dielectric constant would be infinite. Table 1.1 lists several typical materials' dielectric constant values for reference.

1.3.1.1 Dielectric Polarization Mechanisms

A material can be used as a capacitor's dielectric if it can be polarized when an electric field is applied. To have polarization ability, a material should have at least one of the following properties: (1) covalently bonded molecules that possess some natural electron polarities through which they endure electrical stress and orbital deformation from an induced electrical field; (2) polarizable ions where displacement of the ion center occurs under electrical stress; (3) dipole polarity causing dipole rotations; and (4) domain polarity where entire domains undergo rotation induced by an electric field. This property is strongly dependent upon frequency, voltage, and temperature, and can produce dielectric constants as large as 400,000 observed in ceramics. Figure 1.4 represents the polarization mechanism induced by an electric field [5].

1.3.1.2 Ceramic Dielectrics and Their Capacitors

A ceramic is a material composed of non-metal bonding inorganic compounds. Ceramic dielectric materials are insulating and include titanates, porcelains, and metal oxides (see Table 1.1). A ceramic dielectric can be designed according to the shape of a capacitor such as a disc or a rectangle. Since 1930, ceramics have been widely used to construct capacitors for radio receivers and in modern electronic equipment for decoupling and bypass applications. For example, some non-polar ceramic capacitors consist of alternating layers of metal and the several geometric designs of dielectric ceramics can provide high capacitance within a small design.

Different applications of current ceramic capacitors have different requirements, such as accuracy, stability over a temperature and voltage range, and volumetric efficiency. Ceramic-based capacitors are recognized to have a typical accuracy to their nominal capacitance rating of 5% to 10% with a deviation as low as 1%, offering high stability with a low dissipation factor. However, they have the lowest volumetric efficiency and are primarily used in applications requiring frequency filtering. Ceramic-based capacitors have operating temperatures between 10°C and 200°C. Exceptions are the negative–positive–zero ceramic capacitors, as they do not vary with temperature and offer a capacitance range between 1.0 picofarads to a few microfarads. In addition, depending on the application, the accuracies can offset each other. For example, a decrease in the accuracy of the capacitance can be offset by the progressive improvements of volumetric efficiency.

1.3.1.3 Electrolytic Dielectrics and Their Capacitors

The use of electrolytes as dielectrics was initially developed by Charles Pollak in 1886 as a result of his investigations of the anodization of metals. However, electrolytic capacitors have faced numerous difficulties in applications primarily due to their low reliability that hindered their use despite several patents filed for alternative designs. During World War II, the low reliability of electrolyte capacitors was improved and their dependability was increased in numerous applications. Key processes leading to their success were the etching and pre-anodizing of the metal foils prior to assembly. Etching can increase a surface 100 times the area of an unetched metal foil and yields a larger capacitance. However, this kind of capacitor needs to use less corrosive electrolyte solutions; otherwise the electrode will corrode, causing performance deterioration under discharge conditions [6].

Electrolytic dielectrics include metal oxides, aqueous-based liquid electrolytes, and non-aqueous-based liquid electrolytes. The most commonly used metal oxides are aluminum oxide, tantalum pentoxide, and niobium oxide. A typical aqueous-based liquid electrolyte contains boric acid or sodium borate in aqueous solution with various sugars or ethylene glycol to retard evaporation. For example, there are three major types of water-based electrolytes for aluminum electrolytic capacitors: standard water-based (with 40% to 70% water), those containing ethylene glycol (with less than 25% water), and dipropyl ketone (with less than 25% water).

A typical non-aqueous-based liquid electrolyte is generally composed of a weak acid, a salt derived from a weak acid, a solvent, an optional thickening agent, and other additives. The electrolyte is usually soaked into an electrode separator that serves as the dielectric. Weak acids are organic and include glacial acetic acid, lactic acid, propionic acid, butyric acid, crotonic acid, acrylic acid, phenol, and cresol. The salts are ammonium or metal salts of organic acids, including ammonium acetate, ammonium citrate, aluminum acetate, calcium lactate, and ammonium oxalate; or weak inorganic acids such as sodium perborate and trisodium phosphate. Electrolyte solvents are based on alkanolamines (monoethanolamine, diethanolamine, and triethanolamine) or polyols (diethylene glycol, and glycerol).

In practice, the two major types of electrolytic capacitors are (1) aluminum electrolytic capacitors and (2) tantalum electrolytic capacitors. Aluminum electrolytic capacitors are fabricated from two conducting aluminum foils, one of which is coated with an insulating oxide layer and a paper spacer soaked in electrolyte. The foil insulated by the oxide layer is the anode while the liquid electrolyte and the second foil act as the cathode. Tantalum electrolytic capacitors are subdivided into wet and dry types based on whether their counter electrodes are served by sulfuric acid or a manganese dioxide film. Dry tantalum electrolytic capacitors possess a greater capacitance-to-volume ratio relative to aluminum counterparts and are utilized in computer and miniaturized electronic devices that require high reliability and extended lifetimes.

Electrolytic capacitors present some advantages over other types of capacitors: (1) larger capacitance per unit volume due to their high dielectric constants and storage capabilities; (2) suitability for use in relatively high-current and low-frequency electrical circuits, especially in power supply filters to moderate output voltage and current fluctuations in rectifier output; (3) widespread use as coupling capacitors to allow conduction of AC and halt DC. One drawback of electrolytic capacitors is that the standard design requires the applied voltage to be polarized; that is, one specified terminal must always have a positive potential with respect to the other. Therefore, electrolytic capacitors need a DC polarizing bias to be used with AC signals. Some other drawbacks include relatively low breakdown voltages, higher leakage current and inductance, poorer tolerances and temperature ranges, and shorter lifetimes than other types of capacitors.

1.3.1.4 Paper and Polymer Dielectrics and Their Capacitors

In the past, paper impregnated with wax or oil was used as a dielectric material as an alternative to glass and mica dielectrics. A major drawback to using paper was its moisture-absorbing nature because moisture degrades the capacitive ability and increases internal resistances. Polymer dielectrics are used in applications with low frequency and high stability. A list of polymers along with their respective features and shortcomings can be found in Table 1.2. Polymer dielectric-based capacitors include polymerized organic semiconductor solid electrolyte capacitors and conductive polymer capacitors. A typical polymer capacitor has an aluminum foil cathode, a solid state polymer electrolyte, and aluminum foil with an oxide layer as the anode. In general, they last longer but cost more than standard (liquid) electrolytic capacitors.

1.3.1.5 Classification of Dielectric Materials

Dielectric materials fall into two classes: Class I used for linear capacitors and Class II used for non-linear capacitors. In general, Class I materials are natural dielectrics such as glass and mica and have capacitances within the range of a couple to several hundred picofarads. When capacitors use Class I materials, the capacitance will not change with a dynamic operating voltage and frequency. Class I dielectric materials are more costly.

Typical Class II materials are ferrodielectric ceramics and polymer electrolytes with high dielectric constants. They are used to develop high capacitance capacitors. For example, metalized polymer foils can achieve a few thousand picofarads to a few microfarads; an electrolytic capacitor can offer a capacitance in the range of a few to several thousand microfarads; some

TABLE 1.2
Polymer Dielectrics and Their Advantages and Disadvantages

Polymer	Advantages	Disadvantages
Paper	Inexpensive, applicable for high voltage use	Hygroscopic, degradation of insulation and quality factor over time
Polyester	Low moisture absorption, inexpensive, high operating voltage (60,000 V)	Low operating temperature (125°C) and poor stability; dielectric heats quickly, thus limiting use to low frequency AC applications
Polyimide	Similar advantages to polyester, with increased operating temperature (250°C)	Less temperature stability than paper yielding an increase in power factor, higher cost than polyester
Polystyrene	Excellent stability and low moisture absorption	Low operating temperature limited to 85°C
Polycarbonate	Higher stability, lower dissipation factor, and less moisture absorption than polystyrene, operating voltage available across temperature range of –55 to 125°C	Limited operating temperature (125°C)
Polypropylene	Lower dissipation factor, moisture absorption, and power factor, with higher dielectric strength than polyester and polycarbonate; self-healing possible after minor breakdown	Susceptible to damage from over or reverse potential for pulse power applications
Polysulfone	Operates at full voltage at ~125°C	High cost and limited production
Polytetrafluoroethylene	Lowest power factor for solid dielectric and good stability at high operating temperature (250°C)	High cost and limited production
Polyamide	Low dissipation factor, good stability, and high operating temperature of 200°C	High cost

Note: See References 3 and 4.

dielectric polymers such as polystyrene, polyethylene terephthalate, and polytetrafluoroethylene can be used to obtain hundreds of picofarads to a few microfarads. However, capacitors using a Class II material as a dielectric will experience a change in capacitance with changing operating voltage and frequency.

In the development and manufacturing of capacitors and their applications, several properties must be considered for dielectric materials including dielectric strength, dielectric constant, leakage current, power and quality factors, operating temperature, and breakdown voltage.

1.4 Capacitor Charging and Recharging Processes

Charging a capacitor is done by integrating it into an electrical circuit containing an external power source (i.e., a battery). When the circuit is a closed loop, the battery will provide an electromotive force (*emf*) to generate a flow of electrons through the circuit. During this process, the positive capacitor plate loses electrons to the positive terminal of the battery, thus making the plate positively charged. Simultaneously, electrons flowing from the negative terminal of the battery accumulate on the capacitor's negative plate, making it negatively charged. The number of electrons accumulated on the negative electrode is exactly equal to the number of positive charges accumulated on the positive electrode. The charging process continues until the potential between the plates that was initially zero is equal to the potential difference between the battery terminals. The capacitor is then said to be fully charged in reference to the driving potential.

1.4.1 DC and AC Currents

Due to charging and discharging, a capacitor needs a charge flow within the plate through an external wire. This electric flow is called electric current, defined as the flow of electric charge across a defined point or area and measurable in units of Coulombs per second [C/s^1, also called amperes (A)]. The perception is that a current is produced by the electrons traveling through a conducting wire medium, but ions transported through a medium (electrolyte) can also produce current. This discussion will focus on the flow of electrons through solid metallic conductors.

In metal, electrons are negative and singularly mobile charges that flow in the direction of an electric potential gradient from a lower potential to a higher one. This potential gradient, commonly called electromotive or driving force, is primarily responsible for the work necessary for the travel of electrons through a closed electric circuit or more importantly, with a connected electric load. In electric circuit analysis, the current direction is the direction of the positive charge flow rather than the electron charge flow. In a metal electric wire, only electrons are mobile to produce current. Therefore, the electron flow direction is opposite to the current direction. The current can be constant or momentary and classifying a current is necessary. The two types are (1) direct current (DC) and (2) alternating current (AC).

Direct current (DC) in an electric circuit describes a unidirectional flow of electrons traveling continuously from a low to high potential area. The relationship between the direct current I and the difference between high and low potentials V are described as follows. If there are two points designated 1 and 2 along a circuit wire loop, the potential of Point 1 is V_1 and that of Point 2 is V_2 (i.e., the potential difference or voltage difference between these two points is $V = |V_2 - V_1|$); when a current (I) flows from Point 1 to Point 2, the relationship between I and V can be expressed as

$$R = \frac{V}{I} \tag{1.21}$$

This relationship is called Ohm's Law, where R is the electric resistance expressed in ohm (Ω) units if the unit of current is the ampere (A) and the potential difference is voltage (V). According to Ohm's law, in practice, some specifically designed resistors using appropriate materials are fabricated and connected inside the electric loop to control current flow.

Alternating current (AC) is distinct from direct current in that it describes the directional change of the current flow when the electromotive force continually reverses its direction [3]. Normally, AC is generated by the forced rotation of a loop conductor through a magnetic field. During rotation, alteration of the polarity takes place in a continuous oscillating frequency varying sinusoidally with time. An induced potential $\tilde{V}$ through the loop can be expressed as

$$\tilde{V} = \tilde{V}_m \sin \omega_d t \tag{1.22}$$

where $\tilde{V}_m$ is the amplitude of the oscillation (i.e., maximum value), ω_d is the angular frequency of the rotating loop, and t is the time. As a result of this periodic potential change, the alternating current $\tilde{I}$ sinusoidally oscillates over time at the same angular frequency and can be written as

$$\tilde{I} = \tilde{I}_m \sin(\omega_d t - \varphi) \tag{1.23}$$

where $\tilde{I}_m$ is the maximum amplitude of current oscillation and φ is the phase constant and describes a situation in which the current is out of phase with the potential. Note that Ohm's law, described by Equation (1.21), is also applicable to the case of AC. The electrical resistance expressed by the equation is in opposition to the passage of electric charge and is dependent on the intrinsic resistivity of the material through which the current is passed. In general, the resistance (R) of an object can be derived from the

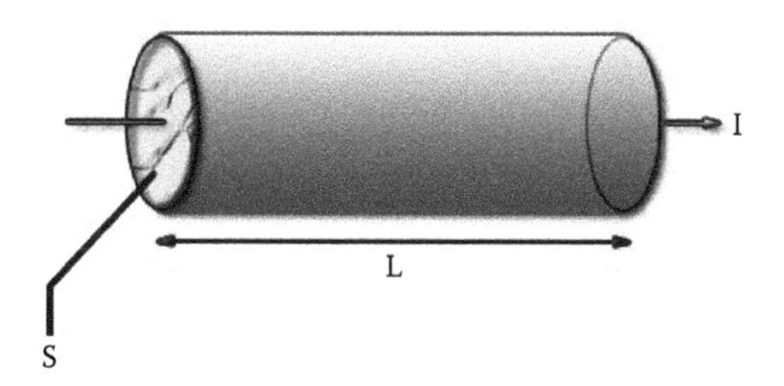

FIGURE 1.5
Length of wire l with cross-sectional area S through which an established current travels.

relation $R = l/\sigma S$ where l is the length of the object, σ the intrinsic conductivity, and S the cross-sectional area for current flow as shown in Figure 1.5.

If a material has a pure electric resistance property, it is considered an ohmic material and has a constant resistance R largely independent of the potential applied or the current passed through. Other materials that do not comply with Ohm's law have non-linear resistances. Ideal resistors are considered to have no function in storing energy via an electric or magnetic field. However, in AC applications, this is hardly the case because an equivalent inductance or capacitance in series with the resistor element is often considered. The use of AC circuits requires the consideration of additional opposition to current flow due to electrical and magnetic fields treated as electrical reactance effects. An electrical circuit's impedance is defined by the sum effect of resistance and resistance [3].

1.4.2 Charging of Capacitor: *RC* Time

The connection of a single capacitor in series with a potential or current source can quickly charge the plates of a capacitor over time. If a resistor is placed between the capacitor and the potential or current source, the charge time will be increased. However, this will yield an important relation, called the *RC* time constant that provides a method for measuring the charge and relaxation times for capacitor charging [3]. Depending on the application, the *RC* time constant can be very important in circuit design. Figure 1.6 shows a circuit design that can be used to obtain the *RC* constant using Kirchoff's law:

$$V^0 - IR - \frac{q}{C} = 0 \tag{1.24}$$

where V^0 is the battery's voltage. By substituting $I = dq/dt$ into Equation (1.24) and rearranging the equation into a first order differential, the function

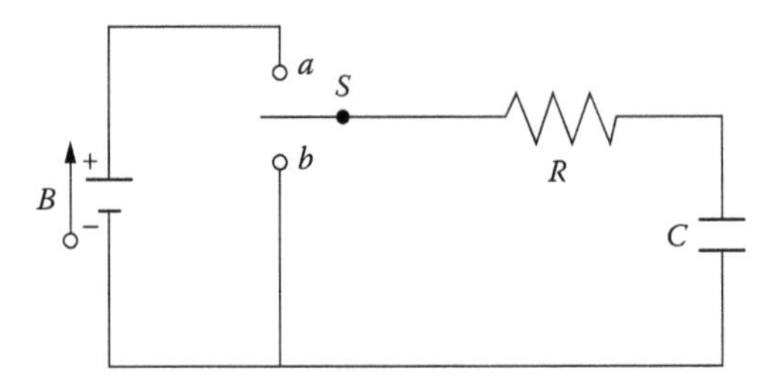

FIGURE 1.6
Electric circuit of resistor and capacitor connected in series with switch (S), elemental resistor (R), capacitor (C), and battery (B).

satisfying the equation with the initial condition of $q = 0$ at $t = 0$ is then described as

$$q = CV^0(1 - e^{-t/RC}) \tag{1.25}$$

According to Equation (1.25), when time approaches infinity, a charge $q = CV^0$ will be obtained, which agrees with the previous definition of capacitance in Equation (1.11). [Note the e in Equation (1.25) denotes the natural logarithm constant, and is not an electron charge]. Using Equation (1.11) to substitute charge for potential, Equation (1.25) can be written in terms of the potential V_p across the capacitor plates as a result of the driving voltage by

$$V_p = V^0(1 - e^{-t/RC}) \tag{1.26}$$

In addition, a relation of the charge time to the current traveling through the circuit can also be derived by taking the derivative of q with respect to t in Equation (1.25) as shown by

$$\frac{dq}{dt} = I = \frac{V^0}{R} e^{-t/RC} \tag{1.27}$$

In all these equations, the RC product is the capacitive time constant and is represented by τ_c (= RC). For example, when $t = \tau_c = RC$, Equation (1.25) is reduced to $q = 0.63CV^0$, indicating that at $t = \tau_c = RC$, only 63% of the total charge can be achieved given the driving potential. For general purposes, a capacitor is considered to have a full charge after five time constants.

1.4.3 Discharge of Capacitor

After a complete charging of a capacitor to a potential equivalent to that of the power source V^0, a discharge can be started by connecting the charged capacitor to a loop circuit with a resistor R. In this discharging loop, the only power source is the fully charged capacitor [3]. The sum of the potentials in this discharge loop is

$$\frac{q}{C}+\frac{dq}{dt}R=0 \tag{1.28}$$

The solution to this first-order differential in terms of charge q is

$$q = q_0e^{-t/RC} \tag{1.29}$$

where the initial charge q_0 of the capacitor is equal to V^0C. Additionally, a function describing the discharge current is derived by differentiating Equation (1.29) with respect to t

$$I=-\frac{q_0}{RC}e^{-t/RC} \tag{1.30}$$

As an example, Figure 1.7 shows the charge and discharge curves to illustrate a capacitor's behavior. Ideally, the energy used to charge the capacitor and capacitive charge it stores do not leak or dissipate, and are retained indefinitely until discharged [1].

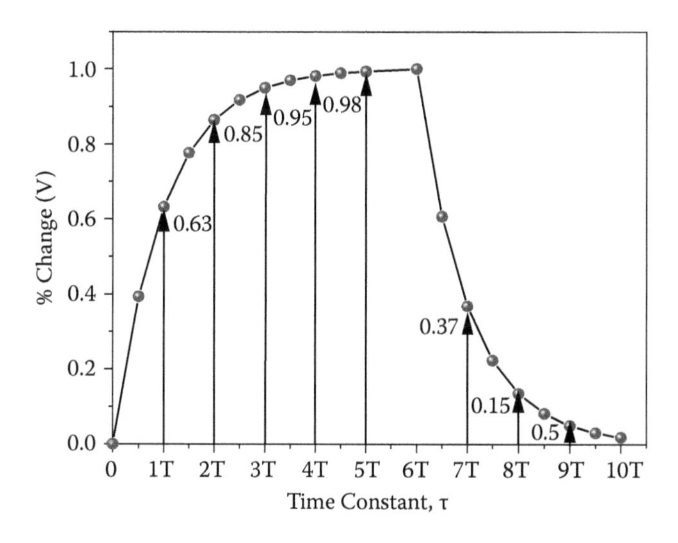

FIGURE 1.7
Graph of time constants for capacitor charging and discharging.

1.5 Energy Storage in Capacitor

For an electrostatic charge to develop along the plates of a capacitor, work must be done by an external driving force. At the beginning of the charging, the net charge between the plates of the capacitor is zero. When a potential is applied, the charge will accumulate on the conducting plates. However, as an electric field develops between the plates it becomes more difficult to accumulate like charges and subsequently the process requires more work. This work, done by an external power source such as a battery, transfers energy into electric potential energy E and is stored in an electrical field within the dielectric material. Recovery of this stored energy is achieved by discharging the capacitor into a circuit. To calculate the work done by a potential difference V^o for charge transfer on a capacitor, Equation (1.31) is used where the incremental change (dq) in charge requires incremental work (dW):

$$dW = Vdq = \frac{q}{C}dq \tag{1.31}$$

The total work performed, and thus the total potential energy stored in the capacitor, is

$$E = \int dW = \frac{1}{C}\int_0^q q\,dq = \frac{q^2}{2C} \tag{1.32a}$$

Note that the capacitance C in Equation (1.32a) is independent of charge and can be taken out of the integral [2]. Combining Equation (1.11) with (1.32a), a more familiar form of the energy stored in a capacitor can be obtained:

$$E = \frac{q^2}{2C} = \frac{1}{2}C(V^o)^2 \tag{1.32b}$$

Ideally, the energy stored in a capacitor and capacitive charge stored by the capacitor do not leak or dissipate and are retained indefinitely until discharged [1]. However, in practice, due to the leaking of dielectric material, the self-discharge rate of the capacitor is faster relative to batteries.

1.6 Capacitor Containing Electrical Circuits and Corresponding Calculation

In general, all electric circuits are driven by an external power source such as a portable battery, supercapacitor, stationary electric generator, solar cell, or thermopile. Despite their distinct modes of operation, they all have the same principal function: performing work on charge carriers while keeping a potential difference between their connected terminals.

1.6.1 Circuit Resistors

For a single loop circuit with a passive resistance element R connected to an external power source with a voltage of V^0, the sum of the potentials within this loop should equate to zero according to Kerchoff's voltage law:

$$V^0 - IR = 0 \tag{1.33}$$

From Equation (1.33), the current flow through the resistor

$$I = \frac{V^0}{R}$$

and the corresponding power I^2R can be obtained. Series resistances in a single loop experience an identical current passing through them. By applying Kerchoff's voltage law, the sum of the series resistances can be equated to a single equivalent resistance R_{eq}. In contrast, resistances in parallel experience the same potential derived from the external power source, and a reciprocal R_{eq} is equal to the sum of the reciprocal resistances.

1.6.2 Circuit Capacitors

Figure 1.8a and Figure 1.8b show an arrangement of capacitor elements in series and/or parallel within a single-loop circuit. For simplifying the circuit, a replacement equivalent capacitor can be considered to have the same capacitance as that of all the actual capacitors combined. If the capacitors are arranged in parallel to an applied potential V^o, each of them will experience the same equivalent potential difference V^o. The charge in each individual capacitor can be expressed as the following, according to Equation (1.11):

$$Q_1 = C_1V^o; Q_2 = C_2V^o; Q_3 = C_3V^o; \text{etc.} \tag{1.34}$$

The total charge of the combination can then be calculated as

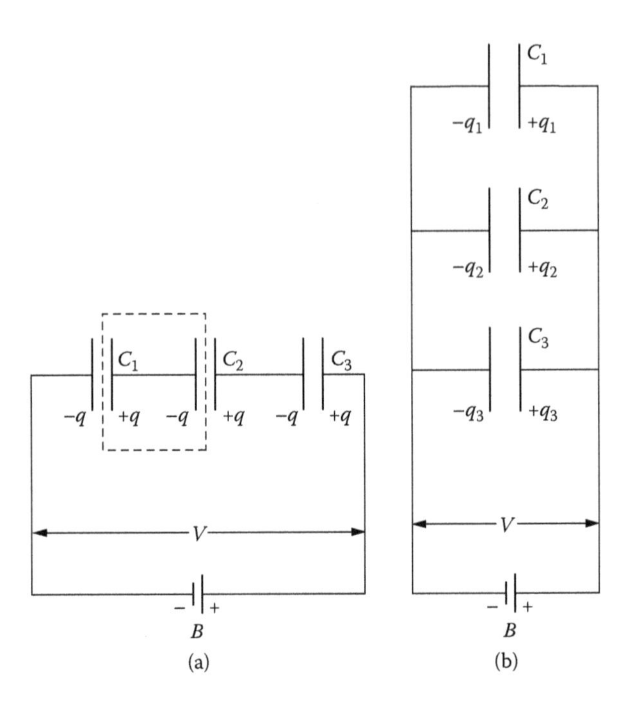

FIGURE 1.8
Electric circuit of single-loop capacitor connected in (a) series and (b) parallel.

$$Q = Q_1 + Q_2 + Q_3 \ldots = (C_1 + C_2 + C_3 \ldots)V^o \tag{1.35}$$

Under a given potential and known total charge storage, the equivalent capacitance ($C_{Eq,\ parallel}$) can be expressed as

$$C_{Eq,parallel} = \frac{Q}{V^o} = \sum_{i=1}^{n} C_i \tag{1.36}$$

Capacitors connected in series to a voltage source with a potential difference V^o will experience an equivalent chain-reaction charging of each plate rather than an equivalent potential across each capacitor. Therefore, each capacitor will theoretically store an equal amount of charge. The potential difference across the plates of each capacitor can be expressed as

$$V_1 = \frac{Q}{C_1}; V_2 = \frac{Q}{C_2}; V_3 = \frac{Q}{C_3}; \text{etc} \tag{1.37}$$

From this, the sum of the potential across the series-wired capacitors should be

$$V^0 = \sum_{i=1}^{n} V_i = Q\left(\sum_{i=1}^{n} \frac{1}{C_i}\right) = \frac{Q}{C_{Eq,series}} \tag{1.38}$$

The equivalent capacitance of capacitors connected in series, which is the sum of their reciprocal, is always smaller than any individual capacitor and is more approximate to the smallest one.

1.6.3 Inductors

Inductors are analogous circuit elements used to generate a controlled magnetic field. An inductor is made from a continuously winded conducting wire that results in a coil that is significantly shorter than the overall length of the wire used. A current I passing through a coil can produce a magnetic flux Φ_B. The inductance, L is defined as

$$L = \frac{N\Phi_B}{I} \tag{1.39}$$

where N is the number of turns used in winding the coil. With the use of an ideal inductor (i.e., negligible resistance through the conductor), an induced potential difference V_L can develop across the element following a change in the current:

$$V_L = -L\frac{dI}{dt} \tag{1.40}$$

Equation (1.40) indicates that the generated inductance can oppose a change in the flow of current. This principle is of more significance in AC applications.

1.6.4 Resistor–Inductor Circuits

Inductors are similar to capacitors in that they store an electric charge through an induced magnetic field rather than an electric field. The storage process can be described using an analysis of a single-loop series resistor–inductor (RL) circuit as shown in Figure 1.9. When an RL circuit is closed at time $t = 0$, a current is generated by the external power source with a potential V^0 and will flow through the resistor. However, a rapid increase of the current to a stable value

$$\frac{V^0}{R}$$

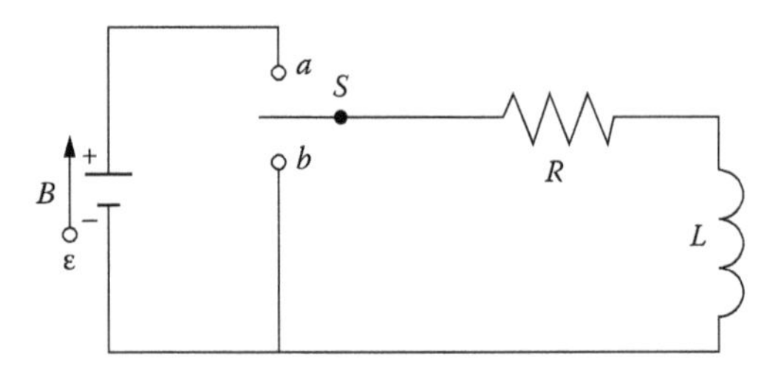

FIGURE 1.9
Electric circuit for single-loop resistor and inductor in series with inductor element L.

across the resistor cannot be observed due to the presence of the conductor. The loop analysis of the potentials provides the following equation:

$$V^0 - IR - L\frac{di}{dt} = 0 \tag{1.41}$$

Solving this first-order differential equation with respect to I gives the solution:

$$I = \frac{V^0}{R}(1 - e^{-Rt/L}) \tag{1.42}$$

The inductor element also has an inductive time constant $\tau_L = L/R$. After time $t = \tau_L$ has passed, the current through the resistor will be

$$0.63\frac{V^o}{R}$$

and will approach the equilibrium value

$$\frac{V^o}{R}$$

after five time constants. If the external power source is removed from the circuit and the circuit is closed, the current through the circuit will gradually decay. To solve for the decay function of the current, the potential sum of the loop is

$$IR + L\frac{dI}{dt} = 0 \tag{1.43}$$

With initial conditions of $t = 0$ and $I = V^0/R$, Equation (1.43) can be modified to give the expression of I:

$$I = \frac{V^0}{R} e^{-t/\tau_L} \tag{1.44}$$

1.6.5 Inductor–Capacitor Circuits

As discussed earlier, the two-element series combinations such as *RC* and *RL* circuits showed the exponential functions for the growth or decay of charge, potential, and current over time with a time scale measured by their respective time constants τ_C and τ_L. However, an inductor–capacitor (*LC*) circuit demonstrates different behavior from that of *RC* and *RL* circuits. There are two new parameters introduced to describe *LC* circuit behavior: (1) a sinusoidal oscillation period *T* and (2) an angular frequency ω. The *LC* circuit's capacitor is initially charged to a maximum potential V^0 and the magnetic energy stored in the inductor is zero.

Upon removal of the external power source, the capacitor discharges through the inductor and the magnetic field stores the charge and then releases it to oppositely charge the capacitor in an oscillating manner. For an ideal situation, this reverse process will then occur and operate indefinitely in the circuit. This is referred to as a *tank circuit*, similar to an analogous ideal block spring or flywheel design [6–7]. If the total energy of an ideal system is U (i.e., no loss dissipated as thermal energy), the *LC* circuit energy can be defined as

$$E = E_B + E_E = \frac{LI^2}{2} + \frac{q^2}{2C} \tag{1.45}$$

where E_E is the energy stored in the capacitor's electric field and E_B is the energy stored in the inductor's magnetic field. In an ideal circuit with no energy loss,

$$\frac{dE}{dt} = 0$$

and

$$I = \frac{dq}{dt}$$

Thus,

$$\frac{dI}{dt} = \frac{d^2q}{dt^2}$$

and Equation (1.45) can be simplified to the differential equation:

$$\frac{dE}{dt} = L\frac{d^2q}{dt^2} + \frac{1}{C}q = 0 \tag{1.46}$$

The solution to this differential equation with a maximum charge Q (i.e., amplitude of oscillation) at time $t = 0$ can be written as:

$$q = Qcos(\omega t + \varphi) \tag{1.47}$$

where ω is the angular frequency of electromagnetic oscillation and φ is the phase angle constant. By taking the derivative of Equation (1.46) with respect to t, a function describing the current through an LC circuit can be obtained:

$$I = -\omega Qsin(\omega t + \varphi) \tag{1.48}$$

The amplitude ωQ is equivalent to the maximum current I passing through the circuit. Solving for ω is done by substituting Equations (1.47) and (1.48) into Equation (1.46):

$$\omega = \frac{1}{\sqrt{LC}} \tag{1.49}$$

This ω is often called the natural angular frequency by which the LC circuit will oscillate devoid of any external power source.

1.6.5 Resistor–Inductor–Capacitor Circuits

Circuits containing the elements of resistance, inductance, and capacitance are called RLC circuits [6–7]. By integrating a resistor into an LC circuit, the oscillating electromagnetic energy is no longer perpetual and will dissipate as thermal energy. Thus, the differential equation relating the change in electromagnetic energy becomes

$$\frac{dE}{dt} = L\frac{d^2q}{dt^2} + R\frac{dq}{dt} + \frac{1}{C}q = 0 \tag{1.50}$$

Solving this second-order differential results in the following function:

$$q = Qe^{-Rt/2L}\cos(\omega' t + \varphi) \tag{1.51}$$

Because the resistor is present, the angular frequency is modified to

$$\omega' = \sqrt{\omega^2 - \left(\frac{R}{2L}\right)^2} \tag{1.52}$$

If oscillation of the *RLC* circuit needs to continue, an external power source is needed to apply a constant potential driving the current in compensating the dissipation of thermal energy though the resistor. With an external power source, the oscillations of the charge, potential, and current can then be described as forced oscillations, replacing the natural frequency ω with ω_d. Normally, the amplitudes of the oscillating quantities are highly reliant on how similar the two frequencies are to each other. A resonating frequency can produce maximum amplitudes.

1.6.6 Resistive, Capacitive, and Inductive Loads for AC Circuits

A single resistive load *R* in an AC circuit can be analyzed in the same manner as previous resistor elements through a loop analysis of the potentials to give an expression:

$$\tilde{V}^0 - \tilde{V}_R = 0 \tag{1.53}$$

where $\tilde{V}^0$ is the external AC power source and $\tilde{V}_R$ is the potential drop across the resistor and is equivalent to

$$\tilde{V}_R = \tilde{V}^0 \sin \omega_d t \tag{1.54}$$

where the amplitude is equivalent to the maximum value of the driving potential. Using the relation

$$R = \frac{\tilde{V}_R}{\tilde{I}_R}$$

the current can be derived as

$$\tilde{I}_R = \frac{\tilde{V}_R}{R} \sin \omega_d t = \tilde{I}_R \sin \omega_d t \tag{1.55}$$

Equation (1.55) indicates that the phase constant is zero ($\varphi = 0$); thus the current through the resistor is in phase with the driving potential [3]. For a capacitive load, the potential difference across the capacitor is

$$\tilde{V}_c = \tilde{V}^0 \sin \omega_d t \tag{1.56}$$

The current passing through the capacitor can be expressed as

$$\tilde{I}_c = C \frac{d\tilde{V}_c}{dt} \tag{1.57}$$

Combining Equations (1.56) and (1.57), the following current expression is obtained:

$$\tilde{I}_c = \omega_d C \tilde{V}^0 \cos \omega_d t = \omega_d C \tilde{V}^0 \sin(\omega_d t + 90^0) \tag{1.58}$$

A capacitive reactance X_C can be defined and expressed as

$$X_C = \frac{1}{\omega_d C} \tag{1.59}$$

This capacitive reactance has the same unit (ohm) as that of a resistor. Substituting Equation (1.59) into Equation (1.58) yields

$$\tilde{I}_c = \frac{\tilde{V}^0}{X_C} \sin(\omega_d t + 90°) = \tilde{I}^0 \sin(\omega_d t + 90°) \tag{1.60}$$

Thus, the AC current has a phase constant of 90° ahead of the driving potential. Similar to capacitive circuits, a phase constant of –90° for an inductive circuit can be obtained, indicating the current lags the potential by one quarter cycle. An equivalent inductive reactance X_L in a circuit only containing an inductor can be defined as

$$X_L = \omega_d L \tag{1.61}$$

1.6.6.1 Series Resistor–Inductor–Capacitor Circuit

With the use of brief derivations of current through AC circuit elements, more explicit evaluation of the current amplitude and equivalent resistances in an *RLC* circuit can be performed [2]. By applying symmetry of notation and the Pythagorean theorem, the amplitude of these oscillating functions shown by phasor notation in Figure 1.10 gives a relationship as

$$(\tilde{V}^o)^2 = \tilde{V}_R^2 + (\tilde{V}_L - \tilde{V}_C) \tag{1.62}$$

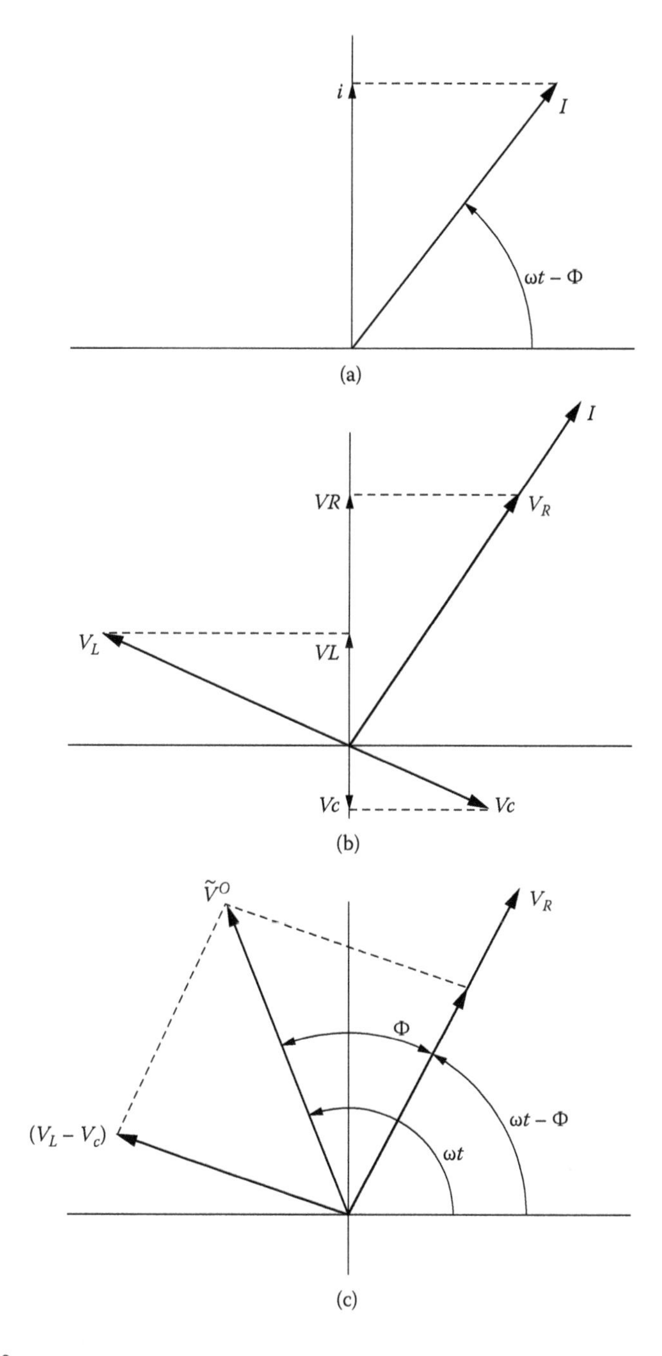

FIGURE 1.10
Phasor diagram of voltage through inductance (*VL*), capacitance (*VC*), and resistance (*VR*) for alternating current circuit. (*Source:* Halliday, D., R. Resnick, and J. Walker. 2008. Fundamentals of Physics, 8th ed. New York: John Wiley & Sons, pp. 600–900. With permission.)

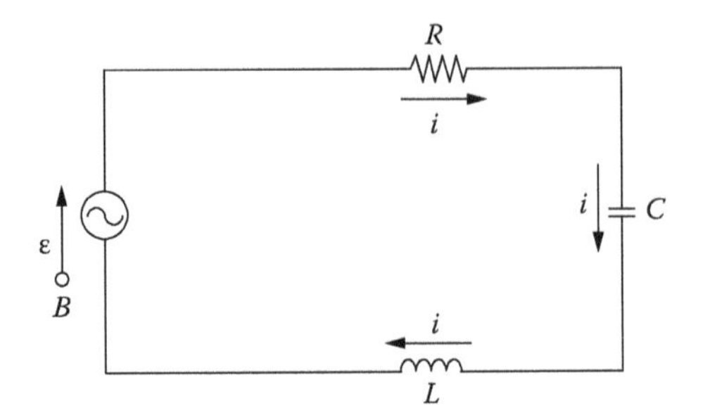

FIGURE 1.11
Electric circuit schematic of resistor, inductor, and capacitor (*RLC*) connected in series.

According to the circuit loop rule, the sum of the potentials through each element must be equivalent to the driving potential applied:

$$\tilde{V}^o - \tilde{V}_{\bar{R}} - \tilde{V}_L - \tilde{V}_C = 0 \tag{1.63}$$

If the *R*, *C*, and *L* are in series in the circuit, as depicted in Figure 1.11, and a current $\tilde{I}$ passes through it, Equation (1.63) can be rewritten as

$$(\tilde{V}^o)^2 = \left(\tilde{I}R\right)^2 + (\tilde{I}X_L - \tilde{I}X_C) = 0 \tag{1.64}$$

By rearranging Equation (1.64), a function for the current is then obtained:

$$\tilde{I} = \frac{\tilde{V}^o}{\sqrt{R^2 + \left(X_L - X_C\right)^2}} \tag{1.65}$$

The current is equal to the driving potential divided by a denominator term, hereafter known as impedance $Z_{R\text{-}L\text{-}C}$ of an AC *RLC* circuit:

$$Z_{R\text{-}L\text{-}C} = \sqrt{R^2 + (X_L - X_C)^2} = \sqrt{R^2 + (\omega_d L - \frac{1}{\omega_d C})^2} \tag{1.66}$$

Equation (1.67) represents equivalent resistance or AC impedance of the whole series *RLC* circuit. It indicates that when a circuit is operating at a resonant frequency (i.e., $X_L = X_C$), the overall impedance is equal to a real resistance *R*. If the frequency is increased to exceed the resonance, the capacitive reactance will be reduced and the inductive reactance will be increased. The situation is likewise reversed at frequencies below resonance.

1.6.6.2 RLC Circuits Having Other R, L, and C Combinations

There are different *RLC* circuits with various *R*, *L*, and *C* combinations. Other *RLC* circuits include parallel *RLC* (*R/L/C*), parallel *LC* series with *R* ((*L/C*)-*R*), series *CL* parallel with *R* ((*C-L*)/*R*), parallel *RL* series with *C* ((*R/L*)-*C*), series *RL* parallel with *C* ((*RL*)-*C*), parallel *RC* series with *L* ((*R/C*)-*L*), and series *RC* parallel with *L* ((*R-C*)/*L*) [3]. The expressions of AC impedance for these *RLC* circuits can be obtained and summarized as follows:
Series *RLC* (*R-L-C*) circuit:

$$Z_{R\text{-}L\text{-}C} = \sqrt{R^2 + (\omega_d L - \frac{1}{\omega_d C})^2} \tag{1.67}$$

Parallel *RLC* circuit (*R/L/C*):

$$Z_{R/L/C} = \sqrt{\left(\frac{\frac{1}{R}}{(\frac{1}{R})^2 + (\omega_d C - \frac{1}{\omega_d L})^2}\right)^2 + \left(\frac{\omega_d C - \frac{1}{\omega_d L}}{(\frac{1}{R})^2 + (\omega_d C - \frac{1}{\omega_d L})^2}\right)^2} \tag{1.68}$$

Parallel *LC* series with *R* ((*L/C*)-*R*):

$$Z_{(L/C)\text{-}R} = \sqrt{R^2 + \left(\frac{1}{\frac{1}{\omega_d L} - \omega_d C}\right)^2} \tag{1.69}$$

Series *CL* parallel with *R* ((*C-L*)/*R*):

$$Z_{(C\text{-}L)/R} = \sqrt{\left(\frac{\frac{1}{R}}{(\frac{1}{R})^2 + (\frac{\omega_d C}{\omega_d^2 CL - 1})^2}\right)^2 + \left(\frac{\frac{\omega_d C}{\omega_d^2 CL - 1}}{(\frac{1}{R})^2 + (\frac{\omega_d C}{\omega_d^2 CL - 1})^2}\right)^2} \tag{1.70}$$

Parallel *RL* series with *C* ((*R/L*)-*C*):

$$Z_{(R/L)\text{-}C} = \sqrt{\left(\frac{\omega_d R^2 L}{R^2 + \omega_d^2 L^2}\right)^2 + \left(\frac{\omega_d R^2 L}{R^2 + \omega_d^2 L^2} - \frac{1}{\omega_d C}\right)^2} \tag{1.71}$$

Series *RL* parallel with *C* ((*RL*)-*C*):

$$Z_{(R-L)/C}=\sqrt{\left(\frac{R}{(1-\omega_d^2LC)^2+(\omega_dRC)^2}\right)^2+\left(\frac{\omega_dL(1-\omega_d^2LC)-\omega_dR^2C}{(1-\omega_d^2LC)^2+(\omega_dRC)^2}\right)^2} \tag{1.72}$$

Parallel RC series with L ((*R*/*C*)-*L*):

$$Z_{(R/C)-L}=\sqrt{\left(\frac{R}{1+\omega_d^2R^2C^2}\right)^2+\left(\omega_dL-\frac{\omega_dR^2C}{1+\omega_d^2R^2C^2}\right)^2} \tag{1.73}$$

Series RC parallel with L ((*R*-*C*)/*L*):

$$Z_{(R-C)/L}=\sqrt{\left(\frac{R}{(1-\frac{1}{\omega_d^2LC})^2+(\frac{R}{\omega_dL})^2}\right)^2+\left(\frac{\frac{R^2}{\omega_dL}-\frac{1}{\omega_dC}(1-\frac{1}{\omega_d^2LC})}{(1-\frac{1}{\omega_d^2LC})^2+(\frac{R}{\omega_dL})^2}\right)^2} \tag{1.74}$$

1.7 Types and Structures of Capacitors

Depending on the dielectric materials and applications, various types of capacitors have been designed, fabricated, and commercialized. These capacitors are categorized as fixed, variable, power, high voltage, interference suppression, ferrodielectric, polar polymer dielectric, linear, and nonlinear capacitors [7]. The following sections will briefly discuss each type.

1.7.1 Fixed Capacitors

A fixed capacitor is constructed to possess a fixed value of capacitance that cannot be adjusted. The schematic diagrams of fixed capacitors show various dielectrics in both axial and radial designs. The capacitance of a fixed capacitor dielectric should not be changed by changes of charge quantity, cell voltage, or frequency. This behavior is called the linear property of the dielectric material and the corresponding capacitor is called a linear capacitor.

1.7.2 Variable Capacitors

A variable capacitor is constructed to allow a user to adjust its capacitance mechanically. For example, in a radio application, using a variable capacitor coupled with an inductor allows the capacitance to be manipulated to resonate a desired radio station frequency. These tuner capacitors have low capacitance values in a range of 10 to 500 pF and are used primarily for frequency filtering. The dielectric materials used in these devices include air, plastic foil, and inert sulfur hexafluoride gas. A typical variable device is a trimmer capacitor, which is a miniaturized version of a variable capacitor. In variable capacitors, several dielectric materials including air, ceramic, mica, and polymers are used.

1.7.3 Power Capacitors

Power capacitors are generally large and heavy in comparison to fixed capacitors. A typical weight range is 10 to 50 kg. Smaller power capacitors are also available and have capacitance ranges from 2.5 to 20 μF with an AC circuit working voltage of 150 V and an RMS voltage as high as 550 V. These devices are normally used in electrical engines as starting motor capacitors.

The dielectric materials used in power capacitors include paper, polypropylene, and mixed vacuums impregnated with mineral or synthetic oil. The most reliable dielectric material is polypropylene. The metal electrodes can be aluminum foils or thin metal films deposited onto the surfaces of the dielectrics. The reliabilities and lifetimes of power capacitors are always concerns because of the possibility of overheating.

1.7.4 High-Voltage Capacitors

For DC voltage or low frequency applications, polymer foil-based capacitors are used. Ceramic capacitors are more suitable for high frequency applications. Capacitors shaped like disks and tubes have been fabricated. The tube-shaped types are easier to protect against hazardous flashovers and can be used with voltage ratings up to 6 kV. A series connection of several individual elementary capacitors with matching capacitances is a way to reduce the risk of high voltage when they are used on some AC circuits.

When using high-voltage capacitors, it is important to be aware of dielectric absorption (the percentage ratio of the regained potential with respect to the charging potential). For purposes of accident prevention, high-voltage capacitors are required to be shunted to high-rated resistors when they are switched from the main intended circuit.

1.7.5 Interference-Suppression Capacitors

In general, interference-suppression capacitors have low inductances. These capacitors can be separated into two classes depending on their use: Class X and Class Y. Class X types are used in applications with lower safety requirements, for example, where a collapse would not pose a danger. Class Y types are used in environments or applications that have strict safety requirements. These two classes have different test conditions. For example, an 1100 V test voltage is used for testing Class X capacitors such as a 4 μF/250 V capacitor; a

2250 V test is used for testing Class Y capacitors. Both types are used in high (ceramic) and low (polymer) radio frequency and automotive applications.

1.7.6 Ferrodielectric Capacitors

Ferrodielectric capacitors are shaped as tubes, monoliths, multilayers, or disks. They normally use ceramic dielectrics because of their high dielectric constants ($1000 < \varepsilon_r < 30{,}000$). However, the capacitance of a ferrodielectric ceramic capacitor is strongly dependent on the operating temperature, frequency, and voltage, where the effect of instability on the relative capacitance with respect to dielectric strength is demonstrated using two different ε_r values. These types are called non-linear capacitors and will be discussed briefly below.

1.7.7 Polar Polymer Dielectric Capacitors

Polar polymer dielectric capacitors are miniaturized devices that are widely used in miniaturized circuits in modern electronics. In fabricating polar polymer dielectric capacitors, a physical vapor deposition process is used to coat polycarbonate foil with an Al layer (≤0.5 μm) to act as the electrode plates. Melted metal is then deposited by an airbrush method on either a cylindrical or flattened roll to provide contacts for the terminal connections. By changing the dielectric thickness, the capacitor's voltage can be changed from 60 to 250 V.

1.7.8 Linear and Nonlinear Capacitors

Capacitor types can also be divided into linear and nonlinear capacitors according to the dielectric material used. In general, if the capacitance does not change when the operating voltage and frequency are changed, the capacitor is a linear type. If the capacitance changes when the operating voltage and frequency are changed, the capacitor is a non-linear type. Generally, linear capacitors are the most widely used; nonlinear capacitors have specific niche applications.

1.8 Summary

As essential elements in electrical circuits, capacitors have been used in the operation of countless systems and devices ranging from high power engines to microcircuits. The two classes of capacitors are: (1) Class I linear capacitors and (2) Class II nonlinear capacitors. In this chapter, a concise introduction of the principles and fundamental operations of capacitors from a science and engineering perspective is given with the intention to further the understanding of the important factors facing electrochemical supercapacitors.

References

1. Hickey, H. V. and W. M. Villines. 1970. *Elements of Electronics*. New York: McGraw Hill.
2. Halliday, D., R. Resnick, and J. Walker. 2008. *Fundamentals of Physics*, 8th ed. New York: John Wiley & Sons, pp. 600–900.
3. Ballou, G. 1993. *The Electrical Engineering Handbook*. Boca Raton: CRC Press.
4. K-Tek, *Dielectric Constants Chart* [online]. http://www.asiinstruments.com/technical.asp [accessed May 22, 2012].
5. Von Hippel, A. R. 1954. *Dielectrics and Waves*. Cambridge, MA: MIT Press. pp. 90–100.
6. Dummer, G. W. A. and H. M. Nordenberg. 1960. *Fixed and Variable Capacitors*. New York: McGraw Hill.
7. Nowak, S. et al. 2009. *Fundamentals of Circuits and Filters*. Boca Raton: Taylor & Francis.

2

Fundamentals of Electrochemical Double-Layer Supercapacitors

2.1 Introduction

The early concept of an electrochemical supercapacitor (ES) was based on the electric double-layer existing at the interface between a conductor and its contacting electrolyte solution. The electric double-layer theory was first proposed by Hermann von Helmholtz and further developed by Gouy, Chapman, Grahame, and Stern. The electric double-layer theory is the foundation of electrochemistry from which the electrochemical processes occurring at an electrostatic interface
between a charged electrode material and an electrolyte are investigated. Based on this knowledge, many electrochemical theories and technologies including electrochemical supercapacitors, batteries, and fuel cells have been invented and established since the double-layer theory was put forward.

As discussed in Chapter 1, electrostatic and electrolytic capacitors are considered the first and second generation capacitors. These early capacitors were developed for use as primary circuit elements in holding microfarad to picofarad charges of direct current or to filter the frequencies for alternating current circuits. With the rapid developments in materials, the third generation known as the supercapacitor was invented. The earliest supercapacitor with unusually high capacitance was invented by Becker at SOHIO in 1957, and intended to serve as an electrolytic capacitor for low voltage operation. Carbon material served as its electrodes. With further modifications, the first practical supercapacitor was developed by Boos [1],[2] as described in a 1970 patent.

The rapid growth of mobile electronics and alternative energy vehicles created a need for advanced electrochemical energy storage devices with high power capabilities. This need led to substantial research and development of supercapacitors. In the early 1990s, the United States Department of Energy (DOE) strongly advocated funding for battery and supercapacitor research, creating international awareness of the potential of the supercapacitor. Since then, great effort has focused on supercapacitor research and development in terms of electrode materials, composites, hybridizations, and suitable electrolytes to improve performance and reduce costs.

At the same time, a fundamental understanding of supercapacitor design, operation, performance, and component optimization led to improvements of supercapacitor performance, particularly increasing their energy density. To further increase energy density, more advanced supercapacitors called *pseudocapacitors*, in which the electroactive materials are composited with carbon particles to form composite electrode materials, were developed. The electrochemical reaction of the electroactive material in a *pseudocapacitor* takes place at the interface between the electrode and electrolyte via adsorption, intercalation, or reduction–oxidation (redox) mechanisms. In this way, the capacitance of the electrode and the energy density can be increased significantly.

This chapter will provide a comprehensive discussion of the fundamentals of double-layer supercapacitors, including the electric double-layer charging and discharging mechanism, the theoretical principles that govern their operation, and their structural designs.

2.2 Electrode and Electrolyte Interfaces and Their Capacitances

An electrochemical device consists of two electrodes with an electrolyte between them. The electrolyte can be a solid or a solution. Solid state electrolytes serve two functions. They conduct ions and separate the positive electrode from the negative electrode. For liquid state electrolytes such as electrolyte solutions, an inert porous separator sheet allows the ions to pass through, creating a conducting current. The structure of an electrochemical capacitor is very similar to that of an electrochemical cell but there is no electron transfer across the interface.

Figure 2.1 shows a typical double-layer capacitor. On the positive electrode, an accumulation of positive charges attracts an equal number of negative charges around the electrode in the electrolyte side due to Coulomb's force. However, due to heat fluctuation in the electrolyte, the charges carried by the ions have a scattering distribution, leading to some net negative charges in the electrolyte zone near the electrode. The charge balance between the electrode and the electrolyte represents an electric double-layer.

To maintain the electric neutrality of the system, an equal number of negative charges accumulates at the negative electrode near which is an equal number of net positive charges in the adjacent electrolyte, forming another double-layer. Therefore, a complete double-layer capacitor has two electric double-layers, one at the positive electrode–electrolyte interface and the other at the negative electrode–electrolyte interface. These two double-layers constitute the capacitor's "heart" and determine its performance. The following sections will present a theoretical description of the electric double-layer in terms of capacitance for charge storage.

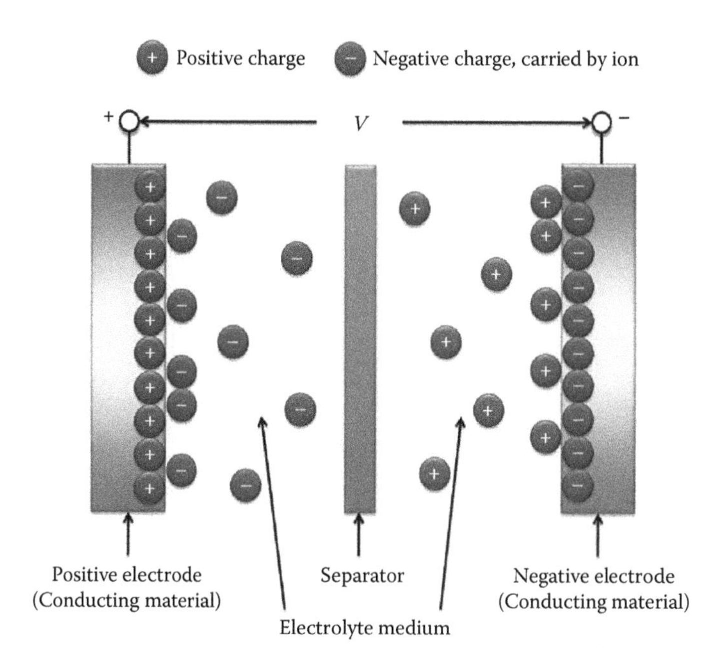

FIGURE 2.1
(See color insert.) Electric double-layer supercapacitor.

2.2.1 Electric Double-Layer at Interface of Electrode and Electrolyte Solution

The electric double-layer models at the interface between an electrode and an electrolyte solution were initially developed using aqueous solutions. Later models were extended into both non-aqueous electrolyte solutions and ionic liquids with some modifications. Therefore, the electric double-layer discussed here applies to all three of these electrolyte solutions. As shown in Figure 2.2a, the positive (or negative charge) developed along the interface of the electrode and electrolyte solution can be balanced by an induced accumulation of oppositely charged solution ions near the electrode surface in the solution through Coulomb's force, forming the electric double-layer. Due to thermal fluctuation in the solution, the net negative ions are scattered with a higher concentration near the electrode surface and a lower concentration in the solution.

This scattered layer plus the electrode positive charge array is called the diffuse double-layer or diffuse layer. In the literature, this diffuse layer is also called the Gouy point charge layer or model or the Gouy–Chapman model. We will use the *diffuse layer* term in this chapter. The thickness of the diffuse layer is dependent on the temperature, the concentration of the electrolyte, the charge number carried by the ion, and the dielectric constant of the electrolyte solution.

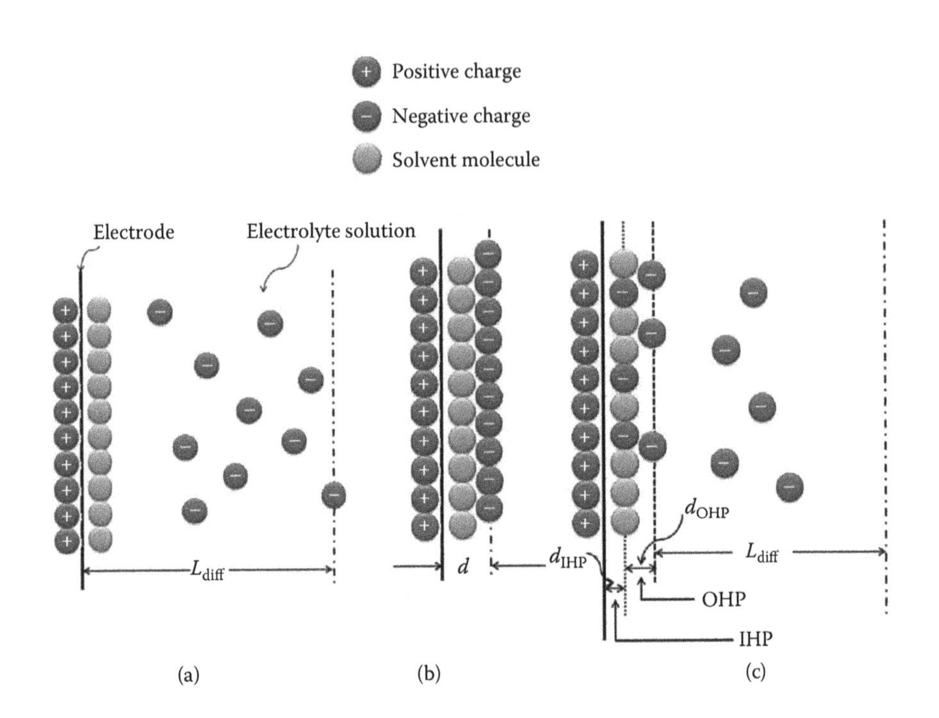

FIGURE 2.2
(See color insert.) Electric double-layer models at interface of electrode and electrolyte solution. (a) Diffuse layer or Gouy-Chapman model. (b) Helmholtz layer or model; the *d* represents the double-layer thickness. (c) Stern-Grahame layer or model in which the IHP represents the inner Helmholtz plane and the OHP represents the outer Helmholtz plane.

In general, the higher the temperature, the thicker the diffuse layer; the higher the electrolyte concentration, charge number carried by the ion, and dielectric constant, the thinner the diffuse layer. However, if the temperature is low, the concentration of the electrolyte, the charge number carried by the ion, and the dielectric constant of the electrolyte solution are high, the diffuse layer will become very thin, forming an array of negative ions near the electrode surface with a distance of the size of the solvent such as water in aqueous solution, leading to a compact electric double-layer as shown in Figure 2.2b. This compact layer is called the Helmholtz layer [4]. In reality, these two layers coexist as shown in Figure 2.2c and form the Stern–Grahame model. To take care of the specific adsorption of the ions on the electrode surface, the Helmholtz layer can also be divided into two layers called the inner plane and the outer plane, and will be discussed more in a later section of this chapter.

In electrochemistry, the potential drop across the double-layer is described by the potential difference ($\Delta\phi_{M/S}$) between the inner potential at a point in the metal electrode (ϕ_M) and the inner potential at the end point of the diffuse layer in electrolyte solution (ϕ_S): ($\Delta\phi_{M/S}$) = (ϕ_M) − (ϕ_S). This is called the absolute potential difference. The other expression of the potential drop is the outer potential difference ($\Delta\psi_{M/S}$), and is simplified as ($\Delta\psi$) in Figure 2.3. This outer potential difference is also called the Volta potential (also Volta potential difference, contact potential difference, or outer potential difference). It represents the potential difference between one point close to the electrode outer surface and the end point of the diffuse layer in the solution. The relationship between the outer and inner potentials can be expressed as:

$$\phi = \psi + \chi \tag{2.1}$$

where ϕ is the inner potential and χ is the surface potential determined by short range effects of adsorbed ions and oriented water molecules. In general, the outer potential can be measured directly, but the surface potential cannot. Therefore, this inner potential drop is not experimentally measurable. Regarding the potential distribution within the double-layer, Figure 2.3 shows how the double-layer potential drop can be expressed as either $\Delta\phi_{M/S}$ or $\Delta\psi_{M/S}$, that is, $\Delta\phi_{M/S} = \Delta\psi_{M/S}$. If we separate the double-layer into several phases as shown in Figure 2.4, this relationship can be derived. According to Equation (2.1), the overall double-layer potential drop in Figure 2.4 can be expressed as:

$$\Delta\phi_{M/S} = \Delta\chi_{O/M} + \Delta\chi_{M/O} + \Delta\psi_{M/H} + \Delta\chi_{O/H} + \Delta\chi_{H/O} + \Delta\psi_{H/S} + \Delta\chi_{O/S} + \Delta\chi_{S/O} \tag{2.2}$$

Due to $\Delta\chi_{O/M} = -\Delta\chi_{M/O}$, $\Delta\chi_{O/H} = -\Delta\chi_{H/O}$, and $\Delta\chi_{O/S} = -\Delta\chi_{S/O}$, Equation 2.2 becomes: $\Delta\phi_{M/S} = \Delta\psi_{M/H} + \Delta\psi_{H/S} = \Delta\psi_{M/S}$.

The potential drop across the diffuse layer is normally expressed as the outer potential drop rather than the inner potential drop. The outer potential drop ($\Delta\psi_{H/S}$) is expressed as ψ_1. As shown in Figure 2.3, $\Delta\psi_{M/S}$ and ψ_1 are the potential drops across the entire double-layer and the potential drop across the diffuse layer, respectively. Therefore the potential drop across the Helmholtz layer should be ($\Delta\psi_{M/S} - \psi_1$), and the potential across the entire double-layer can be expressed as:

$$\Delta\psi = (\Delta\psi_{M/S} - \psi_1) + \psi_1 \tag{2.3}$$

If the unit charge quantity accumulated on the electrode side or in the electrolyte solution side can be expressed as q on a unit area

$$\left(q = \frac{Q}{A}\right),$$

A is the real rather than geometric planar electrode surface area touched with the electrolyte solution. For instance, with a unit of $\mu C.cm^{-2}$, the reciprocal of the overall double-layer differential capacitance with a unit of $\mu F.cm^{-2}$ can be expressed as:

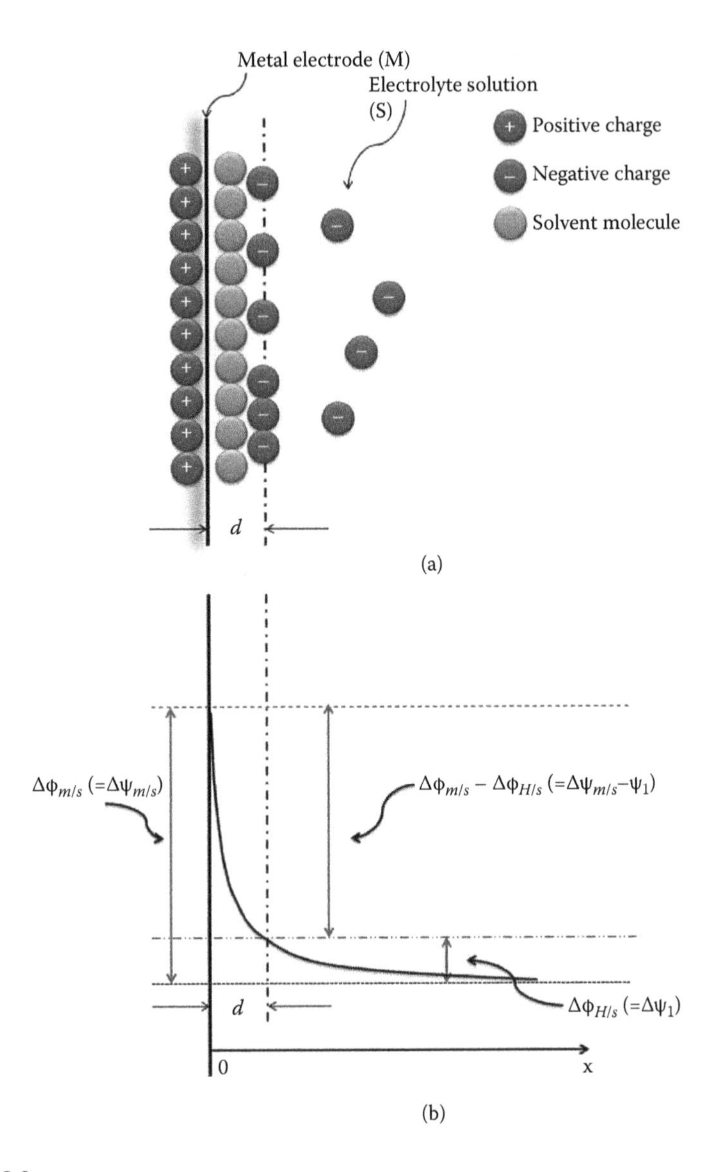

FIGURE 2.3
(See color insert.) A schematic description where the double-layer potential drops.

$$\frac{1}{C_{dl}} = \frac{d\Delta\psi}{dq} = \frac{d(\Delta\psi - \psi_1)}{dq} + \frac{d\psi_1}{dq} = \frac{1}{C_H} + \frac{1}{C_{diff}} \tag{2.4}$$

In Equation (2.4), C_H is the differential capacitance of the Helmholtz layer [Equation (2.5)] and C_{diff} is the differential capacitance of the diffuse layer [Equation (2.6)]:

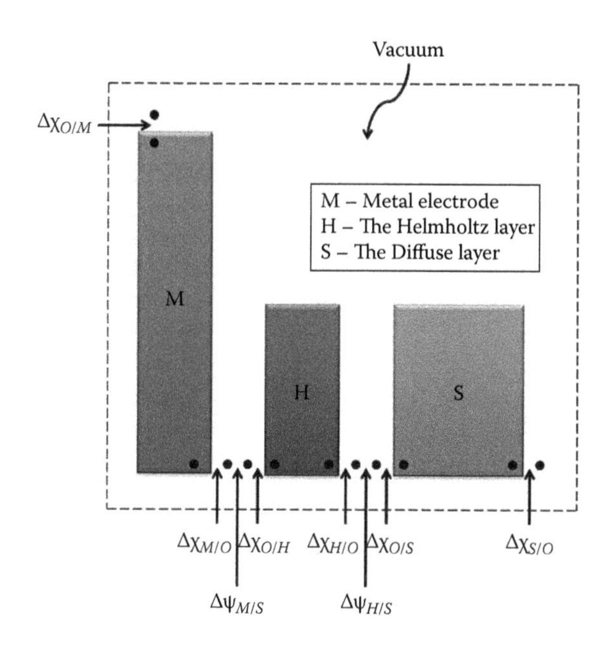

FIGURE 2.4
Separated electric double-layer.

$$C_H = \frac{dq}{d(\Delta\psi_{M/S} - \psi_1)} \tag{2.5}$$

$$C_{diff} = \frac{dq}{d\psi_1} \tag{2.6}$$

Equation (2.4) indicates that the equivalent circuit of the entire double-layer can be treated as C_H and C_{diff} in series as shown in Figure 2.5 [5].

C_{dl} of the electrode–electrolyte interface can be easily measured using electrochemical methods in the potential range where there is no electron transfer across the interface (the ideal non-polarizable potential range).Unfortunately, it is not easier to measure the C_H and C_{diff} independently. However, using the Gouy–Chapman–Stern (GCS) model, the individual differential capacitances can be theoretically treated.

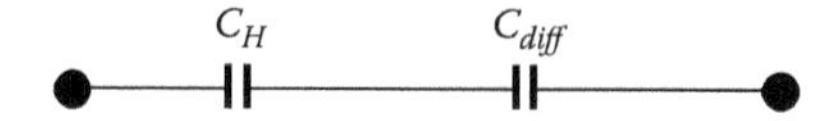

FIGURE 2.5
Helmholtz and diffuse layer differential capacitances (C_H and C_{diff}, respectively) connected in series.

In general, the Helmholtz layer can be treated as a linear capacitor. In a theoretical model of the electric double-layer, the compact Helmholtz layer is generally treated as an ideal capacitor with a fixed thickness (*d*), and its capacitance is considered unchanging with the potential drop across it. Therefore, the capacitance of the Helmholtz layer can be treated as a constant if the temperature, the dielectric constant of the electrolyte solution inside the compact layer, and its thickness are fixed. However, if the specific ion adsorption happened on the electrode surface, the dielectric constant of the electrolyte solution inside the compact layer may be affected, leading to non-linear behavior of the Helmholtz layer. This will be discussed more in a later section.

In a theoretical model, the diffuse layer appears more complicated. According to the definition in Chapter 1, the diffuse layer capacitor should be treated as a nonlinear rather than linear capacitor because its capacitance is dependent on the electrode potential. Figure 2.6 shows the differential capacitance of a graphite electrode as a function of an electrode

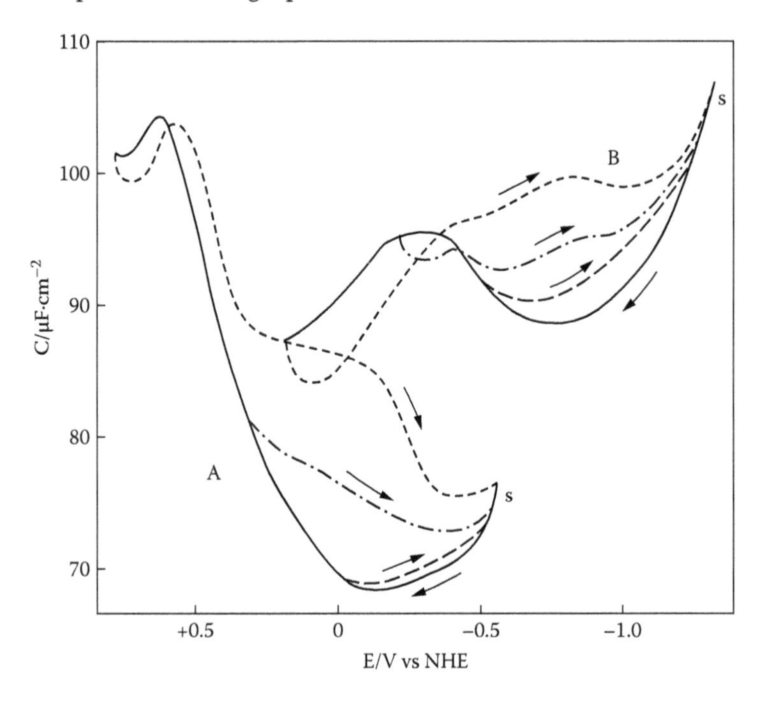

FIGURE 2.6
Capacity-potential curves for edge orientation of stress-annealed pyrolytic graphite in 0.5 *M* H_2SO_4 and 1 *M* NaOH at 25°C and 1000 Hz without hood. (A) 0.5 *M* H_2SO_4. (—) potentials going positive, (— —) potentials going negative from +0.1 V, (—·——) potentials going negative from +0.4 V, (·········) potentials going negative from +0.8 V. (B) 1 *M* NaOH. (—) potentials going positive, (— —) potentials going negative from −0.4 V, (— ·— ·—) potentials going negative from −0.2 V, (·········) potentials going negative from +0.2 V. (*Source:* Randin, J. P. and E. Yeager. 2001. *Electroanalytical Chemistry and Interfacial Electrochemistry*, 58, 313–322. With permission.)

potential in two different electrolyte solutions [6]. It can be seen that the differential capacitance is strongly dependent on the electrode potential as well as the type of electrolyte used. Therefore, the theoretical treatment of the diffuse layer seems complicated. In the next subsection, we will discuss how to theoretically calculate the differential capacitance of the double-layer.

2.2.2 Double-Layer Net Charge Density by Gouy–Chapman–Stern (GCS) Modeling

To provide a theoretical description of the diffuse layer, the potential distribution shown in Figure 2.3 is used. According to the Gouy–Chapman–Stern (GCS) model [7], the ion concentration ($C_{i(x)}$) within the diffuse layer at point x can be expressed as:

$$C_{i(x)} = C^0 \exp\left(-\frac{z_i F \psi_x}{RT}\right) \tag{2.7}$$

This equation is called the Boltzmann distribution, where C^0 is the bulk concentration of ion *i*, F is Faraday's constant, ψ_x is the potential at x point, *R* is the universal gas constant, and *T* is the temperature. Thus the net charge in a unit volume, ρ (μC.cm^{-3}), can be expressed as:

$$\rho = \sum z_i F C_{i(x)} = \sum z_i F C^o \exp\left(-\frac{F\psi_x}{RT}\right) \tag{2.8}$$

The following Poisson formula can be used to connect volume charge with the potential:

$$\frac{d^2\psi_x}{dx^2} = -\frac{\rho}{\varepsilon_r \varepsilon_o} \tag{2.9}$$

where ε_r is the relative dielectric constant of the electrolyte solution and ε_o is the dielectric constant of the vacuum. Then Equation (2.8) can be written as:

$$\frac{d^2\psi_x}{dx^2} = -\frac{1}{\varepsilon_r \varepsilon_o} \sum z_i F C^o \exp\left(-\frac{z_i F \psi_x}{RT}\right) \tag{2.10}$$

According to

$$\frac{d^2\psi_x}{dx^2} = \frac{1}{2}\frac{d}{d\psi_x}\left(\frac{d\psi_x}{dx}\right)^2,$$

Equation (2.10) can be rewritten as:

$$d\left(\frac{d\psi_x}{dx}\right)^2 = -\frac{2}{\varepsilon_r\varepsilon_o}\sum z_i FC^o \exp\left(-\frac{z_i F\psi_x}{RT}\right)d\psi_x \tag{2.11}$$

When integrating the conditions $x \to \infty$, $\psi_x = 0$ and $\frac{d\psi_x}{dx} = 0$

into Equation (2.10), Equation (2.12) is obtained:

$$\left(\frac{d\psi_x}{dx}\right)^2 = -\frac{2RT}{\varepsilon_r\varepsilon_o}\sum C^o\left[\exp\left(-\frac{z_i F\psi_x}{RT}\right) - 1\right] = \frac{2RTC^o}{\varepsilon_r\varepsilon_o}\left[\exp\left(\frac{z_i F\psi_x}{RT}\right) - \exp\left(\frac{z_i F\psi_x}{RT}\right)\right] \tag{2.12}$$

At the point of $x = d$, $\psi_x = \psi_1$, Equation (2.12) can be alternatively expressed as:

$$\left(\frac{d\psi_x}{dx}\right)^2_{x=d} = \frac{2RTC^o}{\varepsilon_r\varepsilon_o}\left[\exp\left(\frac{z_i F\psi_1}{RT}\right) - \exp\left(-\frac{z_i F\psi_1}{RT}\right)\right] \tag{2.13}$$

In order to obtain the total net charge within the diffuse layer, Equation (2.9) can be integrated from $x = d$ to $x = \infty$, resulting in:

$$\left(\frac{d\psi_x}{dx}\right)_{x=d} = \frac{1}{\varepsilon_r\varepsilon_o}\int_{x=d}^{x=\infty}\rho dx = -\frac{q}{\varepsilon_r\varepsilon_o} \tag{2.14}$$

Combining Equations (2.13) and (2.14), the total net charge within the diffuse layer becomes:

$$q = \sqrt{2\varepsilon_r\varepsilon_o RTC^o}\left[\exp\left(\frac{z_i F\psi_1}{RT}\right) - \exp\left(-\frac{z_i F\psi_1}{RT}\right)\right] \tag{2.15}$$

Because ρ is expressed as $\mu C.m^{-3}$, q in Equation (2.15) should be in $\mu C.m^{-2}$ units.

2.2.3 Theoretical Differential Capacitance of Electric Double-Layer

To determine differential capacitance of the diffuse layer, Equation (2.15) can be differentiated with respect to the potential drop, leading to:

$$C_{diff} = \frac{dq}{d\psi_{H-S}} = \frac{z_i F}{2RT}\sqrt{2\varepsilon_r \varepsilon_o RTC^o}\left[\exp\left(\frac{z_i F\psi_1}{RT}\right) + \exp\left(-\frac{z_i F\psi_1}{RT}\right)\right] \tag{2.16}$$

Equation (2.16) shows the differential capacitance of the diffuse layer as a function of the potential drop across the layer at various electrolyte concentrations. The data in Figure 2.7 were obtained by using Equation (2.16). It can be seen that the differential capacitance can be significantly decreased when decreasing the electrolyte concentration. In addition, the differential capacitance can be significantly increased when increasing the potential drop at both positive and negative potential directions. When the potential drop across the diffuse double-layer is zero, the differential capacitance has a minimum value. This zero potential point is called the electrode potential at zero charge (pzc).

From Equation (2.16), when the potential drop across the diffuse layer is near zero, the equation becomes:

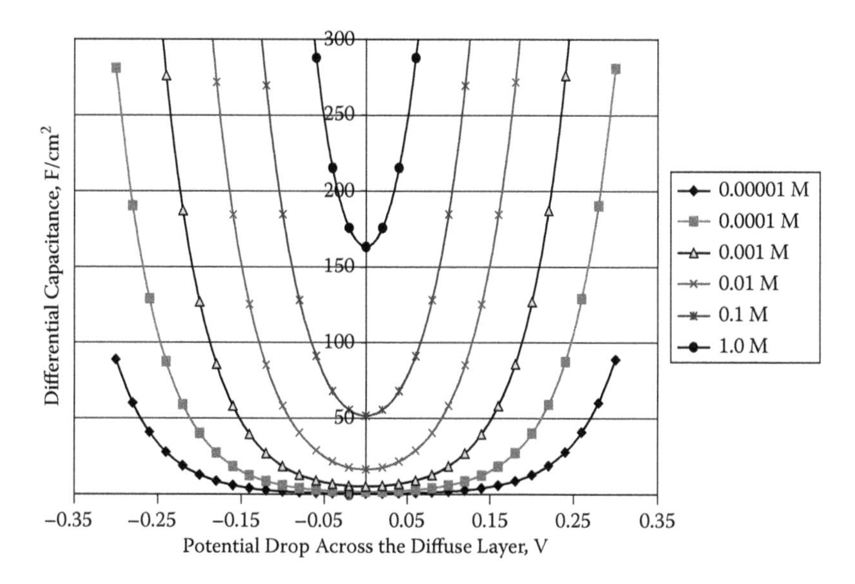

FIGURE 2.7
Differential capacitance as function of potential drop cross diffuse layer at several electrolyte concentrations calculated according to Equation (2.16) by assuming $\varepsilon = 40$, $\varepsilon_o = 8.854 \times 10^{-12}$ F.m^{-1}, z = 1, 25°C and 1.0 atm.

$$C_{diff} \approx \frac{z_i F}{RT}\sqrt{2\varepsilon_r \varepsilon_o RTC^o} \tag{2.17}$$

Equation (2.17) indicates that when $\psi_1 = 0$, the differential capacitance of the diffuse layer is proportional to the square roots of both the electrolyte concentration and the dielectric constant. Furthermore, according to the definition of capacitance, if the C_{diff} is expressed as

$$C_{diff} \approx \frac{\varepsilon_r \varepsilon_o}{L_{diff}},$$

the rearrangement of Equation (2.17) will give the equivalent thickness of the diffuse layer (L_{diff}):

$$L_{diff} = \frac{1}{z_i F}\sqrt{\frac{\varepsilon_r \varepsilon_o RT}{2C^o}} \tag{2.18}$$

For a 1-1 type electrolyte ($|z_i| = 1$) with concentration less than 0.001 *M* in an aqueous solution, the value of L_{diff} could be as high as several hundred angstrom (Å), however, if the electrolyte concentration is higher than 0.1 *M*, the value of L_{diff} could fall below 10 Å.

2.2.4 Differential Capacitance of Entire Double-Layer

The thickness of the Helmholtz layer has a fixed value. Therefore, its differential capacitance should be constant and not change with its potential drop. The differential capacitance of the Helmholtz layer is in the range of 18 to 100 $\mu F.cm^{-2}$ and depends on the electrolyte concentration, type of solution solvent, and status of the electrode surface. Combining Equations (2.4) and (2.16) will produce the differential capacitance of the entire double-layer:

$$C_{dl} = \frac{C_H C_{diff}}{C_H + C_{diff}} = \frac{C_H \frac{|z|F}{2RT}\sqrt{2\varepsilon_r \varepsilon_o RTC^o}\left[\exp\frac{|z|F\psi_l}{2RT} + \exp\left(-\frac{|z|F\psi_l}{2RT}\right)\right]}{C_H + \frac{|z|F}{2RT}\sqrt{2\varepsilon_r \varepsilon_o RTC^o}\left[\exp\frac{|z|F\psi_l}{2RT} + \exp\left(-\frac{|z|F\psi_l}{2RT}\right)\right]} \tag{2.19}$$

In capacitors, the differential capacitance of the Helmholtz layer can be expressed as

$$C_H = \frac{\varepsilon_r \varepsilon_o}{d},$$

in which the thickness of the Helmholtz layer (d) is about the diameter of a water molecule, ~1.9 ×10^{-10} m. The high electric field within the Helmholtz layer (>10^9 V.m^{-1}) can make all polarities of the water molecules in order, leading to a smaller relative dielectric constant (~ 6) than that of randomly distributed water, which is about 40 in the presence of an electrolyte. The calculated C_H value is about 28 μF.cm^{-2}. If C_H = 28 μF.cm^{-2}, ε = 40 (aqueous electrolyte solution), z = 1, and T = 25°C, then C_{dl} can be calculated according to Equation (2.19). The obtained differential capacitance data at different electrolyte concentrations can be plotted as a function of the potential drop across the diffuse layer, as shown in Figure 2.8.

Comparing Figure 2.8 with Figure 2.7, we see that in a very dilute electrolyte solution, the differential capacitance of the diffuse layer is much smaller than that of the Helmholtz layer, meaning that the differential capacitance of the entire double-layer is close to that of the diffuse layer. However, with a highly concentrated electrolyte solution, the differential capacitance of the diffuse layer is much larger than that of the Helmholtz layer, signifying that the differential capacitance of the entire double-layer is gradually

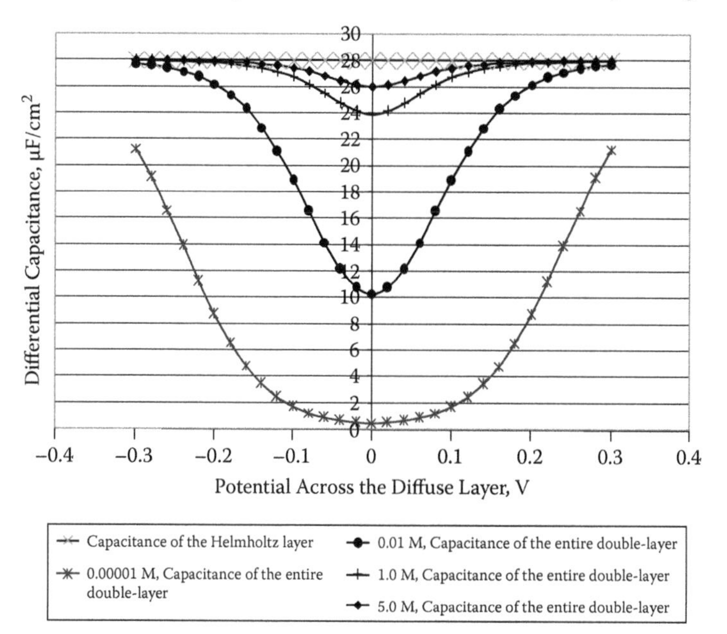

FIGURE 2.8
Differential capacitance as function of potential drop across diffuse layer at three electrolyte concentrations calculated according to Equation (2.9) by assuming C_H = 28 μF.cm^{-2}, ε = 40, ε_o = 8.854 × 10^{-12} F.m^{-1}, z = 1, 25°C and 1.0 atm.

approaching that of the Helmholtz layer. Therefore, if a very dilute electrolyte solution is used, the capacitance obtained should represent that of the diffuse layer; if a highly concentrated electrolyte solution is used, the capacitance obtained should represent that of the Helmholtz layer, except for values near the potential of the point of zero charge.

In addition, Equation (2.19) demonstrates that the dielectric constant has the same weight as that of the electrolyte concentration, meaning that the differential capacitance of the diffuse layer is also proportional to the square root of the dielectric constant. Therefore, using different electrolyte solutions such as aqueous, non-aqueous, and ion liquid solutions can produce different capacitances of the double-layer.

2.2.5 Potential Drop Distribution within Electric Double-Layer

The relationship between the Helmholtz layer and the diffuse layer potential drops can be obtained by combining Equations(2.5) and (2.15). The resulting expression is:

$$\psi - \psi_1 = \frac{1}{C_H}\sqrt{2\varepsilon_r\varepsilon_o RTC^o}\left[\exp\left(\frac{|z|F\psi_1}{2RT}\right) - \exp\left(-\frac{|z|F\psi_1}{2RT}\right)\right] \quad (2.20)$$

For example, assuming $C_H = 28\ \mu\text{F.cm}^{-2}$, $\varepsilon = 6$ in an aqueous electrolyte solution, $\varepsilon_o = 8.854 \times 10^{-12}\ \text{F.m}^{-1}$, $z = 1$, and $T = 25°\text{C}$, the relationship between the Helmholtz layer potential drop ($\psi - \psi_1$) and the diffuse layer potential drop (ψ_1) can be calculated according to Equation (2.20), and is plotted in Figure 2.9 at different electrolyte concentrations. Note that Figure 2.9 represents only one case at a fixed differential capacitance of the Helmholtz layer (C_H), and does not fully reflect the situations at other C_H values.

Figure 2.9 indicates that the potential of the Helmholtz layer is smaller than that of the diffuse layer with diluted electrolyte solutions. However, when increasing the electrolyte concentration, the potential of the diffuse layer becomes much smaller than that of the Helmholtz layer. This observation reinforces the notion that at dilute electrolyte concentrations the potential drop of the entire double-layer is dominated by that of the diffuse layer, and at high electrolyte concentrations the dominating potential drop will be that of the Helmholtz layer. Furthermore, Equation (2.20) also indicates that the potential drop across the Helmholtz layer is not only a function of the square root of the electrolyte concentration, but also a function of the square root of the dielectric constant ($\varepsilon_r\varepsilon_o$), suggesting that different electrolyte solutions can cause different potential drops across the diffuse and Helmholtz layers, and the dielectric constant has the same effect on the potential drop distribution.

From Figure 2.9 it seems that the potential drops can go very high. In practice, this is impossible because the magnitudes of these potential drops are

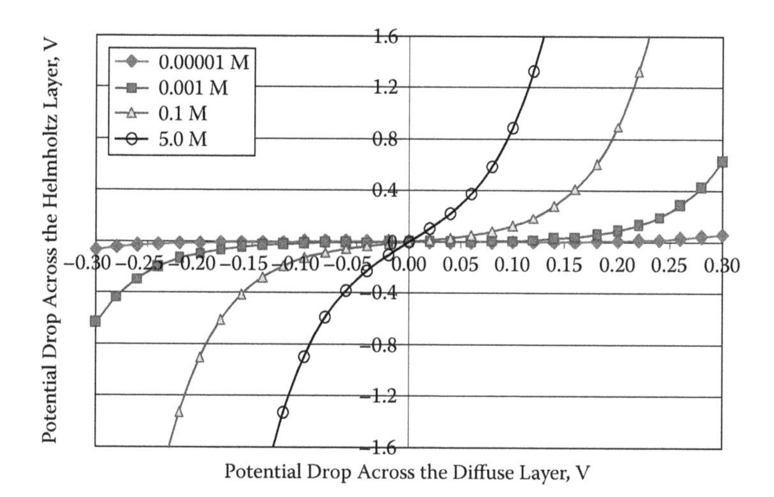

FIGURE 2.9
Potential of Helmholtz layer as function of potential of diffuse layer calculated according to Equation (2.20) by assuming $C_H = 28\ \mu F.cm^{-2}$, $\varepsilon = 6$, $\varepsilon_o = 8.854 \times 10^{-12}\ F.m^{-1}$, z = 1, 25°C and 1.0 atm.

limited by the stable windows of both the electrode material and the electrolyte. Therefore, if the potential drops go too high above the electrode or electrolyte reduction–oxidation, these reactions will stop the further increase of the potential drops.

2.2.6 Factors Affecting Double-Layer Capacitance

The discussion about Equations (2.16) and (2.19) shows that the differential capacitance of the double-layer is mainly dependent on the charge (z_i), the electrolyte concentration (C^o), the solvent used (ε_r), and the temperature (T), but does not depend on the types of electrolytes or electrode materials and their structures. It may be true that as long as an electrode is electrically conductive, the differential capacitance should be similar if other conditions are the same. However, if the electrode is a semiconductive material, the net charge accumulated at the electrode will have a diffuse distribution near the interface at the electrode side.

In fact, due to the electrode surface status, electrolyte structure, and their interaction at the interface, different electrode materials and different electrolytes have different double-layer differential capacitances. In particular, when electrolyte ions (both inorganic and organic) are strongly adsorbed on the electrode surface, the differential capacitance of the double-layer is significantly affected.

For a chosen electrode material, different types and sizes of electrolyte ions have different interactions with the electrode surface, resulting in different strengths of adsorption; or for a chosen electrolyte, different electrode materials have different affinities to the electrolyte ion, leading to different

TABLE 2.1

Capacitances of Carbon Electrode Materials and Electrolytes at Room Temperature

			Aqueous Electrolyte		Organic Electrolyte	
Materials	**Specific Surface Area ($m^2.g^{-1}$)**	**Density ($g.cm^{-3}$)**	$F.g^{-1}$	$F.cm^{-3}$	$F.g^{-1}$	$F.cm^{-3}$
Carbon Materials						
Commercial activated carbons (ACs)	1000 to 3500	0.4 to 0.7	< 200	< 80	< 100	< 50
Particulate carbon from SiC/TiC	1000 to 2000	0.5 to 0.7	170 to 220	< 120	100 to 120	< 70
Functionalized porous carbons	300 to 2200	0.5 to 0.9	150 to 300	< 180	100 to 150	< 90
Carbon nanotube (CNT)	120 to 500	0.6	50 to 100	< 60	< 60	< 30
Templated porous carbons (TC)	500 to 3000	0.5 to 1	120 to 350	< 200	60 to 140	< 100
Activated carbon fibers (ACF)	1000 to 3000	0.3 to 0.8	120 to 370	< 150	80 to 200	< 120
Carbon cloths	2500	0.4	100 to 200	40 to 80	60 to 100	24 to 40
Carbon aerogels	400 to 1000	0.5 to 0.7	100 to 125	< 80	< 80	40
Carbon-Based Composites						
TC-RuO_2	600	1	630	630	—	—
CNT-MnO_2	234	1.5	199	300	—	—
AC-polyaniline	1000	—	300	—	—	—

Source: Zhang, L. L. and X. S. Zhao. 2009. *Chemical Society Reviews*, 38, 2520–2531. With permission.

strengths of adsorption. As a result, different differential capacitances can be observed when using different electrode materials and electrolytes. The most popular electrode materials used to construct electrochemical supercapacitors are carbon-based nanoparticles and related composite materials that have high surface areas, giving high capacitances. The metal is normally used as the current collector on which a layer of carbon or composite particles acts as the electrode layer. This will be discussed in detail in a later section of this chapter. Table 2.1 shows some differential capacitances of typical carbon-based electrode materials.

2.2.7 Specific Adsorption of Ions and Effect on Double-Layer

The specific adsorptions of ions can affect the Helmholtz layer. For example, if the total charge of anions adsorbed on the electrode surface is more than the

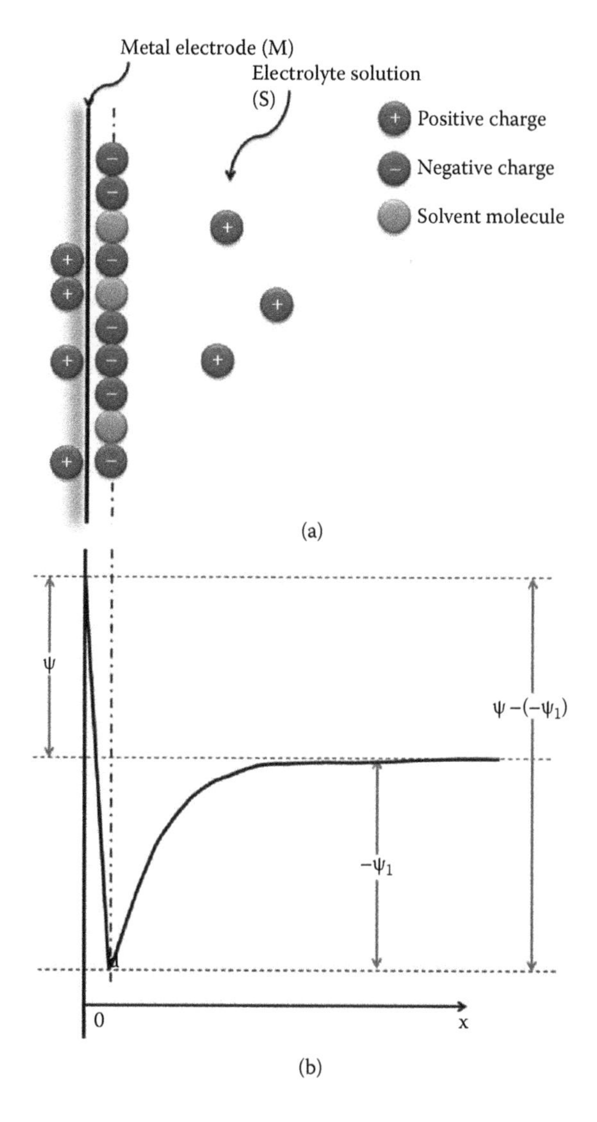

FIGURE 2.10
(See color insert.) Double-layer with specific ion adsorption and its corresponding potential distribution.

total positive charge on the electrode surface, the potential drop of the diffuse layer would be negative ($\psi_1 < 0$), resulting in $\psi - \psi_1 > \psi$, as shown in Figure 2.10.

To address the effect of specific adsorption of ions on the electrode surface, two models were developed to divide the Helmholtz layer into: (1) the inner Helmholtz plane (IHP) with a thickness of d_{IHP} and (2) the outer Helmholtz plane (OHP) with a thickness of d_{OHP}, as shown in Figure 2.10. The capacitance of IHP (C_{IHP}) is induced by the net charge near the electrode surface and its adsorbed counter ions, and the capacitance of OHP (C_{OHP}) is induced by the rest of the net charge layer outside the IHP. In this case, the differential capacitance of the entire Helmholtz layer can be treated as the serially connected capacitances of the IHP (C_{IHP}) and OHP (C_{OHP}):

$$\frac{1}{C_H} = \frac{1}{C_{IHP}} + \frac{1}{C_{OHP}} = \frac{d_{IHP}}{\varepsilon_{IHP}\varepsilon_o} + \frac{d_{OHP}}{\varepsilon_{OHP}\varepsilon_o} \tag{2.21}$$

where ε_{IHP} and ε_{OHP} are the relative dielectric constants of IHP and OHP, respectively. Both the values of d_{IHP} and d_{OHP} are about the diameter of a water molecule if the electrolyte solution is aqueous. However, the waters inside the IHP are in order due to the strong electric field and the waters in the OHP are randomly-distributed, so the ε_{IHP} value (~6) is much smaller than that of ε_{OHP} (~40). In this case, Equation (2.21) will become:

$$C_H \approx \frac{\varepsilon_{IHP}\varepsilon_o}{d_{IHP}} \tag{2.21a}$$

Equation (2.21a) suggests that the capacitance of the Helmholtz layer is mainly determined by the IHP's capacitance in the presence of a specific ion adsorption. For deeper understanding about this model, please refer to the related electrochemistry books [7].

Equation (2.21) suggests that the capacitance of the Helmholtz layer can be changed when the dielectric constant or the thickness of the Helmholtz layer is changed by a specific ion adsorption. The ion adsorption is dependent on the electrode potential, and the capacitance of the Helmholtz layer changes with the electrode potential. In this case, the Helmholtz capacitor may not be considered a linear capacitor.

It is often observed that in practical systems, even using high concentration of electrolytes, the measured differential capacitances that are supposed to be those of the Helmholtz layers show some degree of electrode potential dependency. This may be explained by ion adsorption at high electrolyte concentrations.

2.3 Electrode Potential and Double-Layer Potential Windows Using Different Electrode Materials and Electrolytes

2.3.1 Electrode Potential

In electrochemistry, the theoretical electrode potential is defined as the potential drop at the interface of the electrode–electrolyte solution, shown as $\Delta\psi_{M/S}$

in Figure 2.3. This electrode potential is composed of the potential drops of both the Helmholtz and diffuse layers. However, this potential drop is not experimentally measurable. This can be illustrated and explained by the electrochemical measurement system shown in Figure 2.11. The measurement for the potential drop across an interface such as the interface between metal phase M^I and the solution phase S must involve at least four interfaces such as between the metal phase M^I and the metal phase M^{II} ($\Delta\phi_{M^I/M^{II}}$), M^{II} and solution phase S ($\Delta\phi_{M^{II}/S}$), S and M^{III} ($\Delta\phi_{S/M^{II}}$), and M^{III} and $M^{I'}$ ($\Delta\phi_{M^{III}/M^{I'}}$). If the reading in the voltage meter is V (note V is the cell voltage), then V can be expressed as:

$$V = \Delta\phi_{M^I/M^{II}} + \Delta\phi_{M^{II}/S} + \Delta\phi_{S/M^{III}} + \Delta\phi_{M^{III}/M^{I'}} \tag{2.22}$$

Equation (2.22) suggests that it is impossible to measure $\Delta\phi_{M^{II}/S}$ in the presence of three uncertain interface potential drops unless they are fixed. Equation (2.22) can be simplified to:

$$V = \phi_{M^I} - \phi_{M^{I'}} \tag{2.23}$$

where both ϕ_{M^I} and $\phi_{M^{I'}}$ are the inner potentials of the two lead metals. According to Equation (2.1), Equation (2.23) can be alternatively expressed as:

$$V = \left(\psi_{M^I} + \chi_{M^I}\right) - \left(\psi_{M^{I'}} - \chi_{M^{I'}}\right) \tag{2.24}$$

For the same metal lead M^I, $\chi_{M^I} = \chi_{M^{I'}}$, Equation (2.24) can be converted into:

$$V = \psi_{M^I} - \psi_{M^{I'}} = \Delta\psi_{M^I/M^{II}} + \Delta\psi_{M^{II}/S} + \Delta\psi_{S/M^{III}} + \Delta\psi_{M^{III}/M^{I'}} \tag{2.25}$$

Comparing Equations (2.22) and (2.25), it can be seen that the sum difference in the inner potential differences of the two electrodes is equal to that of the outer potential differences.

In Equation (2.25), both $\Delta\psi_{M^I/M^{II}}$ and $\Delta\psi_{M^{III}/M^{I'}}$ are metal–metal interfaces, so their potential drops should be fixed as long as the types of metal are fixed. If the value of $\Delta\psi_{S/M^{III}}$ is fixed, the relative value of $\Delta\psi_{M^{II}/S}$ can be measured:

$$\Delta\psi_{M^{II}/S} = V - \left(\Delta\psi_{S/M^{III}} + \Delta\psi_{M^I/M^{II}} + \Delta\psi_{M^{III}/M^I}\right) = V - V_{ref} \tag{2.26}$$

where V is the cell voltage, and V_{ref} is the electrode potential of the reference electrode. Therefore, although the absolute potential drop across the interface is not measureable, its relative value can be measured. In electrochemistry, the relative potential drop is called the relative electrode potential or electrode potential. A commonly used reference is the normal hydrogen electrode (NHE) with a potential defined as zero V that does not change with temperature. This NHE contains a Pt or Pt black electrode immersed into a H^+ solution (1.0 *M*) bubbled with 1.0 atm of hydrogen gas.

Another commonly used reference electrode is the saturated Calomel electrode whose electrode potential is 0.245 V versus NHE at 25°C and 1.0 atm. The electrode reaction kinetics for these reference electrodes are fairly fast. Even with current passing through the electrode–electrolyte interface, the electrode potential will show an insignificant shift. This device is called a non-polarizable electrode and will be discussed in the following section. Any change in cell voltage will truly reflect the potential change of the targeted electrode–electrolyte interface ($\Delta\psi_{M^{II}/S}$):

$$d\left(\Delta\psi_{M^{II}/S}\right) = dV \tag{2.27}$$

Therefore, the differential capacitance of the double-layer can be rewritten as:

$$C_{dl} = \frac{dq}{d\left(\Delta\psi_{M^{II}/S}\right)} = \frac{dq}{dV} \tag{2.28}$$

Equation (2.28) suggests that the double-layer differential capacitance can be directly measured by measuring the electrode charge change when the relative electrode potential is altered.

In supercapacitor literature, the reported electrode potentials are all relative, as expressed in Figure 2.11 in which an NHE reference electrode was used. For the remainder of this book, all mentions of electrode potentials refer to relative electrode potentials.

2.3.2 Double-Layer Potential Ranges or Windows

In Figure 2.1, the net charge (q) can be forced to accumulate on the electrode to form an electrode–electrolyte double-layer using a power source such as a battery connected in the circuit loop. Consequently, the potential drop across the double-layer (V) forms and is expressed as Equation (2.29) if the concentration of the electrolyte is high enough to form the Helmholtz layer:

$$V = \frac{q}{C_{dl}} \tag{2.29}$$

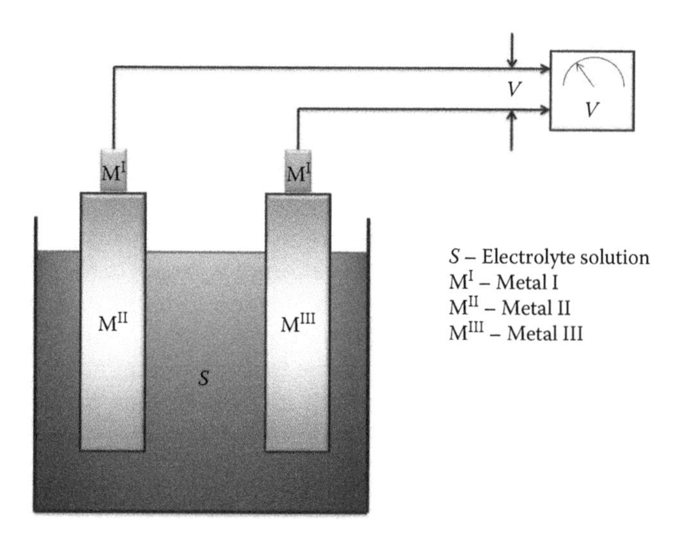

FIGURE 2.11
Electrochemical cell containing two electrode–electrolyte interfaces. M^I = metal lead I. M^{II} = metal electrode II. M^{III} = metal electrode III. $M^{I'}$ = metal lead I′. S = electrolyte solution.

Equation (2.29) seems to suggest that as long as enough net charge can be put on the electrode surface, the potential drop could be increased to the desired value. Unfortunately, it is not possible for V to go very high because some factors limit its increase. For example, if an electrode–electrolyte interface contains a graphite carbon electrode and 1.0 M NaI aqueous solution [carbon/1.0 M NaCl (aq) interface] and the electrode potential is positively moved from 0 V versus NHE to around 0.6 V versus NHE. The oxidation of two I^- ions occurs through a one electron transfer from I^- to the electrode to form I^o (two I^o could combine to form I_2), limiting the increase of electrode potential.

After the surface I^- is exhausted, the electrode potential could further increase until water oxidation produces O_2 at ~0.8 V versus NHE. In the potential range from 0 to 0.8 V, the surface oxidation of graphite may also happen when the potential is greater than 0.2 V to form surface groups that may also compress the potential increase. For the same interface, if the potential is negatively moved to about −0.6 V versus NHE, the water reduction would produce H_2, compressing further decrease of the electrode potential. If the potential could be further moved to −3.0 V, Na^+ reduction could occur. Therefore, only the electrode potential range from −0.6 to 0.2 V versus NHE is free of electrochemical reactions. This is called the double-layer range or window, and only in this potential range can the electrode be charged or discharged without interference from electrochemical reactions.

In electrochemistry, when no electrode reactions can occur within a fairly wide electrode potential range, the result is called an ideal polarizable electrode, completely polarizable electrode, or totally polarized electrode.

Consequently, the electrode behaves like a capacitor and only capacitive current (no faradic current) flows upon a change of potential in a certain electrode potential range. The electrode [carbon/1.0 *M* NaCl (aq) interface] mentioned above can behave as an ideal polarized electrode but only within the double-layer range from –0.6 to 0.2 V versus NHE.

Likewise, ideal non-polarizable electrodes are not polarizable. The potential of the ideal non-polarizable electrode will not change from its equilibrium potential even with the application of a large current density. The electrode reaction is extremely fast with an almost infinite exchange current density. For example, using an NHE electrode with a large Pt surface (Pt black) and an electrochemical current as high as several amperes per geometric square centimeter will not result in a significant change in its electrode potential. However, most electrode–electrolyte interfaces are between the ideal polarizable and non-polarizable electrodes.

In a practical application of a double-layer supercapacitor, the wider the double-layer, the higher the capacity for charge storage. However, the double-layer potential range or window is strongly dependent on the electrode material, the electrolyte, and the solvent used. The most practical electrode materials in supercapacitors are carbon-based and have almost ideal polarizable potential windows in an electrolyte solution. However, their surface reversible redox reactions to produce pseudocapacitance are actually beneficial by contributing to the capacity of charge storage.

It is important to choose an electrolyte with a wide electrochemically stable range. For a solvent, the selection seems difficult due to its intrinsic electrochemical stability. For example, for an aqueous solution, the electrochemical disassociation window of water is around 1.23 V at room temperature. If water is used as a supercapacitor electrolyte solvent, the maximum cell voltage will be around 1.23 V; if acetonitrile is the solvent, the electrode potential window is around 2.0 V; with an ion liquid, the electrode potential window can be as high as 4.0 V. Therefore, different solvents have different potential windows. Table 2.2 lists several common solvents and their potential windows for supercapacitors.

2.4 Capacitance of Porous Carbon Materials

The differential capacitance for a smooth electrode–electrolyte interface was discussed earlier in this chapter. The unit of differential capacitance is expressed as farads per square meter ($F.m^{-2}$) or microfarads per square meter ($\mu F.cm^{-2}$). Strictly speaking, this is called the capacitance density. If the capacitance density is C_{dl} in $F.m^{-2}$ and the electrode surface area is defined as A in square meters, the capacitance of the entire electrode is $C_{dl}A$ in F. If

TABLE 2.2

Typical Electrolyte Solvents and Their Potential Windows for Supercapacitors

Solvent	Electrolyte Salt	Temperature (°C)	Potential Range (V)[c]
Water	KOH, 4*M*	25	1
	H_2SO_4, 2*M*	25	1
	KCl[b], 2*M*	25	1
	Na_2SO_4, 1*M*	25	1
	K_2SO_4, 1*M*	25	1
Propylene carbonate	Et_4NBF_4, 1*M*	25	2.7
Acetonitrile	Et_4NBF_4, 1*M*	25	2.7
Ionic liquid[a]	$[EtMeIm]^+[BF_4]^-$	25	4
	$[EtMeIm]^+[BF_4]^-$	100	3.25

[a] Ionic liquids are molten salts that at higher temperatures display stability and significantly increase conductivity.
[b] Chloride ions are corrosive to metal current collectors during charging.
[c] Potential range can be shifted due to variations in electrode material stability.
Note: See References 9 through 12.

the capacitor is used for energy storage, the capacitance is one of the most important parameters for evaluating performance.

For example, when storing energy in a fixed size capacitor, the greater the capacitance, the greater the performance. However, when using a smooth electrode surface, the differential capacitance density is in the range of less than 1.0 $F.m^{-2}$, and needs 100 m^2 of the total electrode area to obtain 100 F of charge storage. This is not practical at all. Therefore, it is necessary to enlarge the electrode surface without increasing the device volume when a capacitor needs to store a lot of charge. Fortunately, with the rapid development of new materials, this surface area enlargement has become feasible for practical applications.

2.4.1 Carbon Particles and Their Associated Electrode Layers

Commercially available carbon particles such as active carbon powders are the most common active electrode materials for double-layer supercapacitors. They can have surface areas between 1000 and 3000 $m^2.g^{-1}$. When this kind of carbon material is used to make the electrode layers and no electrochemical reaction is employed for charge generation, the resulting devices are called double-layer supercapacitors.

Figure 2.12 shows an electrode layer in a double-layer supercapacitor. This electrode layer is composed of carbon particles and a binder. In the layer, the sources of the capacitance are the pores on the carbon particles and the porous channels within the matrix layer. It is obvious that only locations that are accessible by the electrolyte ions can form the electrode–electrolyte double-layer for capacitance generation.

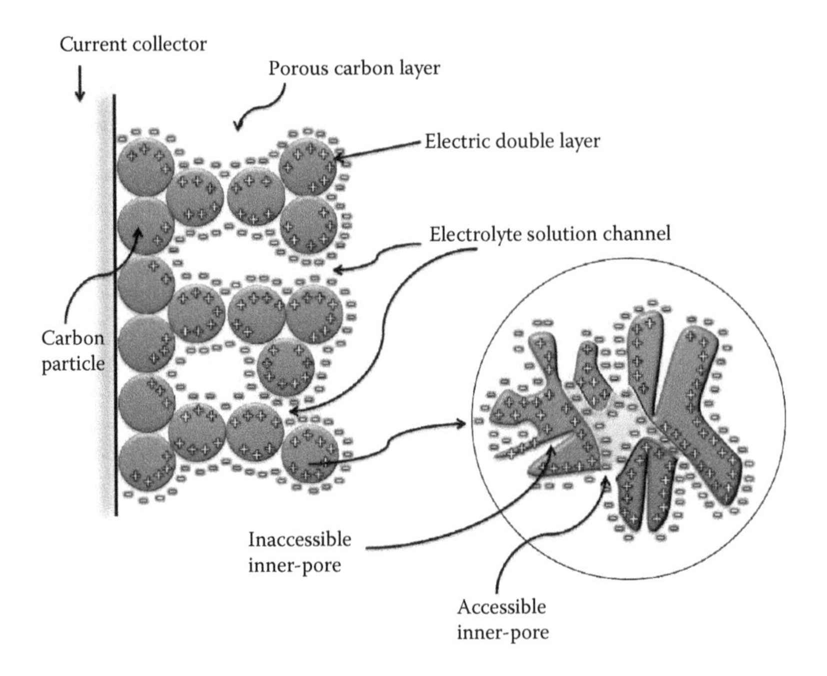

FIGURE 2.12
(See color insert.) Electrode for electrochemical double-layer supercapacitor.

Although places not accessible by the electrolyte ions may also give capacitance, the capacitance is significantly smaller than in places accessible by the electrolyte ions. Each carbon particle (inset in Figure 2.12) has two kinds of pores: (1) wide and shallow and (2) narrow and deep. In an ideal situation, both of kinds of pores can be accessed by the electrolyte to produce capacitance. However, if pores are too narrow and deep, the ions may not be able to penetrate them to contribute their capacitance. Therefore, it is important to match the size of the electrolyte ions with pore size to achieve high capacitance. This will be discussed more in a later section.

Based on Figure 2.12, particle size can also affect the exposed areas of carbons particle to the electrolyte solution. In general, the more porous the matrix layer, the larger the exposed area. Therefore, the carbon particle size should be optimized to yield the best porosity. However, if the porosity of the electrode matrix layer is too high, the electric conductivity of the matrix layer will be reduced, leading to high resistance of the electrode layer and lower power density of the supercapacitor. Therefore, there is a trade-off between the porosities and conductivities of the electrode materials.

2.4.2 Capacitances of Porous Carbon Materials and Their Associated Electrode Layers

Regarding the differential capacitance of such an electrode matrix layer, the Gouy–Chapman–Stern (GCS) double-layer modeling for capacitance is still applicable if the concentration of the electrolyte used is high enough to make the diffuse layer disappear. However, if a very dilute electrolyte solution is used, the situation will become more complicated due to the potential distribution within the electrolyte channels inside the porous layer.

In supercapacitors, the electrolyte concentration is normally high (>1.0 *M*), causing the diffuse layer to disappear and the Helmholtz layer to remain. This increases the capacitance of the electrode layer and allows more charge storage. Furthermore, if the material particle size is larger than the thickness of the Helmholtz layer (<1 nm), conclusions derived from the planar electrode may still be applicable to the situation of particles. For example, the carbon particle materials used in supercapacitors are normally 10 to 50 nm—much larger than the thickness of the Helmholtz layer of <1 nm.

To express the capacitance of such a high surface electrode material, a specific capacitance (C_{sp}, expressed in $F.g^{-1}$) is defined as:

$$C_{sp} = \frac{C_m}{m} \tag{2.30}$$

where C_m is the measured capacitance (F) using the electrode layer constructed from this material and m is the mass of the electrode material (g). Chapter 7 will give a detailed description of capacitance measurements using both cyclic voltammetry and cell charging–discharging techniques.

Theoretically, we should be able to calculate the specific capacitance of an electrode material according to its mass in the matrix layer, its differential capacitance density (C_{dl} in $F.m^{-2}$), and the total specific surface area of the carbon particles. In normal conditions, this surface area can be measured by the Brunauer-Emmett-Teller (BET) technique and expressed as S_{BET} in square meters per gram ($m^2.g^{-1}$):

$$C_{sp} = \frac{S_{BET} C_{dl}}{m} \tag{2.31}$$

From this equation, it seems that the larger the material surface area, the higher the specific capacitance. However, when the specific surface area (SSA) is larger than ~1200 $m^2.g^{-1}$, the specific capacitance will be saturated by further increasing the SSA [13]. At high SSAs, the pore walls become thinner, causing a space constriction for charge accumulation inside the pore walls. It is not accurate to estimate the specific capacitance of an electrode material based on Equation (2.31) due to the large scattering of differential capacitance values for carbon porous materials. The reason is the wide variety of carbon types, such as active carbon powders and fabrics, nanotubes, and aerogels.

Even a single carbon particle has a complex structure with several forms. For example, an active carbon particle (SPECTRACAB) contains randomly oriented single-layers, bilayers, and trilayers of small graphite sheets. These graphite sheets contain two orientations, the basal plane, and the edge plane [14]. Due to the different differential capacitances of the many forms of carbon, it is difficult to choose a differential capacitance value for the calculation. For carbon materials, the differential capacitance is between 0.05 and 1.0 $F.cm^{-2}$ and depends on the carbon form and electrolyte used.

Using Equation (2.31), it is expected that the calculated specific capacitance value based on BET surface area should be larger than that of the measured value. This occurs because a small portion of the entire carbon particle area is ineffective in the matrix layer. Thus the utilization of the particle area does not reach 100%. For example, when using active carbon (Carbon Black BP2000, Carbot Inc.) as the material to construct the electrode layer, the measured specific capacitance of this material is ~90 $F.g^{-1}$ [15]. This BP2000 carbon has a BET surface of ~1500 $m^2.g^{-1}$. If the differential capacitance for carbon material is assumed to be in the range of 0.1 $F.m^{-2}$, the calculated specific capacitance should be 150 $F.g^{-1}$, which is larger than the measured 90 $F.g^{-1}$. This suggests that the calculated value does not practically reflect the real situation.

Rather than calculating the specific capacitance of the electrode material, experimental measurements should be more practical in obtaining both the specific and differential capacitances using the electrode layer constructed from this material. In the experiments, the capacitance of the entire electrode layer can be measured (expressed as ε_0), then the differential capacitance can be obtained using the following equation:

$$C_{dl} = \frac{C_m}{S_{BET}} \tag{2.32}$$

This differential capacitance is called the apparent differential capacitance. Table 2.3 lists both the differential and specific capacitances for several carbon materials [16].

2.5 Electrochemical Double-Layer Supercapacitors

2.5.1 Structure and Capacitance

The structure of an electrochemical double-layer supercapacitor is similar to that of a battery: two electrodes comprised of a carbon material as shown

TABLE 2.3

Typical Differential and Specific Capacitances of Porous Carbon Materials

Carbon Material	Surface Area ($m^2.g^{-1}$)	Electrolyte	Differential Capacitance ($\times 10^{-4}$ $F.cm^{-2}$)	Specific Capacitance ($F.g^{-1}$)
Activated carbon	1200	10% NaCl	0.19	45
Carbon black	80 to 230	1*M* H_2SO_4	0.08	6.6 to 23.7
		31% KOH	0.10	
Carbon fiber cloth	1630	0.51*M* $EtNBF_4$ in propylene carbonate	0.069	30
Graphite	Solid	0.9 *N* NaF	Basal plane: 0.03 Edge plane: 0.5 to 0.7	60
Graphite powder	4	10% NaCl	0.35	
Carbon aerogel	650	4*M* KOH		
Glassy carbon	Solid	0.9 *N* NaF	0.13	
Carbon nanotube	400 to 600	$LiPF_6$/EC:DEC		12 to 120
Graphene	400 to 1500	2 *M* KCl		100 to 200

Note: See References 6, 17–23, and 24–31.

in Figure 2.1. Both porous electrodes are identical and are charged by an external power supply to hold opposite charges, one negative and the other positive. The electrode's active layers are made from carbon particles that are compacted together through a binder such as PTFE and two pressed current collectors. A separator between the two electrodes is made of a porous electrically insulated material used to prevent contact and short circuiting and also provide pathways for electrolyte ions. The electrode layers and the porous separator are filled with an electrolyte solution.

A double-layer is established at each side of the electrode and contains the Helmholtz and diffusion layers along the carbon particle–electrolyte solution interface. The capacitance is expressed in Equation (2.4). The overall capacitance of a supercapacitor can be treated as two differential capacitances connected in series. If the capacitance for the positive electrode can be expressed as

$$C_{dl,p}\left(=\frac{C_{H,p}C_{diff,p}}{C_{H,p}+C_{diff,p}}\right),$$

and the negative electrode as

$$C_{dl,n}\left(=\frac{C_{H,n}C_{diff,n}}{C_{H,n}+C_{diff,n}}\right),$$

the total capacitance of the supercapacitor (C_{dl}^{T}) is:

$$C_{dl}^{T}=\frac{C_{dl,P}C_{dl,n}}{C_{dl,P}+C_{dl,n}} \tag{2.33}$$

For electrochemical double-layer supercapacitors, the two electrodes are identical, meaning that $C_{dl,P} = C_{dl,n}$. Therefore, Equation (2.33) can be converted into:

$$C_{dl}^{T}=\frac{1}{2}C_{dl,P}=\frac{1}{2}C_{dl,n} \tag{2.34}$$

The total capacitance of a double-layer supercapacitor is equal to half of the individual electrode capacitance. Note that Equation (2.34) is applicable only to symmetric supercapacitors whose two electrodes are identical. It is not applicable to asymmetric supercapacitors in which the two electrodes are not identical. However, Equation (2.33) is applicable for both symmetric and asymmetric supercapacitors. In an asymmetric supercapacitor, the electrode with the smaller capacitance will dominate the total capacitance.

2.5.2 Equivalent Series Resistance (ESR)

As discussed in Chapter 1, if a sinusoidal alternative current is applied on an ideal capacitor, the output voltage should be out of phase by 90°, independent of the frequency. However, in a supercapacitor, the output voltage is normally out of phase fewer than 90°, suggesting that an equivalent series ohmic resistor is coupled. This ohmic component is defined as the equivalent series resistance (ESR).

In electrochemical supercapacitors, the ESR is a real series resistance that involves: the contact resistance between the current collector and the electrode layer, the resistance of the electrode layer interparticles due to the porous and particulate nature of the electrode matrix, the resistance of the external lead contact, the resistance of the electrolyte, and the resistance caused by the dielectric loss of the interphasal solvent and ions when the AC frequency is higher than hundreds of megahertz (MHz). Figure 2.13 displays a simple equivalent circuit of a supercapacitor in the presence of ESR.

As described in Chapter 1, the complex AC impedance of this equivalent circuit (Z_{cell}^{j}) in Figure 2.13, which is the complex AC impedance of the

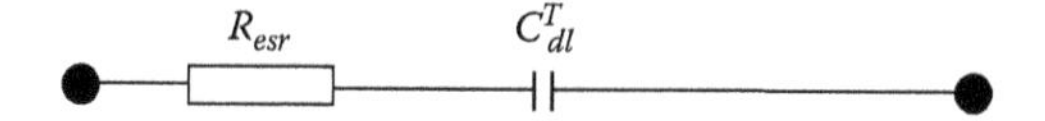

FIGURE 2.13
Equivalent circuit of supercapacitor in presence of equivalent series resistance.

supercapacitor cell, can be expressed as Equation (2.35) if the equivalent series resistance is expressed as R_{esr}:

$$Z_{cell}^{j} = R_{esr} - j\frac{1}{2\pi f C_{dl}^{T}} \tag{2.35}$$

where f is the AC frequency in hertz. The magnitude of this Z_{cell}^{j} can be expressed as:

$$\left|Z_{cell}^{j}\right| = \sqrt{\left(R_{esr}\right)^{2} + \left(\frac{1}{2\pi f C_{dl}^{T}}\right)^{2}} \tag{2.35a}$$

Using AC impedance spectroscopy at the high end of AC frequency, the corresponding impedance would be R_{esr} because the second term on the right side of Equation (2.35a) will be much smaller than the first term. At the low end of the AC frequency, this second term will become much larger than the first one and the obtained AC impedance is dominated by this second term from which C_{dl}^{T} can be obtained. This allows R_{esr} and C_{dl}^{T} to be obtained separately. Another way to measure R_{esr} is to use the charging–discharging curve if its magnitude is large enough. This will be discussed in a later section.

ESR is an important parameter in evaluating a supercapacitor's performance, in particular its power density, because the ESR restricts the rates at which the capacitance can be charged or discharged upon application of a given current or voltage.

2.5.2.1 Thermal Degradation from ESR

The voltage drop created by cell resistance affects both the charge and discharge capacities of a cell. The non-ideal loss limits the effective region for usable charge storage and thereby limits the charge capacity of the cell. The lost charge is primarily dissipated as heat. The non-ideal resistive power losses can generate an unacceptable amount of heat very quickly. Even commercial cells designed to have low resistance, for example, a Maxwell K2 cell (Table 2.4) can accumulate heat very quickly.

If the cell based on the specifications in the table was operated at the maximum usable power of 1 kW, the current passing through the device would

TABLE 2.4
Maxwell K2 Cell Specifications

Specification Detail	Value
Cell capacitance	650 F
ESR	0.8 mΩ
Operating voltage	2.7 V
Maximum usable specific power	6.8 KW/kg
Cell weight	160 g/device

Note: See Reference 32.

be around 400 A (I = P/V) and it would generate a significant amount of heat (130W or 0.81kW/kg; $P_{lost} = R_{esr}I^2$) due to the small amount of resistance.

Such a large amount of heat loss within a small volume can cause rapid degradation to performance, damage electrical components, and cause the electrolyte to swell and melt casing materials if the heat is not safely channeled away from the device. The capacitors can handle this sort of high current for a short burst before failure. However, for normal operation, the damage caused by heat limits the practical maximum current of the K2 supercapacitor to 88 A or only 1.49 kW/kg despite the low ESR present within the device [32]. This illustrates the significance of non-ideality through power loss and shows how active materials with higher resistance effectively limit device performance.

2.5.3 Leakage Resistance

In an ideal double-layer supercapacitor, no charges are considered to cross the double-layer interface when the electrode potential is charged in a certain range. The current density passing through the supercapacitor (i_{dl}) (i_{cell}) is the charging or discharging current density of the double-layer. However, there exists a leakage current density (i_{lk}) caused by several undesired processes and also a faradic leakage current density (i_F) when the electrode potential is expanded beyond the electrochemical decomposition limits of the electrolyte or solvent. This causes faradic reactions to occur, leading to charge transfer across the double-layer. The total current density used to charge the supercapacitor cell (i_{cell}) will become:

$$i_{cell}\,(charging) = i_{dl} + i_{lk} + i_{iF} \text{ or } i_{cell}\,(dicharging) = i_{dl} - i_{lk} - i_{iF} \tag{2.36}$$

Equation (2.36) indicates that due to the leakage current and faradic leakage current, the current used to charge a supercapacitor is larger than expected and the current obtained from it is less than expected. Note that these two leakage current densities cause the self discharging of a supercapacitor, which is not desirable for practical applications. For a detailed discussion please see Reference 33. This faradic leakage current density is an

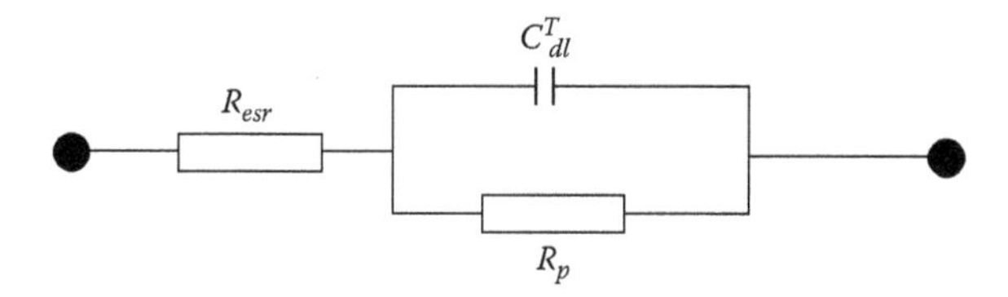

FIGURE 2.14
Equivalent circuit of supercapacitor taking account of both ESR and leakage resistance.

electrochemical reaction current density, so it is strongly dependent on the electrode potential and can be expressed as the combination between the kinetic current density (i_k) and the diffusion limiting current density (i_d):

$$i_F = \frac{i_d i_k}{i_k + i_d} \tag{2.37}$$

For electrolyte and solvent decomposition, i_k may be much larger than i_d due to high concentrations. The faradic leakage current density is mainly dominated by i_k. Regarding the leakage current density, a leakage resistance or leakage parallel resistance can be defined as R_p. When considering the leakage resistance with the equivalent series resistance, the circuit in Figure 2.14 is proposed. Note that the magnitude of R_p is dependent on the electrode potential, while R_{esr} is constant with respect to the potential change.

Furthermore, a R_p value normally has a high magnitude and is much larger than R_{esr} because its effect on supercapacitor charging and discharging is insignificant unless done at a slow rate. In addition, other kinds of non-faradic processes can also cause the self discharging of a supercapacitor, such as non-uniformity of charge acceptance along the surface of pores and possible short circuiting of the anode and cathode from improperly sealed bipolar electrodes.

2.5.3.1 Self Discharge through Leakage Mechanisms

Instabilities in ECs drive charge loss when energy is stored for longer periods. Diffusion of charge and restructuring of ions in pores (charge imbalances) can both lead to loss of charge while a device has no external connections to its terminals [34]. The magnitude of self discharge or internal leakage current is an important indicator of the quality of an EC. The leakage current can be modeled as a resistance in parallel with the capacitor and the different components can be observed by the AC impedance test. However, models of leakage resistance can overly simplify the complex voltage and time-dependent components that contribute to the leakage current behavior seen in real devices.

Alternatively, leakage behavior can be determined by measuring the self-discharge voltage by: (1) charging the device by applying a slow voltage ramp

TABLE 2.5

Overview of Self-Discharge Mechanisms

Controlling Discharge Mechanism	Discharge Type	Linear Graphical Relationship
Activation	Faradic	V versus *logt*
Charge redistribution		
Impurity causing redox reaction		
Diffusion	Faradic	V versus $\sqrt{t}$
Fibril conduction through separator		
Shunt resistances between electrodes	Ohmic	*logV* versus *t*

Note: See References 35 and 36.

(1 to 50 $mV.s^{-1}$), (2) optionally holding voltage to establish steady state, or (3) switching the device to open circuit and monitor voltage over time. Variations in discharge based on hold time and charge time allow the determination of any steady state effects and charge rate effects, providing additional insight into the leakage current mechanisms affecting the device. One example is convolution with time-dependent charge redistribution that occurs within non-uniformly charged pores when an open circuit is initiated. The result allows investigation into discharge trends and dominant mechanisms over long periods. Conway's book provides a thorough discussion of theoretical discharge mechanisms [35]. Experimentally these mechanisms present themselves in the discharge plot shown in Table 2.5.

These faradic leakage measurements are also dependent upon temperature ($\log I$ *vs.* $\frac{1}{t}$) where the slope is determined by the activation energy of the mechanism. As a result faradic leakage currents can have important impacts on high temperature performance.

Alternatively, in industry, leakage current can be measured at constant charge by applying a DC voltage and measuring current required to maintain a full charge. The leakage current decreases quickly with time and stabilizes after a few days. The resulting low leakage current ($\mu A.F^{-1}$) can serve as a baseline for leakage comparison of devices. The low reported value can make the leakage rate of a capacitor seem lower than it is during practical use.

Andreas et al. used modeling and analysis of self-discharge data to determine the discharge mechanism for high surface area carbon in an acidic electrolyte. By varying hold time between 0 and 75 hr they illustrated that charge redistribution has a significant effect on self discharge of a positive electrode, but not on the negative where activation mechanisms are dominant [36]. It is suggested that the small, highly mobile H^+ ions that balance the positive electrode are not rate limiting, while the larger HSO_4^- ions lead to migration limitations on the charge redistribution, making it more dominant. This extremely long charge redistribution time is in opposition to the view based on porous transmission line models that suggests redistribution is complete after only 100 to 600 sec. Further work by Andreas et al. used self discharge to verify that water electrolysis is not a dominant discharge mechanism on either electrode [37]. Their work shows that the discharge is controlled by a combination of activation and charge redistribution mechanisms.

2.5.4 Supercapacitor Charging and Discharging

For a detailed discussion of supercapacitor charging and discharging, see Reference 33. When charging and discharging a supercapacitor, the two options are: (1) charge or discharge at a constant cell voltage to record the cell current changes over time, (2) charge or discharge at a constant current to record cell voltage changes over time, and (3) discharge at constant power, varying current as voltage decreases. The following discussion will use the circuit shown in Figure 2.15 to derive the charging and discharging behaviors of a supercapacitor. It is important to note that a mathematical treatment of the potential dependent leakage resistance (R_p) in Figure 2.15 is difficult and complicated due to its uncertain expression of potential dependence. However, it is important to understand that R_p changes with electrode potential [33].

2.5.4.1 *Charging at Constant Cell Voltage*

As shown in Figure 2.15a, the constant charging voltage is E. When the switch is closed at $t = 0$ (t is the charging time), the charging process is started, assuming that before charging started, the supercapacitor was at a zero charge state, that is, the voltage across the supercapacitor was equal to zero. When $t \geq 0$, the overall supercapacitor cell charging current (i_{cell})can be expressed using

$$i_{cell} = \frac{E}{R_{esr} + R_p} + \frac{R_p E}{R_{esr}\left(R_{esr} + R_p\right)} \exp\left(-\frac{R_{esr} + R_p}{R_{esr} R_p C_{dl}^T} t\right) \quad \text{(Charging process)} \tag{2.38}$$

The voltage drop across the supercapacitor can be expressed as:

$$V_{sc} = \frac{R_p E}{R_{esr} + R_p}\left(1 - \exp\left(-\frac{R_{esr} + R_p}{R_{esr} R_p C_{dl}^T} t\right)\right) \tag{2.39}$$

Equation (2.38) shows that at $t = 0$,

$$i_{cell} = \frac{E}{R_{esr}},$$

which is the maximum charging current. When $t = \infty$ (the moment when the supercapacitor is fully charged),

$$i_{cell} = \frac{E}{R_{esr} + R_p}.$$

This suggests that even if the supercapacitor is fully charged, extra current density is still needed to overcome the self discharging caused by R_p. However, if the leakage reaction does not occur, $R_p \to \infty$, i_{cell} can be simplified to:

$$i_{cell} = \frac{E}{R_{esr}} \exp\left(-\frac{t}{R_{esr} C_{dl}^T}\right) \quad \text{(Charging process)} \tag{2.40}$$

In this case, the charging current will approach zero when $t \to \infty$, that is, $i_{cell} \to 0$, suggesting that no extra current density is needed to overcome self discharging. In practice, R_p is normally much larger than R_{esr}, that is, $R_p >> R_{esr}$, so Equation (2.38) can be simplified to Equation (2.40).

According to Equation (2.38), when charging an ideal supercapacitor ($R_{esr} = 0$, and $R_p = \infty$) using a constant voltage, the current will be infinite as long as voltage is applied to the supercapacitor. Even for the case where $R_{esr} \to 0$ and $R_p \neq \infty$, the supercapacitor voltage will go to E immediately after the constant voltage E is applied, as seen using Equation (2.39).

2.5.4.2 Charging at Constant Cell Current

Figure 2.15b shows supercapacitor charging at a constant current density. The constant charging current density is I_{cell}, and the cell voltage is V_{cell}. Assume that before the charging starts, the supercapacitor is at zero charge state. When the switcher is closed at $t = 0$, the charging process is started. When $t \geq 0$, the supercapacitor charging voltage can be expressed as:

$$V_{cell} = I_{cell} R_{esr} + I_{cell} R_p \left(1 - \exp\left(-\frac{t}{R_p C_{dl}^T}\right)\right) \quad \text{(Charging process)} \tag{2.41}$$

Equation (2.41) shows that when $t = 0$, $V_{cell} = I_{cell} R_{esr}$; when $t = \infty$ (the moment when the supercapacitor is fully charged), $V_{cell} = I_{cell} R_{esr} + I_{cell} R_p$. In this case, the voltage across the supercapacitor (V_{sc}) reaches its maximum value, defined as V_{sc}^o, which is equal to $I_{cell} R_p$. If there is no leakage current, $R_p \to \infty$ and Equation (2.41) can be simplified to:

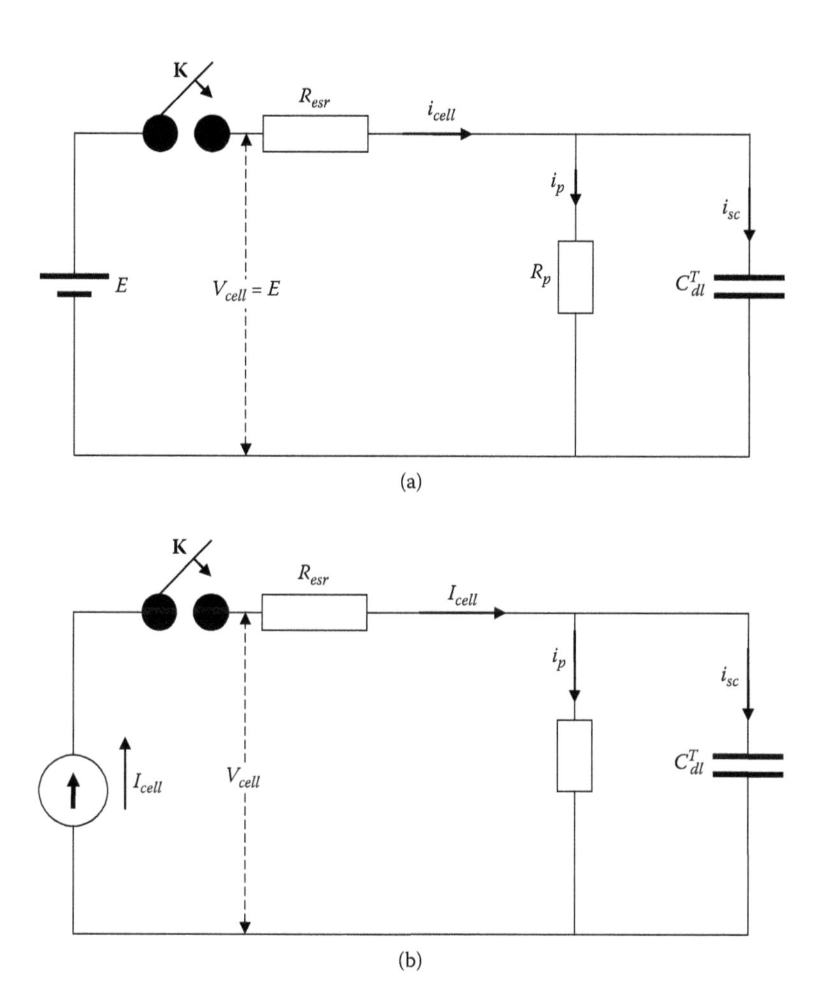

FIGURE 2.15
Equivalent charging circuits of supercapacitor at constant voltage (a) and at constant current (b).

$$V_{cell} = I_{cell}R_{esr} + I_{cell}\frac{t}{C_{dl}} \quad \text{(Charging process)} \tag{2.42}$$

Using Equation (2.42), the time (t_c) required to charge a supercapacitor to a targeted voltage level of V_T can be obtained when

$$V_{cell} = V_T : t_c = C_{dl}^T\left(\frac{V_T - I_{cell}R_{esr}}{I_{cell}}\right).$$

Equation (2.42) suggests that the cell voltage has a linear relationship with charging time. Note that Equation (2.42) can be used to describe the circuit shown in Figure 2.13. If the ESR does not exist ($R_{esr} \to 0$), the cell voltage will become:

$$V_{cell} = I_{cell} R_p \left(1 - \exp\left(-\frac{t}{R_p C_{dl}^T}\right)\right) \tag{2.43}$$

If charging an ideal supercapacitor ($R_{esr} = 0$, and $R_p = \infty$) using a constant current, the cell voltage will become:

$$V_{cell} = I_{cell} \frac{t}{C_{dl}^T} \tag{2.44}$$

2.5.4.3 Discharging Supercapacitor Cell at Constant Resistance

If a supercapacitor's cell is discharged using a load resistance (R_L) after being fully charged, the equivalent circuit is treated as in Figure 2.16a, where V_{sc}^o is the supercapacitor's initial voltage used for discharging. When the switch is closed at $t = 0$, the supercapacitor's voltage (V_{sc}) and current density (i_{sc}) at $t \geq 0$ are expressed by Equations (2.45) and (2.46), respectively:

$$V_{sc} = V_{sc}^o \exp\left(-\frac{R_p + R_{esr} + R_L}{R_p\left(R_{esr} + R_L\right) C_{dl}^T} t\right) \tag{2.45}$$

$$I_{sc} = \frac{V_{sc}^o \left(R_p + R_{esr} + R_L\right)}{R_p\left(R_{esr} + R_L\right)} \exp\left(-\frac{R_p + R_{esr} + R_L}{R_p\left(R_{esr} + R_L\right) C_{dl}^T} t\right) \tag{2.46}$$

The cell current passing through R_L, and the cell voltage, which is the voltage across the load resistance, can be expressed as Equations (2.47) and (2.48), respectively:

$$i_{cell} = \frac{V_{sc}}{R_{esr} + R_L} = \frac{V_{sc}^o}{R_{esr} + R_L} \exp\left(-\frac{R_p + R_{esr} + R_L}{R_p\left(R_{esr} + R_L\right) C_{dl}^T} t\right) \quad \text{(Discharging process)} \tag{2.47}$$

$$V_{cell} = V_{sc} - i_{cell}R_{esr} = V_{sc}^{o}\left(1 - \frac{R_{esr}}{R_{esr} + R_L}\right)\exp\left(-\frac{R_p + R_{esr} + R_L}{R_p\left(R_{esr} + R_L\right)C_{dl}^{T}}t\right) \quad (2.48)$$

(Discharging process)

Equations (2.47) and (2.48) indicate that when $R_L = 0$ (without load), i_{cell} will equal the current that passing through R_{esr} and V_{cell} will equal to zero. Note that when $R_p \rightarrow \infty$ (the case without faradic leakage current, as described by the circuit in Figure 2.13), Equations (2.47) and (2.48) can be simplified to:

$$i_{cell} = \frac{V_{sc}^{o}}{R_{esr} + R_L}\exp\left(-\frac{t}{\left(R_{esr} + R_L\right)C_{dl}^{T}}\right) \quad \text{(Discharging process)} \quad (2.49)$$

$$V_{cell} = V_{sc}^{o}\left(1 - \frac{R_{esr}}{R_{esr} + R_L}\right)\exp\left(-\frac{t}{\left(R_{esr} + R_L\right)C_{dl}^{T}}\right) \quad \text{(Discharging process)} \quad (2.50)$$

2.5.4.4 Discharging Supercapacitor Cell at Constant Voltage

When a supercapacitor cell is discharged using a process with a constant voltage (E) that is smaller than the cell voltage (V_{sc}^{o}), the equivalent circuit can be treated as shown in Figure 2.16b. When the switch is closed at $t = 0$, the supercapacitor's voltage (V_{sc}) and current density (i_{sc}) at $t \geq 0$ are expressed by Equations (2.51) and (2.52), respectively:

$$V_{sc} = V_{sc}^{o} - \frac{V_{sc}^{o}\left(R_p + R_{esr}\right) - ER_p}{R_{esr} + R_p}\left[1 - \exp\left(-\frac{R_{esr} + R_p}{R_{esr}R_pC_{dl}^{T}}t\right)\right] \quad (2.51)$$

$$i_{sc} = \frac{V_{sc}^{o}\left(R_p + R_{esr}\right) - ER_p}{R_{esr} + R_p}\exp\left(-\frac{R_{esr} + R_p}{R_{esr}R_pC_{dl}^{T}}t\right) \quad (2.52)$$

By combining Equations (2.50) and (2.51), the cell voltage (V_{cell}) and cell current density (i_{cell}) can be obtained:

$$V_{cell} = E \quad \text{(Discharging process)} \quad (2.53)$$

$$i_{cell} = -\frac{E}{R_p + R_{esr}} + \left(\frac{V_{sc}^o}{R_{esr}} - \frac{ER_p}{R_{esr}\left(R_{esr} + R_p\right)}\right)\exp\left(-\frac{R_{esr} + R_p}{R_{esr}R_pC_{dl}^T}t\right) \tag{2.54}$$

(Discharging process)

When Rp → ∞, Equation (2.54) can be converted into:

$$i_{cell} = \left(\frac{V_{sc}^o - E}{R_{esr}}\right)\exp\left(-\frac{R_{esr}}{R_{esr}C_{dl}^T}t\right) \quad \text{(Discharging process)} \tag{2.55}$$

2.5.4.5 Discharging Supercapacitor Cell at Constant Current

When discharging a fully charged supercapacitor using a constant current (I_{cell}), the equivalent circuit can be treated as shown in Figure 2.16c. When the switcher is closed at t = 0, the supercapacitor's voltage (V_{sc}) and current density (i_{sc}) at $t \geq 0$ are expressed by Equations (2.56) and (2.57), respectively:

$$V_{sc} = V_{sc}^o - \left(V_{sc}^o + I_{cell}R_p\right)\left[1 - \exp\left(-\frac{t}{R_pC_{dl}^T}\right)\right] \tag{2.56}$$

$$i_{sc} = \frac{\left(V_{sc}^o + I_{cell}R_p\right)}{R_p}\exp\left(-\frac{t}{R_pC_{dl}^T}\right) \tag{2.57}$$

The cell voltage (V_{cell}) and cell current density can be expressed by Equations (2.58) and (2.59), respectively:

$$V_{cell} = -I_{cell}R_{esr} + V_{sc}^o - \left(V_{sc}^o + I_{cell}R_p\right)\left[1 - \exp\left(-\frac{t}{R_pC_{dl}^T}\right)\right] \quad \text{(Discharging process)} \tag{2.58}$$

$$i_{cell} = -I_{cell} \quad \text{(Discharging process)} \tag{2.59}$$

When $R_p \to \infty$, Equation (2.58) can be simplified into:

$$V_{cell} = -I_{cell}R_{esr} + V_{sc}^o - I_{cell}\frac{t}{C_{dl}^T} \tag{2.58a}$$

The time (t_{fd}) required for the supercapacitor to be fully discharged ($V_{cell} = 0$) is deduced from Equation (2.58a):

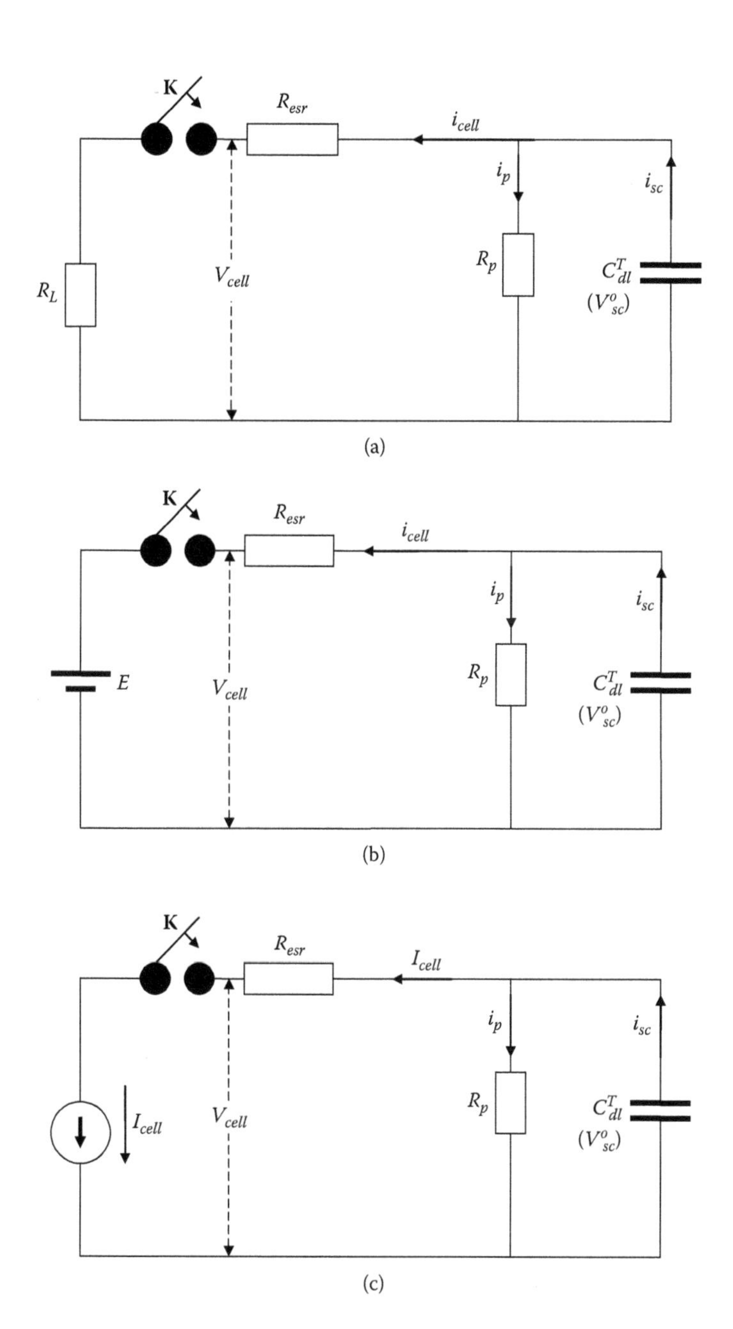

FIGURE 2.16
Equivalent discharging circuits of supercapacitor at constant resistance (a), constant voltage (b), and constant current (c).

$$t_{fd} = C_{dl}^{T} \frac{\left(V_{sc}^{0} - I_{cell} R_{esr}\right)}{I_{cell}}.$$

Equations (2.58) and (2.58a) demonstrate that if the discharging time is longer than t_{fd}, negative cell voltage will result and this is impossible. Therefore, both equations are valid only at $t \leq t_{fd}$.

2.5.4.6 Charging and Discharging Curves at Constant Current

Based on Equation (2.41) for constant charging and Equation (2.58) for constant discharging, the charging and discharging curves can be drawn as shown in Figure 2.17. Note that there are four variable parameters in Equations (2.41) and (2.58): I_{cell}, R_{esr}, R_p and C_{dl}^{T}. In each of the Figure 2.17 charts, one parameter is changed while the other three remain at fixed values.

In Figure 2.17a, R_{esr} is changed and the voltage difference between the charging curves corresponds to the value of $I_{cell}\ R_{esr}$, that is, $\Delta V_{cell} = I_{cell}\ R_{esr}$, from which R_{esr} can be obtained. In Figure 2.17b, the charging and discharging curves at low R_p are both drifted from linearity. This observation suggests that when charging or discharging a supercapacitor at a constant current, the faradic leakage current is large enough to make an impact if the voltage–time curve is not linear. In Figure 2.17b, when R_p is changed from $R_{p,1}$ to $R_{p,2}$, the voltage difference between the two charging curves at any time can be expressed as:

$$\Delta V_{cell} = I_{cell} \left[R_{p,2} \left(1 - \exp\left(-\frac{t}{R_{p,2} C_{dl}^{T}} \right) \right) - R_{p,1} \left(1 - \exp\left(-\frac{t}{R_{p,1} C_{dl}^{T}} \right) \right) \right] \tag{2.60}$$

In Figure 2.17c, I_{cell} is changed from $I_{cell,1}$ to $I_{cell,2}$, and the voltage difference at any time can be expressed as:

$$\Delta V_{cell} = \left(I_{cell,2} - I_{cell,1}\right) \left[R_{esr} + R_p \left(1 - \exp\left(-\frac{t}{R_p C_{dl}^{T}} \right) \right) \right] \tag{2.61}$$

In Figure 2.17d, C_{dl}^{T} is changed from $C_{dl,1}^{T}$ to $C_{dl,2}^{T}$, and the voltage difference at any time can be expressed as:

$$\Delta V_{cell} = I_{cell} R_p \left[\exp\left(-\frac{t}{R_p C_{dl,1}^{T}} \right) - \exp\left(-\frac{t}{R_p C_{dl,2}^{T}} \right) \right] \tag{2.62}$$

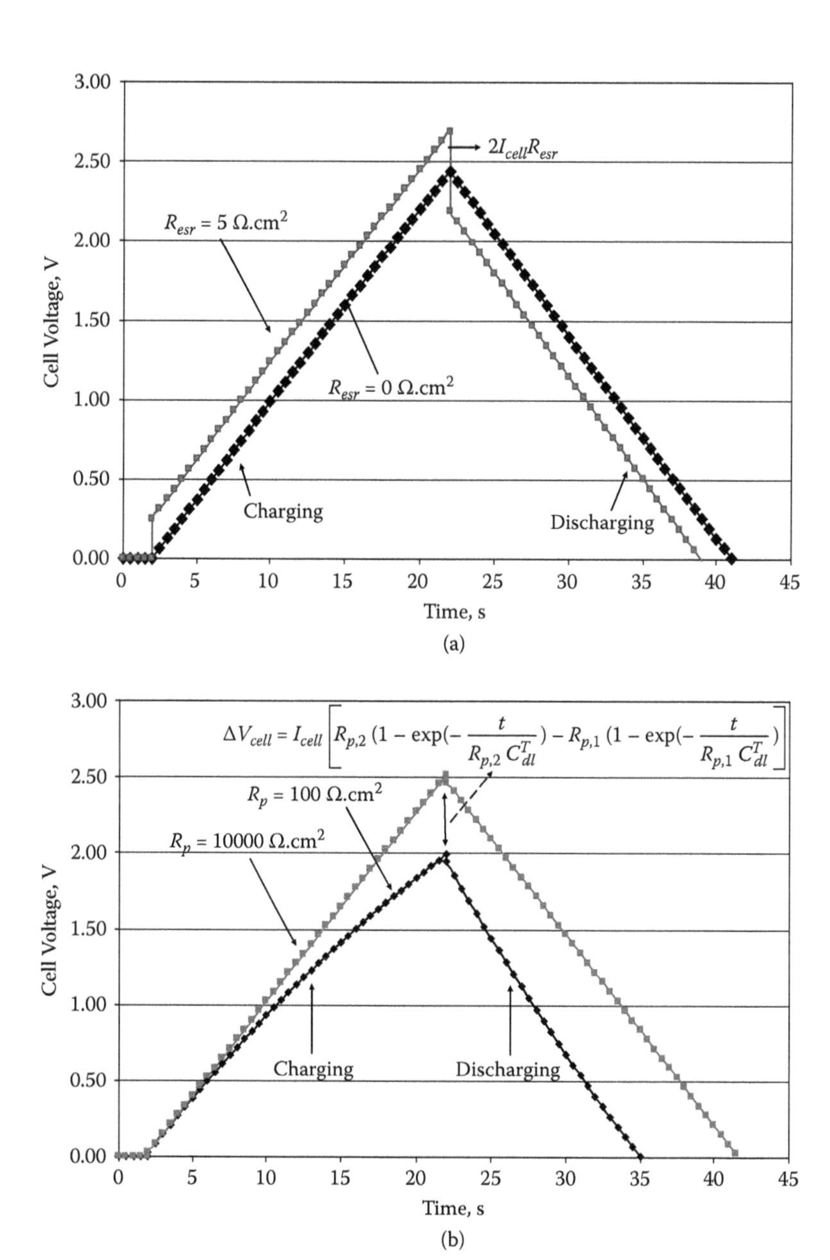

FIGURE 2.17
Calculated charging-discharging curves. (a) R_{esr} is changed from 0.4 to 1.0 Ω.cm² with fixed R_p = 1000 Ω.cm², C^T_{dl} = 0.4 F.cm⁻², and I_{cell} = 0.05 A.cm⁻². (b) R_p is changed from 100 to 10000 Ω.cm² with fixed R_{esr} = 0.5 Ω.cm², C^T_{dl} = 0.4 F.cm⁻², and I_{cell} = 0.05 A.cm⁻². (c) I_{cell} is changed from 0.05 to 0.1 A.cm⁻² with fixed R_p = 100 Ω.cm², C^T_{dl} = 0.4 F.cm⁻², and R_{esr} = 0.5 Ω.cm². (d) C^T_{dl} is changed from 0.4 to 1.0 F.cm⁻² with fixed R_p = 10000 Ω.cm², I_{cell} = 0.05 A.cm⁻², and R_{esr} = 0.05 Ω.cm². (continued)

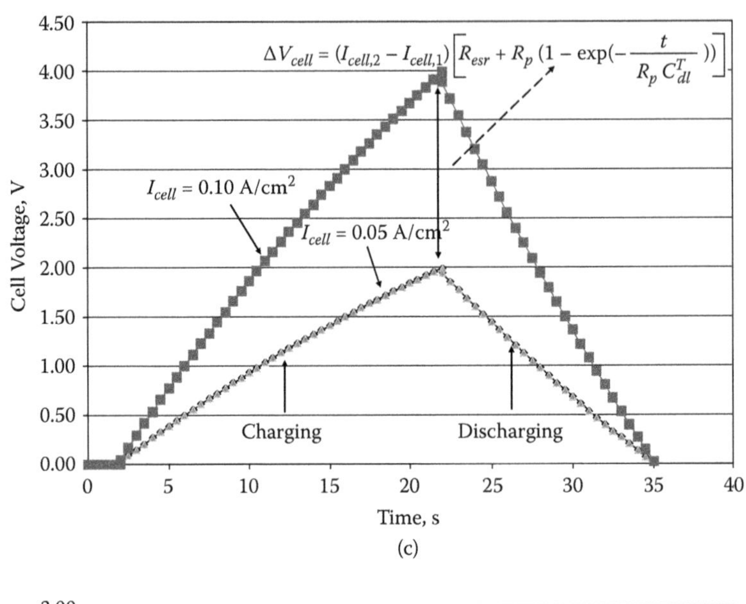

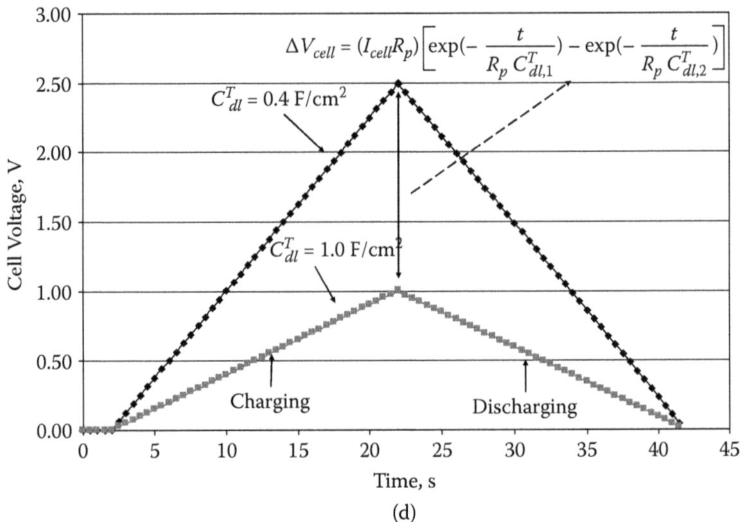

FIGURE 2.17 (CONTINUED)
Calculated charging-discharging curves. (a) R_{esr} is changed from 0.4 to 1.0 Ω.cm² with fixed R_p = 1000 Ω.cm², C^T_{dl} = 0.4 F.cm⁻², and I_{cell} = 0.05 A.cm⁻². (b) R_p is changed from 100 to 10000 Ω.cm² with fixed R_{esr} = 0.5 Ω.cm², C^T_{dl} = 0.4 F.cm⁻², and I_{cell} = 0.05 A.cm⁻². (c) I_{cell} is changed from 0.05 to 0.1 A.cm⁻² with fixed R_p = 100 Ω.cm², C^T_{dl} = 0.4 F.cm⁻², and R_{esr} = 0.5 Ω.cm². (d) C^T_{dl} is changed from 0.4 to 1.0 F.cm⁻² with fixed R_p = 10000 Ω.cm², I_{cell} = 0.05 A.cm⁻², and R_{esr} = 0.05 Ω.cm².

2.5.4.7 AC Impedance Equivalent Circuit

The complex impedance of the equivalent circuit (Z_{cell}^{j}) shown in Figure 2.14 can be written as:

$$Z_{cell}^{j} = R_{esr} + \frac{R_p}{1+\left(2\pi f R_p C_{dl}^{T}\right)^2} - j\frac{2\pi f \left(R_p\right)^2 C_{dl}^{T}}{1+\left(2\pi f R_p C_{dl}^{T}\right)^2} \tag{2.63}$$

The magnitude of this complex impedance can be expressed as:

$$\left|Z_{cell}\right| = \sqrt{\left(R_{esr} + \frac{R_p}{1+\left(2\pi f R_p C_{dl}^{T}\right)^2}\right)^2 + \left(\frac{2\pi f \left(R_p\right)^2 C_{dl}^{T}}{1+\left(2\pi f R_p C_{dl}^{T}\right)^2}\right)^2} \tag{2.63a}$$

Using AC impedance spectroscopy, R_{esr}, R_p, and C_{dl}^{T} can be obtained simultaneously from the complex plane (Nyquist) plot. Chapter 7 will provide a more detailed discussion of supercapacitor measurements including the AC impedance spectroscopic method.

2.6 Energy and Power Densities of Electrochemical Supercapacitors

2.6.1 Energy Densities

Energy density is one of the most important parameters for evaluating an electrochemical supercapacitor. In a double-layer supercapacitor, the energy density can be expressed as:

$$E = \int_0^q V_{sc}\,dq = \int_0^q \frac{q}{C_{dl}^{T}}\,dq = \frac{1}{2}\frac{q^2}{C_{dl}^{T}} = \frac{1}{2}\frac{\left(C_{dl}^{T} V_{sc}\right)^2}{C_{dl}^{T}} = \frac{1}{2} C_{dl}^{T} V_{sc}^2 \tag{2.64}$$

where q is the total charge quantity stored in the supercapacitor (C.cm^{-2}) and the double-layer capacitance of the cell (C_{dl}^{T}) is expressed as F.cm^{-2}. In practical applications, the specific energy density is more popular and useful and is defined as:

$$E_m = \frac{1}{2}\frac{C_m}{m} V_{sc}^2 = \frac{1}{2} C_{sp} V_{sc}^2 \tag{2.65}$$

$$E_m = \frac{1}{2}\frac{C_m}{M}V_{sc}^2 \tag{2.66}$$

In Equation (2.65), E_m is the specific energy density of the active material in an electrode layer in Wh.kg^{-1}. In Equation (2.66), E_m is the specific energy density of the supercapacitor device and M is its mass. Note that when a supercapacitor is fully charged, it will reach a maximum voltage (V_{sc}^o). Therefore, its maximum specific energy densities can be expressed as Equations (2.65a) and (2.66a), respectively:

$$\left(E_m\right)_{\max} = \frac{1}{2}\frac{C_m}{m}\left(V_{sc}^o\right)^2 = \frac{1}{2}C_{sp}\left(V_{sc}^o\right)^2 \tag{2.65a}$$

$$\left(E_M\right)_{\max} = \frac{1}{2}\frac{C_m}{M}\left(V_{sc}^o\right)^2 \tag{2.66a}$$

However, in practical application, the linear voltage drop during discharge creates additional circuitry limitations on the usable voltage range. The quadratic potential drop means that 75% of the stored energy is depleted before voltage reaches the usable range of 50%. To utilize the last 25% energy stored in the device, the circuitry becomes more complex and expensive because of the need to up-convert and regulate the voltage to a useful level for the circuit or load to function efficiently. Therefore, in practical design applications, the maximum usable energy for a capacitor is commonly calculated for a voltage window of V_{sc}^o to half V_{sc}^o resulting in [38]:

$$\left(E_M\right)_{usable} = \frac{3}{8}\frac{C_m}{M}\left(V_{sc}^o\right)^2 \tag{2.66b}$$

The specific energy density is strongly dependent on the materials used. For example, different electrolytes have different voltage windows that directly affect cell voltage; different electrode materials have different particle sizes and porosities and can result in different capacitances. Different current collector materials have different densities. Lighter, highly conductive, and more stable current collector materials are always wanted. Furthermore, the interaction of the electrolyte ion and the electrode layer can also play a role in altering the energy density of a supercapacitor by altering the differential capacitance.

If an electrolyte solution is chosen, the effect of capacitance of the electrode material on the specific energy density can be expressed as:

$$\left(\Delta E_m\right)_{\max} = \left(E_m\right)_{\max} \frac{\Delta C_m}{C_m}$$

and

$$\left(\Delta E_M\right)_{\max} = \left(E_M\right)_{\max} \frac{\Delta C_m}{C_m}.$$

If an electrode material is chosen, the effect of the electrolyte solution on the specific energy density can be expressed as:

$$\left(\Delta E_m\right)_{\max} = 2\left(E_m\right)_{\max} \frac{\Delta V_{sc}^o}{V_{sc}^o},$$

and

$$\left(\Delta E_M\right)_{\max} = 2\left(E_M\right)_{\max} \frac{\Delta V_{sc}^o}{V_{sc}^o}.$$

For example, if the capacitance (C_m) is increased from 100 to 150 F ($\Delta C_m = 50$ F) at a voltage of 1.2 V and the specific energy density ($(E_m)_{\max}$) is 10 Wh.kg^{-1}, the increase in specific energy density will be 5 Wh.kg^{-1}.

However, if C_m is fixed at 100 F and the voltage is increased from 1.2 V to 1.8 V, the increase in specific density will be 10 Wh.kg^{-1}. This suggests that increasing a cell's voltage by using electrolyte solutions with high voltage windows is more effective than using electrode materials with high capacitances.

In theory, the ESR should have no effect on energy density, but it can affect the power density by slowing the discharge rate. Leakage resistance exerts a negative effect through self discharging to reduce the energy density. Table 2.6 lists typical specific energy densities obtained using different electrode materials and electrolytes.

As reported by Liu et al. [54], the graphene material used for double-layer supercapacitors can give specific energy densities as high as 85.6 Wh.kg^{-1} at room temperature and 136 Wh.kg^{-1} at 80°C. It was claimed to be a breakthrough in supercapacitor technology. These specific energy densities are comparable to that of a Ni metal hydride battery, but a supercapacitor can be charged or discharged in seconds or minutes.

2.6.2 Power Densities

Another important performance indicator of supercapacitors is power density because it describes how quickly the energy stored in the device can be

TABLE 2.6

Typical Specific Energy and Power Densities Obtained Using Various Electrode Materials, Electrolytes, and Solvents

Carbon Material	Energy Density ($Wh.kg^{-1}$)	Power Density ($kW.kg^{-1}$)	Electrode Capacitance ($F.g^{-1}$)	References
Carbon fiber cloth	2–36	5 to 11	3.5 to 60	8, 39–42
Activated carbon	5 to 25	10 to 40	50 to 125	10, 24, 38
Carbon aerogel	–	–	5 to 80	8, 43–45
Carbon nanofiber	10 to 20	5 to 20	50 to 100	46–48
Templated carbon	5 to 60	5 to 40	30 to 150	49–53
Carbon nanotube	0.5 to 40	30 to 1000	12 to 120	24–28
Graphene	20 to 70	40 to 250	100 to 200	29–31

Note: Performance parameters vary strongly depending on electrode configuration, electrolyte type, and circuit load used during testing. The range is also subject to variation in material durability and electrode size. Performance is shown for weight of active material only. For packaged devices, active material performance is estimated by assuming 25% active mass within cell. Electrolyte data are restricted to performance in organic solvents.

delivered to an external load. The definition of the specific power density (P_m) is the cell voltage multiplied by the cell current density:

$$P_m = \frac{I_{cell} V_{cell}}{m} \tag{2.67}$$

Note because I_{cell} in Equation (2.67) is in units of $A.cm^{-2}$, m should be expressed in $kg.cm^{-2}$ in order for P_m units to be $W.kg^{-1}$. During a constant current (I_{cell}) discharging process, power density can be expressed as Equation (2.68) by combining Equation (2.67) with (2.58):

$$P_m = \frac{I_{cell}}{m}\left[-I_{cell}R_{esr} + V_{sc}^{o} - \left(V_{sc}^{o} + I_{cell}R_p\right)\left(I - \exp\left(-\frac{t}{R_p C_{dl}^{T}}\right)\right)\right] \tag{2.68}$$

Equation (2.68) indicates that by increasing the time of discharge, the power density of the supercapacitor will gradually decrease. According to Equation (2.69), the maximum specific power density can be expressed as:

$$\frac{\partial P_m}{\partial I_{cell}} = \frac{1}{m}\left[-2\left(I_{cell}\right)_{max}\left(R_{esr} + R_p\left(1 - \exp\left(-\frac{t}{R_p C_{dl}^{T}}\right)\right)\right) + V_{sc}^{o}\exp\left(-\frac{t}{R_p C_{dl}^{T}}\right)\right] = 0 \tag{2.69}$$

where $(I_{cell})_{\max}$ is the maximum current density at the maximum power density, which can be solved through:

$$\left(I_{cell}\right)_{\max}=\frac{V_{sc}^{o}\exp\left(-\frac{t}{R_{p}C_{dl}^{T}}\right)}{2\left(R_{esr}+R_{p}\left(1-\exp\left(-\frac{t}{R_{p}C_{dl}^{T}}\right)\right)\right)} \tag{2.70}$$

Substituting Equation (2.70) into (2.58), Equation (2.71) is obtained:

$$\left(V_{cell}\right)_{\max}=\frac{1}{2}V_{sc}^{o}\exp\left(-\frac{t}{R_{p}C_{dl}^{T}}\right) \tag{2.71}$$

This allows the maximum specific power density to be obtained:

$$\left(P_{m}\right)_{\max}=\frac{1}{4m}\frac{\left(V_{sc}^{o}\right)^{2}\left(\exp\left(-\frac{t}{R_{p}C_{dl}^{T}}\right)\right)^{2}}{R_{esr}+R_{p}\left(1-\exp\left(-\frac{t}{R_{p}C_{dl}^{T}}\right)\right)} \tag{2.72}$$

According to this equation, at the initial discharge process ($t = 0$), the maximum power density should be:

$$\left(P_{m}\right)_{\max}^{t=0}=\frac{1}{4m}\frac{\left(V_{sc}^{o}\right)^{2}}{R_{esr}} \tag{2.73}$$

Equation (2.73) is the most popular for calculating the maximum specific power density. If $R_p \rightarrow \infty$, Equation (2.73) can be simplified to:

$$\left(P_{m}\right)_{\max}=\frac{1}{4m}\frac{\left(V_{sc}^{o}\right)^{2}}{R_{esr}+\frac{t}{C_{dl}^{T}}} \tag{2.74}$$

Figure 2.18 shows specific power density and cell voltage as a function of cell current at the moment of discharging time (10 sec). It can be seen that the

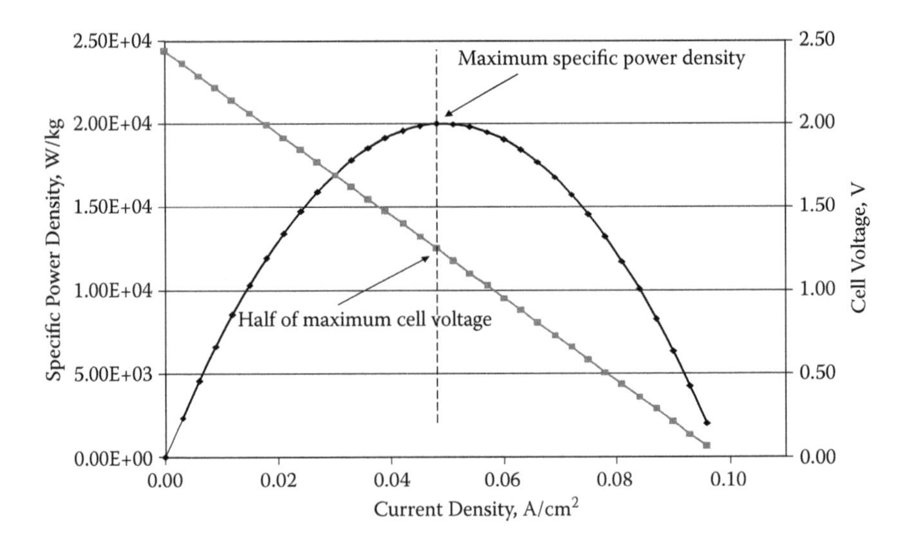

FIGURE 2.18
Calculated specific power density and cell voltage as function of cell current density at discharging moment at 10 sec. $R_p = 10000\ \Omega.cm^2$, $C^T_{dl} = 0.4\ F.cm^{-2}$, $V^0_{SC} = 2.5$ V, and $R_{esr} = 0.05\ \Omega.cm^2$, respectively.

position of the maximum power density is half way through the discharging cell voltage.

Figure 2.19 shows the effects of ESR, leakage resistance, and discharging time on maximum specific power density. Note that ESR has a much stronger effect on maximum power density, in particular at the starting moment of discharge. Leakage resistance has an insignificant effect unless its magnitude is small. In practice, leakage is normally large, so its effect on maximum power density may be negligible during discharging. However, to achieve a long shelf-life for a device, this leakage may be significant.

The matched impedance case for power density assumes that one half the discharge energy is electricity and the other half is lost through resistive heating. The 50% efficiency makes the operating condition unsuitable for most applications. As a result, the matched impedance case largely overestimates the usable maximum power density, which is a much more important consideration for industry in applying ES technologies than it is for materials research. The United States Advanced Battery Consortium (USABC) specifies operating efficiency for electrochemical capacitors at 95% [38]. This value is a more useful estimate of the practical output power available. By using the adjusted efficiency (EF), peak power density becomes:

$$\left(P_m\right)_{\max} = \frac{9}{16}\left(1 - EF\right)\left(V^o_{sc}\right)^2 / R_{esr} \tag{2.75}$$

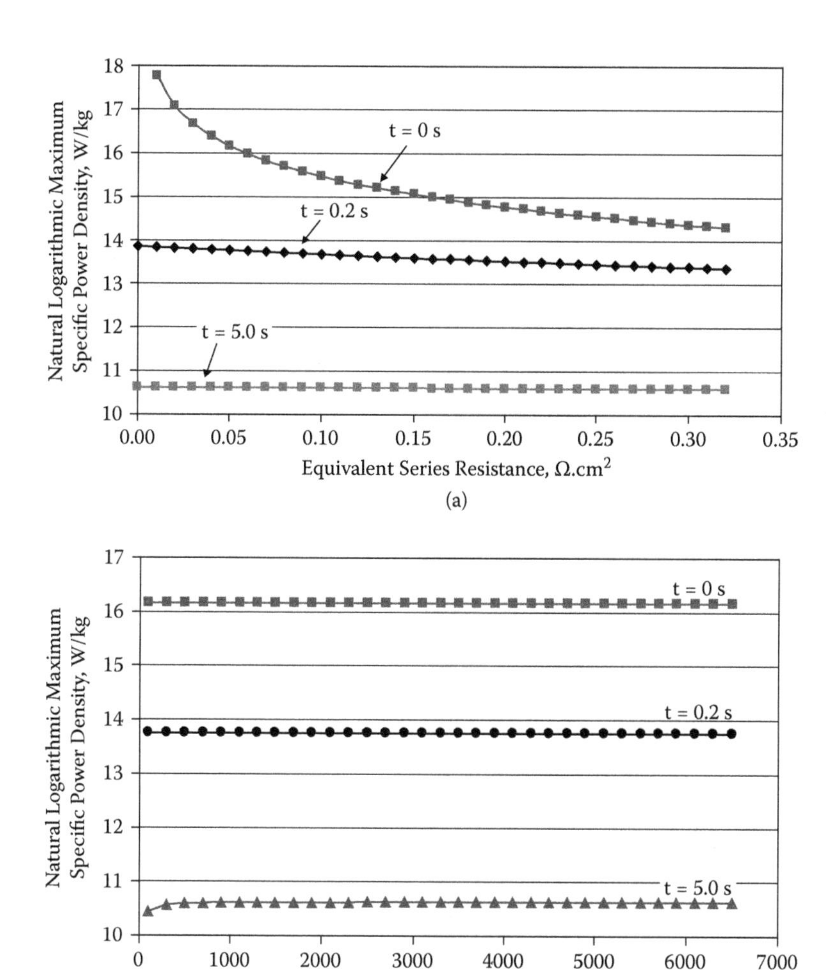

FIGURE 2.19
Calculated maximum specific power density (a) as function of equivalent series resistance and (b) as function of leakage resistance at discharging moments of three times as marked on curves. (a) R_p = 10000 Ω.cm², C^T_{dl} = 0.4 F.cm⁻², V^o_{SC} = 2.5 V, and I_{cell} = 0.05 A.cm⁻², respectively. (b) R_{esr}= 0.05 Ω.cm², C^T_{dl} = 0.4 F.cm⁻², = 2.5 V, and I_{cell} = 0.05 A.cm⁻², respectively.

Ideally, the most accurate way to determine practical and realistic power and energy performance is to use constant power measurements. This is done by monitoring voltage decay (to usable voltage, $V^o_{sc}/2$) over time as constant power is pulled from the cell. A feedback loop controls the current to maintain power at a preset level. The real average power density and discharge time are used to calculate the corresponding energy density of the cell [38].

$$\left(P_m\right)_{average} = \frac{E_M}{t} \tag{2.76}$$

Limitations in cell performance will manifest as a reduction in discharge time (lower energy output) and reduced effective V_{sc}^o (due to IR drop) as power levels are increased. It is possible for a cell to produce higher power for a short burst at the cost of irreversible cell damage and increased ESR. Using multiple tests, devices are monitored for sustainable average output power and average energy density to determine a stable peak power rating and identify failure modes. In practice, the constant power method is underutilized because it requires a capacitor test system that can output a high level of amperage (high power materials) and the response rate of the current must be fast enough to maintain constant power as the voltage quickly drops (low capacitance prototypes).

2.6.3 Ragone Plot: Relationship of Energy Density and Power Density

In applications, the specific energy density and power density are the most important factors determining the performance of electrochemical supercapacitors. The higher the densities, the better a device should perform. Unfortunately, for all electrochemical devices including batteries, fuel cells, and supercapacitors, higher energy densities do not necessarily mean high power densities. The relationship of energy and power densities is illustrated using a Ragone plot [55] on which the specific power density is plotted against the specific energy density. These plots are often used to evaluate and compare the performances of electrochemical energy storage devices.

Figure 2.20 shows Ragone plots for various electrochemical energy storage and conversion devices including batteries, fuel cells, and supercapacitors. The figure demonstrates that supercapacitors have higher power densities and lower energy densities than other types of devices. Therefore, overcoming the challenge of low energy density is currently the major focus in supercapacitor research and development.

To obtain the relationship shown in the Ragone plot, assuming the external load of a supercapacitor cell is R_L, the entire system is shown schematically in Figure 2.21. It is assumed that the leakage current does not exist ($r \to \infty$) in mathematical treatment.

In Figure 2.21, due to the presence of an ESR, the specific energy density (E_L) available for the load can be expressed as:

$$E_L = \frac{1}{2} C_{sp} \left(V_{sc}^o\right)^2 \frac{R_L}{R_L + R_{esr}} = \left(E_m\right)_{\max} \frac{R_L}{R_L + R_{esr}} \tag{2.77}$$

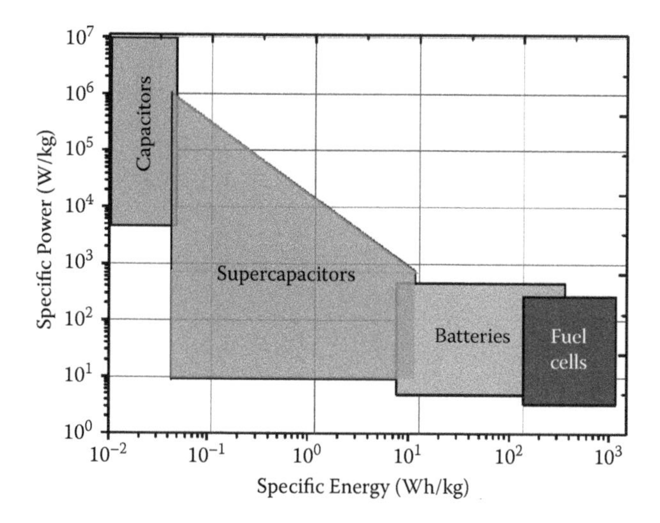

FIGURE 2.20
Ragone plots for electrochemical energy storage and conversion devices including batteries, fuel cells, and supercapacitors along with conversional capacitors. (*Source:* Winter, M. 2004. *Chemical Reviews*, 104(10). With permission.)

The specific power density (P_L) available to the external load is:

$$P_L = \frac{V_L^2}{mR_L} = \frac{\left(V_{sc}^o\right)^2 R_L}{m\left(R_L + R_{esr}\right)} = \frac{\left(V_{sc}^o\right)^2}{4R_{esr}m} \frac{4R_L R_{esr}}{\left(R_L + R_{esr}\right)^2} = \left(P_m\right)_{\max} \frac{4R_L R_{esr}}{\left(R_L + R_{esr}\right)^2} \quad (2.78)$$

By combining Equations (2.77) and (2.78), the Ragone plot can be obtained.

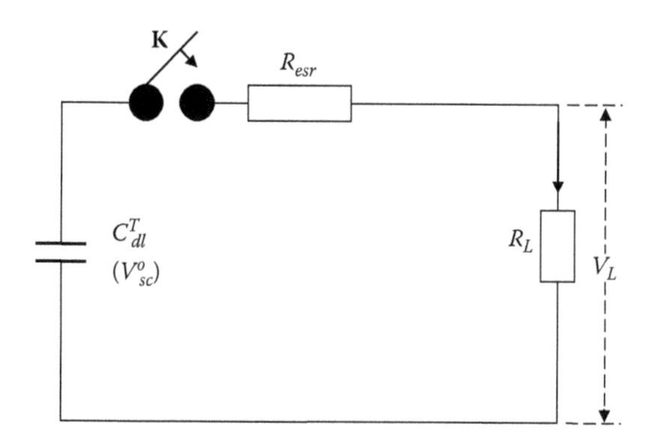

FIGURE 2.21
Electrochemical supercapacitor cell with external load (R_l) and output voltage V_l.

$$E_L = \frac{1}{2}(E_m)_{\max}\left(1+\sqrt{1-\frac{P_L}{(P_m)_{\max}}}\right) \tag{2.79}$$

The plot of

$$\frac{1}{2}(E_m)_{\max}\left(1+\sqrt{1-\frac{P_L}{(P_m)_{\max}}}\right)$$

versus P_L creates the Ragone plot shown in Figure 2.21.

Figure 2.22 indicates that when increasing the load or reducing the load resistance (R_L), the available specific energy will be reduced, leading to a higher specific power density until the load resistance reaches the ESR ($R_L = R_{esr}$). Likewise, further increasing the load or further reducing (R_L) will lead to a decrease in the power density. Therefore, there is a trade-off between power density and energy density.

Note that both the specific power and energy densities shown in Figure 2.22 should be considered the maximum available densities for a double-layer supercapacitor, that is, they are the values at the moment the supercapacitor starts to discharge ($t = 0$). After discharging starts, both their magnitudes will gradually decrease. This situation is only for supercapacitors whose discharge voltages are monotonically decreased by increasing the discharging time. When the Ragone plot is used for supercapacitors, it is necessary to indicate the plot for its maximum energy and power densities.

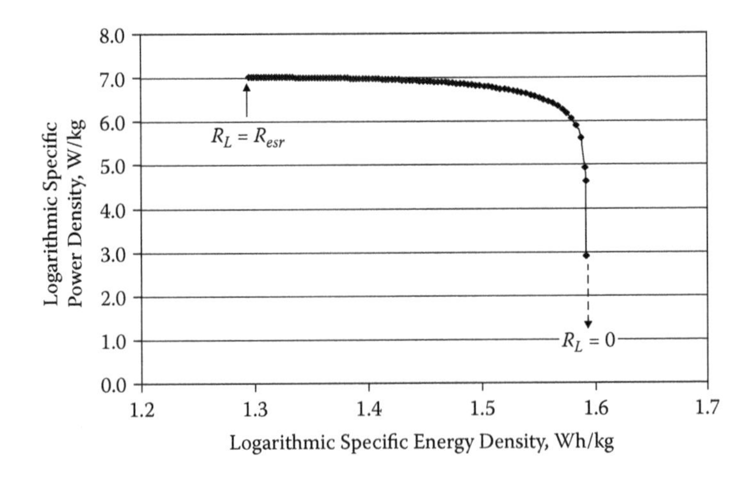

FIGURE 2.22
Calculated Ragone plot according to Equation (2.75) using V^0_{sc} = 2.5 V, C_{sp} = 45 F.g^{-1}, and R_{esr} = 0.05 Ω.cm^2. The magnitude of R_l is set to change from 1.0 × 10^{-6} to 0.91 Ω.cm^2.

2.7 Supercapacitor Stacking

In some practical applications, supercapacitor systems with both high energy and power are necessary. To generate more energy and power, a number of single supercapacitors can be stacked together and several such stacks can be connected to form an energy bank. The two options for stacking are in series and in parallel.

2.7.1 Stacking in Series

In practical applications, the operating voltage of a supercapacitor system must be substantially greater than the 1.2 V or 3.5 V windows provided by aqueous and organic electrolytes, respectively, particularly in transportation applications (e.g., hybrid electric vehicles). High operating voltages are attainable when a number of individual supercapacitor cells are arranged in series to form a supercapacitor stack. A bank of series-stacked capacitors can be connected in parallel to meet the power requirements of the application. The overall stack voltage (V_{stack}) can be expressed as the sum of all individual double-layer supercapacitor voltages:

$$V_{S(stack)} = V_1 + V_2 + V_3 + \ldots.. = \sum_{1}^{n} V_i \tag{2.80}$$

If all individual cells are identical, the stack voltage can be expressed as:

$$V_{S(stack)} = nV_i \tag{2.81}$$

where n is the number of single supercapacitors. The total capacitance of a series of capacitors ($C_{S(stack)}$) can be related to the individual capacitance of each cell C_i by:

$$\frac{1}{C_{S(stack)}} = \frac{1}{C_1} + \frac{1}{C_2} + \frac{1}{C_3} + \ldots.. = \sum_{1}^{n} C_i \tag{2.82}$$

If all individual capacitances are identical, the total capacitance can be expressed as:

$$C_{S(stack)} = \frac{C_i}{n} \tag{2.83}$$

The overall charge ($Q_{S(stack)}$) density accumulated over the stack can be calculated as:

$$Q_{S(stack)} = C_{S(stack)} V_{S(stack)} = \sum_{1}^{n} C_i \sum_{1}^{n} V_i \tag{2.84}$$

If all individual supercapacitors are identical, the total charge can be expressed as:

$$Q_{S(stack)} = \frac{C_i}{n} n V_i = C_i V_i \tag{2.85}$$

Therefore, the overall charge in a series stack is equal to that of a single cell's charge. If all individual cells are not identical, the cells having small capacitances would weigh more in total capacitance and charge. As to the specific energy and power densities of a stack, the following equations can be written according to the individual cell's expressions if all individual cells are identical:

$$\left(E_m\right)_{S(stack)} = \frac{1}{2}\left(C_{sp}\right)_i n V_1^2 \tag{2.86}$$

$$\left(P_m\right)_{S(stack)} = \frac{1}{4m} \frac{n V_i^2}{R_{esr}} \quad \text{(Note that } (R_{esr})_{P(stack)} = nR_{esr}) \tag{2.87}$$

Equations (2.86) and (2.87) indicate that the energy and power densities of a stack containing n identical single cells are the sums of all individual cells' energy and power densities.

2.7.2 Stacking in Parallel

If a number of single supercapacitors are connected in parallel, all single cells should be the same, equal to the stack's voltage ($V_{P(stack)}$). The capacitance of the entire stack ($C_{P(stack)}$) can be expressed as:

$$C_{P(stack)} = C_1 + C_2 + C_3 + \ldots.. = \sum_{1}^{n} C_i \tag{2.88}$$

If all individual cells are identical, the stack capacitance should be:

$$C_{P(stack)} = nC_i \tag{2.89}$$

where n is the number of single supercapacitors. The overall charge, Q_{stack}, density accumulated over the stack can be calculated as:

$$Q_{P(stack)} = C_{P(stack)} V_{P(stack)} = \sum_{1}^{n} C_i \sum_{1}^{n} V_i \tag{2.90}$$

If all individual supercapacitors are identical, then the total charge can be expressed as:

$$Q_{S(stack)} = nC_i V_i = C_p V_i \tag{2.91}$$

Therefore, the overall charge in a series stack is the sum of the charges stored in all individual cells. If all individual cells are not identical, the cells having large capacitances would weigh more in total capacitance and charge. Regarding a stack's specific energy and power densities, the following equations can be written according to the individual cell expressions if all individual cells are identical:

$$\left(E_m\right)_{P(stack)} = \frac{1}{2}\left(C_{sp}\right)_i nV_i^2 \tag{2.92}$$

$$\left(P_m\right)_{P(stack)} = \frac{1}{4m} \frac{nV_i^2}{R_{esr}} \quad \left(\text{Note that } (R_{esr})_{P(stack)} = \frac{R_{esr}}{n}\right) \tag{2.93}$$

Equations (2.92) and (2.93) indicate that the energy and power densities of a stack containing n identical single cells are the sums of all individual cells' energy and power density. When comparing Equations (2.92) and (2.93) with Equations (2.86) and (2.87), it is clear that the energy and power densities of supercapacitors stacked in series or parallel are the same. Series stacking is best used in applications requiring high voltage; parallel stacking is best used in applications requiring high current density and low voltage. More discussion on stacking supercapacitors in series and in parallel will be presented in Chapter 7.

2.8 Double-Layer Supercapacitors versus Batteries

The two distinct methods for achieving electrochemical energy storage and conversion are: (1) using double-layer supercapacitors to directly store the energy (charge) in an electrode–electrolyte solution interface and then delivering the energy stored in a fast and highly reversible non-faradic process; (2) using batteries or fuel cells to store energy inside chemical materials and then delivering the energy stored through faradic oxidation and reduction reactions of the chemical materials.

The primary difference between a battery or fuel cell and a supercapacitor in terms of charge storage and conversion is whether the charge (electron or ion) is transferred across the electrode–electrolyte interface. During charging and discharging, a double-layer supercapacitor does not transfer any charge across the interface. In a battery, the major process is the charge transfer across the electrode–electrolyte interface. This fundamental difference means that the properties of these devices vary.

First, double-layer supercapacitors have much higher power densities than batteries or fuel cells. Charges in a supercapacitor are stored in the interface between the carbon particle surface and the electrolyte. No charge storage occurs inside the carbon particles. However, in a battery the charges are mainly stored inside the electrode active material. In a battery, each active atom in the bulk electrode material carries a charge, whereas in supercapacitors, only atoms near the particle surface carry charges. Therefore, a battery's energy density should be much higher than a supercapacitor's energy density if their cell voltages are the same, as shown in Table 2.7.

Second, supercapacitors have much higher power densities than batteries. The charging and discharging of supercapacitors involves physical charge separation and combination processes; the charging and discharging in a battery are achieved by electrochemical oxidation and reduction processes. For physical charge separation and combination processes, the rates are theoretically infinite if the equivalent series resistance does not exist in the supercapacitor. This suggests that power delivery in a supercapacitor should be extremely fast. In a battery, the rates of oxidation–reduction reactions are limited so that power delivery is limited by the electrochemical reaction rates. Therefore, the energy storage (charging) and power delivery (discharging) in supercapacitors are much faster than those processes in batteries, illustrating that supercapacitor have much high power densities than

TABLE 2.7

Comparison of Relative Properties of Batteries and Electrochemical Double-Layer Supercapacitors

Property	Battery	Electrochemical Double-Layer Supercapacitor
Energy density ($Wh.kg^{-1}$)	10 to 300 (1000 demonstrated)	1 to 10 (100 demonstrated)
Charge time	1 to 5 hr	0.3 to 30 sec
Discharge time	0.3 to 3 hr	0.3 to 30 sec
Power density ($W.kg^{-1}$)	50 to 200	1000
Cycle efficiency	0.7 to 0.85	0.85 to 0.98

Note: See References 17 and 57.

batteries (Table 2.7). This can also be reflected by difference in ESRs. The ESR of a battery is normally much higher than that of a supercapacitor due to differences in their electrode layer conductivities.

Third, the charge and discharge times of supercapacitors are much shorter than the times for batteries. As discussed earlier, the physical charge separation and combination processes in a supercapacitor should be much faster than the electrochemical oxidation–reduction steps in a battery.

Fourth, the life cycles of supercapacitors are much longer than those of batteries. In supercapacitors, the structure of the electrode does not change during each cycle, so theoretically a supercapacitor's life cycle should be infinite. In a battery, the structure of the electrode layer will slightly change for each cycle due to side reactions that cause a gradual reduction in active electrode material, leading to fast performance degradation. Therefore, supercapacitors have much longer cycle times than batteries (Table 2.7).

There is an intrinsic cell voltage increase or decrease with the charging or discharging of a supercapacitor, while a battery has a constant cell voltage during charge and discharge. Also, heat management is easier in supercapacitors, so they are much safer than batteries. To further explain supercapacitor technology, Table 2.8 summarizes the advantages and challenges of supercapacitors and batteries.

Both devices will play roles in future energy storage and conversion technologies. Supercapacitors cannot yet replace batteries due to their low energy densities; they are presently used in conjunction with batteries. It is expected that the high energy densities, low power densities, and short life cycles of batteries can be complemented by the high power densities, low energy densities, and long lives of supercapacitors if hybrid electrical energy storage systems are developed. However, with extensive research and development in new electrode materials and new hybrid technologies, the replacement of batteries by hybrid supercapacitors is desired because of their longer life cycles. For example, to improve the energy density of a double-layer supercapacitor, some hybrid strategies involving these supercapacitors and batteries have been developed by introducing electrochemical reactive materials into the carbon electrode layers. Further discussion is presented in Chapters 4 and 7.

2.9 Applications of Supercapacitors

Supercapacitors have several practical applications including:

Transportation — The most promising market for supercapacitors is in the transportation industry. They can be used in automobiles by coupling them with other energy sources, particularly batteries. They can also improve fuel efficiency by storing energy when an automobile is braking

TABLE 2.8

Advantages and Challenges of Supercapacitors

Advantage	Challenge
Long life, with little degradation over hundreds of thousands of charge cycles	Low energy density
Low cost per cycle	High self-discharging rate, considerably higher than that of electrochemical battery
Good reversibility	Unlike practical batteries, voltage across supercapacitor drops significantly as it discharges; effective storage and recovery of energy requires complex electronic control and switching equipment with consequent energy loss
Very high rates of charge and discharge.	Very low internal resistance allows extremely rapid discharge when shorted, resulting in spark hazard similar to other capacitors of similar voltage and capacitance (generally much higher than electrochemical cells)
Extremely low internal resistance (ESR), consequent high cycle efficiency (95% or more), and extremely low heating levels	Raw material costs are significantly high and play important role in pricing of supercapacitor
High output power	Adoption rates increase only gradually as end users realize benefits
High specific power	Power only available for very short duration
Improved safety, no corrosive electrolyte, and low toxicity of materials; environmentally friendly	
Simple charge methods; no full charge detection needed; no danger of overcharging	
When used in conjunction with rechargeable batteries, can supply energy for short time, reducing battery cycling duty and extending life	

and then releasing the energy when it accelerates. Automobiles powered by coupling fuel cells with supercapacitors are ideal choices for stop-and-go traffic where supercapacitors provide sudden bursts of energy during start-up and fuel cells provide sustained energy. Supercapacitors also perform well at extremely low temperatures (–40°C), are small, and have lightweight designs.

Consumer electronics — Supercapacitors are widely used in consumer electronics as back-up energy sources for system memories, microcomputers, system boards and clocks, toys, and mobile phones. They are ideal for devices requiring quick charges. Supercapacitors are cost-effective options because they have extremely long lifetimes and do not need replacement during the lifetimes of the devices they power.

Uninterrupted power supply (UPS) systems — Supercapacitors can be used for temporary back-up power in UPS systems. They can provide instantaneous supplies of power without delays, helping to prevent malfunctions of mission-critical applications. Supercapacitors quickly bridge the power applications for stationary UPS systems that are augmented with fuel cells. In addition, supercapacitors are best suited to provide power for start-up and during peak load buffering.

Other applications — Supercapacitors show promise for critical-load operations such as hospitals, banking centers, airport control towers, and cell phone towers. The critical time between a power outage and the start of a generator can be bridged effectively by supercapacitors because they provide power within milliseconds to a few seconds after an outage. More detailed applications will be addressed in Chapter 8.

2.10 Summary

This chapter described the fundamentals of electrochemical double-layer supercapacitors and serves as a foundation for understanding supercapacitor science and technology. The fundamental principles of supercapacitors include electrode–electrolyte double-layer theory leading to double-layer models and differential capacitances of both the Helmholtz and diffuse layers. The structure and fabrication of supercapacitors with nanocarbon particle electrode materials and aqueous and non-aqueous electrolyte solutions are presented. Energy and power density calculations were presented and their relationships were discussed. The charging and discharging processes of supercapacitors were also addressed by equivalent series resistance and faradic leakage resistance and included detailed mathematical calculations. In addition, the differences between supercapacitors and batteries, the advantages and disadvantages of supercapacitors, and their applications were also covered.

References

1. Becker, H. E. 1957. *Low Voltage Electrolytic Capacitor.*
2. Boos, D. I., and G. Heights. 1970. *Electrolytic Capacitor Having Carbon Paste Electrodes.*
3. Grahame, D. C. 1947. The electrical double-layer and the theory of electrocapillarity. *Chemical Reviews*, 41, 441–501.

4. Helmholtz, H. 1853, On the laws of the distribution of electrical currents in material conductors with application to experiments in animal electricity. 89.
5. Stern, O. 1924. The theory of the electrolytic double shift. *Zeitschrift Fur Elektrochemie Und Angewandte Physikalische Chemie*, 30, 508–516.
6. Randin, J. P. and E. Yeager. 2001. Differential capacitance study on the edge orientation of pyrolytic graphite and glassy carbon electrodes. *Electroanalytical Chemistry and Interfacial Electrochemistry*, 58, 313–322.
7. Bockris, J. O., B. E. Conway, and E. Yeager, Eds. 1980. *Comprehensive Treatise of Electrochemistry*, New York: Plenum Press.
8. Zhang, L. L. and X. S. Zhao. 2009. Carbon-based materials as supercapacitor electrodes. *Chemical Society Reviews*, 38, 2520–2531.
9. Davies, A. and A. Yu. 2011. Material advancements in supercapacitors: From activated carbon to carbon nanotube and graphene. *Canadian Journal of Chemical Engineering*, 89, 1342–1357.
10. Burke, A. 2007. R&D considerations for the performance and application of electrochemical capacitors. *Electrochimica Acta*, 53, 1083–1091.
11. Lide, D., Ed. 2009. *CRC Handbook of Chemistry and Physics*, 89th ed., Boca Raton: CRC Press, 800–900.
12. Lewandowski, A. and M. Galinski. 2007. Practical and theoretical limits for electrochemical double-layer capacitors. *Journal of Power Sources*, 173, 822–828.
13. Barbieri, O. et al. 2005. Capacitance limits of high surface area activated carbons for double-layer capacitors. *Carbon*, 43, 1303–1310.
14. Qu, D. 2002. Studies of the activated carbons used in double-layer supercapacitors. *Journal of Power Sources*, 109, 403–411.
15. Tsay, K.C., L. Zhang, and J. Zhang. 2012. Effects of electrode layer composition and thickness and electrolyte concentration on both specific capacitance and energy density of supercapacitor. *Electrochimica Acta*, 60, 428–436.
16. Conway, B. E. 1999. *Electrochemical Supercapacitors: Scientific Fundamentals and Technological Applications*. New York: Kluwer/Plenum.
17. Tanahashi, I., A. Yoshida, and A. Nishino. 1990. Electrochemical characterization of activated carbon fiber cloth polarizable electrodes for electric double-layer capacitors. 137, 3052–3057.
18. Soffer, A. and M. Folman. 1972. The electrical double-layer of high surface porous material on carbon electrode. *Electroanalytical Chemistry and Interfacial Electrochemistry*, 38, 25–43.
19. Randin, J. P. and E. Yeager. 1971. Differential capacitance study of stress-annealed pyrolytic graphite electrodes, *Journal of the Electrochemical Society*, 118, 711.
20. Evans, S. 1966. Differential capacity measurements at carbon electrodes. *Journal of the Electrochemical Society*, 113, 165–168.
21. Gagnon, E. G. 1975. Triangular voltage sweep method for determining double-layer capacity of porous electrodes. *Journal of the Electrochemical Society*, 122, 521–525.
22. Kinoshita, K. 1988. *Carbon: Electrochemical and Physicochemical Properties*, New York, John Wiley & Sons, 294–295.
23. Kinoshita, K. and J. A. S. Bett. 1973. Potentiodynamic analysis of surface oxides on carbon blacks. *Carbon*, 11, 403–411.

24. Hiraoka, T. et al. 2010. Compact and light supercapacitor electrodes from a surface-only solid by opened carbon nanotubes with 2200 $m^2.g^{-1}$ surface area. *Advanced Functional Materials*, 20, 422–428.
25. Kaempgen, M. et al. 2009. Printable thin film supercapacitors using single-walled carbon nanotubes. *Journal of the American Chemical Society*, 9, 1872–1876.
26. Hu, L. et al. 2009. Highly conductive paper for energy-storage devices. *Proceedings of the National Academy of Sciences of the United States of America*, 106, 21490–21494.
27. Futaba, D. N. et al. 2006. Shape-engineerable and highly densely packed single-walled carbon nanotubes and their application as super-capacitor electrodes. *Nature: Materials*, 5, 987–994.
28. Niu, C. et al. 1997. High power electrochemical capacitors based on carbon nanotube electrodes. *Applied Physics Letters*, 70, 1480.
29. Jeong, H. M. et al. 2011. Nitrogen-doped graphene for high-performance ultracapacitors and the importance of nitrogen-doped sites at basal planes. *Nanoletters*, 11, 2472–2477.
30. Ku, K. et al. 2010. Characterization of graphene-based supercapacitors fabricated on Al foils using Au or Pd thin films as interlayers. *Synthetic Metals*, 160, 2613–2617.
31. Zhu, Y. et al. 2011. Carbon-based supercapacitors produced by activation of graphene. *Science*, 332, 1537–1541.
32. M Technologies, *K2 Series Ultracapacitor*, p. 4.
33. Ban, S. et al. 2013. Charging and discharging electrochemical supercapacitors in the presence of both parallel leakage process and electrochemical decomposition of solvent. *Electrochimica Acta*, 90, 542–549.
34. Ricketts, B. W. 2000. Self-discharge of carbon-based supercapacitors with organic electrolytes. *Journal of Power Sources*, 89, 64–69.
35. Conway, B. E. 1999. *Electrochemical Supercapacitors*. New York: Plenum.
36. Black, J., and H. A. Andreas. 2009. Effects of charge redistribution on self-discharge of electrochemical capacitors. *Electrochimica Acta*, 54, 3568–3574.
37. Oickle, A. M. and H. A. Andreas. 2011. Examination of water electrolysis and oxygen reduction as self-discharge mechanisms for carbon-based aqueous electrolyte electrochemical capacitors. *Journal of Physical Chemistry C*, 115, 4283–4288.
38. Linden, D. and T. Reddy. 2010. *Handbook of Batteries*, 4th ed., New York: McGraw Hill.
39. Lekakou, C. et al. 2011. Carbon-based fibrous EDLC capacitors and supercapacitors. *Journal of Nanotechnology*, 1–8.
40. Nawa, M., T. Nogami, and H. Mikawa. 1984. Application of activated carbon fiber fabrics to electrodes of rechargeable battery and organic electrolyte capacitor. *Journal of the Electrochemical Society*, 131, 1457–1459.
41. Xue, R. et al. 2011. Effect of activation on carbon fibers from phenol formaldehyde resins for electrochemical supercapacitors. *Journal of Applied Electrochemistry*, 41, 1357–1366.
42. Xu, B. et al. 2007. Activated carbon fiber cloths as electrodes for high performance electric double-layer capacitors. *Electrochimica Acta*, 52, 4595–4598.
43. Kalpana, D., N. G. Renganathan, and S. Pitchumani. 2006. A new class of alkaline polymer gel electrolyte for carbon aerogel supercapacitors. *Journal of Power Sources*, 157, 621–623.

44. Pekala, R. W. et al. 1998. Carbon aerogels for electrochemical applications. *Journal of Non-Crystalline Solids*, 225, 74–80.
45. Zhu, Y. et al. 2006. Cresol formaldehyde-based carbon aerogel as electrode material for electrochemical capacitor. *Journal of Power Sources*, 162, 738–742.
46. Barranco, V. et al. 2010. Amorphous carbon nanofibers and their activated carbon nanofibers as supercapacitor electrodes. *Carbon*, 114, 10302–10307.
47. Kim, C. and K. S. Yang. 2003. Electrochemical properties of carbon nanofiber web as an electrode for supercapacitor prepared by electrospinning. *Applied Physics Letters*, 83, 1216.
48. Ra, E. J. et al. 2009. High power supercapacitors using polyacrylonitrile-based carbon nanofiber paper. *Carbon*, 47, 2984–2992.
49. Vixguterl, C. et al., Electrochemical energy storage in ordered porous carbon materials, *Carbon*, May 2005. 43(6), pp. 1293-1302.
50. Portet, C. et al. 2009. Electrical double-layer capacitance of zeolite-templated carbon in organic electrolyte. *Journal of the Electrochemical Society*, 156, A1–A6.
51. Nishihara, H. et al. 2009. Investigation of the ion storage/transfer behavior in an electrical double-layer capacitor by using ordered microporous carbons as model materials. *Chemistry*, 15, 5355–5363.
52. Lei, Z. et al. 2011. Mesoporous carbon nanospheres with an excellent electrocapacitive performance. *Journal of Materials Chemistry*, 21, 2274.
53. Wang, D.W. et al. 2008. A 3D aperiodic hierarchical porous graphitic carbon material for high-rate electrochemical capacitive energy storage. *Angewandte Chemie*, 47, 373–376.
54. Liu, C. et al. 2010. Graphene-based supercapacitor with an ultrahigh energy density. *Nanoletters*, 4863–4868.
55. Ragone, D. V. *Review of Battery Systems for Electrically Powered Vehicles*. SAE Technical Paper 680453.
56. U.S. Department of Energy. *Future Fuel Cells R&D* (online). http://www.fossil.energy.gov/programs/powersystems/fuelcells [accessed April 30, 2012].
57. Schneuwly, A. and R. Gallay. 2000. *Properties and applications of supercapacitors: from the state-of-the-art to future trends*. 1–10.

3

Fundamentals of Electrochemical Pseudocapacitors

3.1 Introduction

As discussed in the previous chapter, the double-layer capacitance developed in electrochemical supercapacitors (ESs) is mainly due to net electrostatic charge accumulation and separation at the electrode–electrolyte interface. The net negative (or negative) charges such as electrons are accumulated near the electrode surface. At the same time, an equal number of positive charges such as cations are accumulated near the electrode surface at the electrolyte side, forming electric double-layers such as the Helmholtz and diffuse layers [1].

Both layers are responsible for and contribute to the magnitude of the capacitance. Because the magnitude of the capacitance is strongly dependent on the electrode surface, materials having large surface areas such as active carbon are needed to construct a layer forming a porous electrode. However, even with a porous electrode configuration, double-layer capacitance is relatively small because only the carbon particle surface can physically store charges and the electrolyte ion accessibility of the porous structure is limited. Most carbon particles in those inaccessible areas are useless for charge storage.

To increase the capacitance of ESs, some electrochemically active materials are explored for electrode use to provide much higher pseudocapacitance than double-layer capacitance. Pseudocapacitive charge storage fundamentally differs from the electrostatic mechanism that governs double-layer capacitance. For pseudocapacitance, a faradic charge transfer in the electrode porous layer occurs through a thermodynamically and kinetically favored electrochemical reduction–oxidation (redox) reaction [1].

Because this redox reaction is strongly dependent on the electrode potential, the change in charge quantity arising from this reaction (dq) has a relationship with the change in electrode potential (dV). The dependency of dq on dV (dq/dV) is called the pseudocapacitance created by the redox reaction.

The behavior of this redox reaction is similar to the electrochemical reactions in rechargeable batteries. The redox reaction in an electrode layer must be electrochemically reversible or semi-reversible to ensure efficient charge and discharge. If the ES electrode layer is entirely composed of a reversible electrochemical active material, the electrode will behave like a rechargeable battery. However, in a pseudocapacitor electrode layer containing both an electrochemically inert material such as active carbon and an electrochemically active substance such as a redox material, the charge or discharge will involve two processes (1) double-layer charging or discharging as described in Chapter 2 and (2) the electrochemical redox process.

For electrochemical redox, each reactant molecule in the bulk phase contributes one or more charges toward the stored energy, unlike the case in a double-layer charging or discharging process where only charges can physically accumulate on the material particle's surface. Therefore, pseudocapacitance is much higher for a double-layer mechanism than the possible capacitance.

The electrochemical reversibility of the employed redox material in a pseudocapacitor normally means that the redox process follows Nerstian behavior [2]. These redox materials include: (1) electrochemically active materials that can be adsorbed strongly on an electrically conductive substrate surface such as a carbon particle and (2) solid-state redox materials that can combine with or intercalate into an electrode substrate to form a hybrid electrode layer. For example, adsorption on an electrode substrate surface is commonly observed as underpotential deposition of protons on the surface of a crystalline metal electrode (Pt, Rh, Pd, Ir, or Ru). In the case of Ru, the protons can pass through the surface into the metal lattice by an absorption process, similar to the transitional behavior seen in lithium battery intercalation electrodes.

Another example is a largely conjugated organic molecule that has a redox group. The molecules can irreversibly adsorb on the carbon particle surface, providing a pseduocapacitance source. The most common pseudocapacitance is derived from redox reactions on metal oxide materials and conducting polymers that exhibit a combination of protonation reactions and absorption into the polymer matrix. Several examples of high pseudocapacitance shown in Table 3.1 will be discussed later in further detail.

Besides the need for reversibility of ES pseudocapacitive materials discussed above, their redox reaction pathway(s) must occur over a practical potential range. Both reversibility and reaction potential range, along with costs of the materials, are common limitations to the usefulness of pseudocapacitive behavior.

Electrochemical reversibility means that the electroreduction of a pseudomaterial's oxidation state can occur at almost the same electrode potential as that of electrooxidation of its reduction state. However, in practice, most redox reactions are not totally reversible due to their limited reaction kinetics, particularly if the reactions are driven at high rates. For a non-ideal reversible redox reaction to maintain a desired reaction rate, an over-electrode

TABLE 3.1

Capacitive Performances of Some Pseudocapacitive Material Samples

Material	Voltage Range (V)	Capacitance (F/g)	References
NiO	0.7	100 to 3000	3
RuO_2	1.4	350 to 1500	4–6
MnO_2	0.9	150 to 1200	7–11
Fe_3O_4	0.9	75	7
V_2O_5	0.8	170	7
PANI	0.8	115 to 1000	12–14
PEDOT	0.8	60 to 250	15, 16
PPy	0.8	150 to 420	14

potential (called overpotential) is needed and leads to reaction energy loss. If this non-ideal redox reaction is used in a pseudocapactior, its reaction overvoltage will lead to irreversible capacitance losses [2]. In normal cases, the reversibility of a redox reaction is also highly dependent on the structural and chemical reversibility of the reaction mechanism.

Furthermore, in order for a pseudocapacitive material to be useful for storage, the redox reactions must occur within the stability region of the electrolyte. For example, even if a reaction is totally reversible, the amount of pseudocapacitance induced by this reaction must occur within the electrolyte voltage window; otherwise, this reaction would be useless.

The largest benefit in the incorporation of pseudocapacitive redox material with double-layer material within an ES device is the enlargement of capacitance for improving energy density. However, several challenges surround coupling pseudocapacitive material with double-layer material in an ES electrode layer. First, all redox reactions have some degree of side reaction, causing degradation in the charging–discharging cycle life. This degradation is similar to that observed in rechargeable batteries and considerably compromises the advantage of the long cycle life of an ES. Second, the redox reactions and physisorption that occur on ES electrodes often compete with or feed the processes that cause electrolyte breakdown, causing significant ES performance degradation. Finally, a redox reaction may react with the separator material, causing lifetime issues. Therefore, the diversity of materials and added challenges make choosing the correct materials important to optimizing results.

For optimum coupling, both double-layer capacitance and pseudocapacitance of the electrode layer should be maximized. To achieve this, composite electrodes seem more effective when using high surface area carbon to increase deposition mass of the pseudocapacitive component. In designing ES devices, asymmetric configuration seems feasible. By carefully analyzing performance characteristics, it is possible to select separate anode and cathode materials that exhibit extended voltage stability and higher capacitance.

For example, in an acidic medium, carbon electrodes exhibit stronger stability in the negative region due to hydrogen electrosorption [17]. In this approach, the carbon anode seems to work well with respect to another pseudocapacitive cathode material with overpotential in the positive regime to boost performance and the potential window. Furthermore, to determine acceptable materials it is important to understand the charge regimes, identify operable characteristics, and be able to quantify resistances due to faradic components in the system.

3.2 Electrochemical Pseudocapacitance of Electrode–Electrolyte Interface

3.2.1 Fundamental Electrochemistry of Pseudocapacitance

As discussed above, in a typical ES electrode layer containing both double-layer and redox materials, two processes happen during charge or discharge. One is the accumulation of static charge within the double-layer, producing double-layer capacitance, and the other is the charge release or storage induced by the redox reaction, producing pseudocapacitance. For double-layer processes, we provided a detailed discussion in Chapter 2. As discussed above, only reversible or quasi-reversible redox materials are desired for pseudocapacitance generation by an ES. In this section, we will focus on the reversible (or quasi-reversible) redox reactions and discuss their fundamental electrochemistry.

Assuming that the redox material particles and/or reaction sites are uniformly distributed in the electrode layer and both the oxidant (O_X) and the reductant (R_d) are insoluble in the electrolyte, the redox process can be expressed as

$$O_X + ne^- \leftrightarrow R_d \tag{3.I}$$

where n is the overall electron transfer number involved in Reaction (3.I). According to the theory of electrochemical thermodynamics, the reversible electrode potential induced by the (3.I) reaction can be expressed as the Nernst form [18].

$$E = E^o_{O_X/R_d} + \frac{RT}{nF}\ln\left(\frac{C_{O_X}}{C_{R_d}}\right) \tag{3.1}$$

where $E^o_{O_X/R_d}$ is the standard electrode potential (25°C, 1.0 atm) of Reaction (3.I), C_{OX} and C_{Rd} are the concentrations of O_X and reductant R_d within the entire electrode layer (mol.cm^{-3}), E is the electrode potential (V), R is the universal gas constant (8.314 J/K·mol), and T is the temperature (K). If the initial state of the electrode layer contains only oxidant with a concentration of $C^o_{O_X}$, the consumption of O_X due to electrochemical reaction will produce R_d with a concentration of. Therefore, Equation (3.1) can be rewritten as

$$E = E^o_{O_X/R_d} + \frac{RT}{nF}\ln\left(\frac{C_{O_X}}{C^o_{O_X} - C_{O_X}}\right) \tag{3.2}$$

or

$$\frac{C_{O_X}}{C^o_{O_X} - C_{O_X}} = \exp\left(\frac{nF}{RT}(E - E^o_{O_X/R_d})\right) \tag{3.2a}$$

Equation (3.2a) can be alternatively expressed as

$$C_{O_X} = C^o_{O_X}\left(\frac{1}{1+\exp\left(\frac{nF}{RT}(E^o_{O_X/R_d} - E)\right)}\right) \tag{3.3}$$

Equation (3.3) indicates that when the electrode potential E changes over time, for example, in a linear potential scan experiment, the concentration of oxidant will change accordingly, resulting in a current flow through the electrode [19]. The current density (i, A/cm^2) passing through the electrode can be expressed as

$$i = \frac{nFAd}{A}\frac{dC_{O_X}}{dt} = nFd\frac{dC_{O_X}}{dt} \tag{3.4}$$

where A is the geometric area of the electrode layer (cm^2) and d is the thickness of the electrode layer (cm). By differentiating Equation (3.3) with respect to time, then substituting the result into Equation (3.4), the current density expression can be obtained:

$$i = \frac{n^2F^2}{RT} dC^o_{O_X} \frac{\exp\left(\frac{nF}{RT}(E^o_{O_X/R_d} - E\right)}{\left[1+\exp\left(\frac{nF}{RT}(E^o_{O_X/R_d} - E\right)\right]^2} \frac{dE}{dt} \tag{3.5}$$

Note that

$$\frac{dE}{dt}$$

is the potential scan rate (v), which is a controlled constant in electrochemical methods such as linear scan voltammetry and cyclic voltammetry. According to the definition of capacitance discussed in Chapter 2, the electrode potential-dependent pseudocapacitance induced by a redox reaction within the ES electrode layer (C_{pc} (E), F/cm^2) can be expressed as Equation (3.6) by combining Equation (3.5):

$$C_{pc}(E) = \frac{idt}{dE} = \frac{i}{v} = \frac{n^2F^2}{RT} dC^o_{O_X} \frac{\exp\left(\frac{nF}{RT}(E^o_{O_X/R_d} - E\right)}{\left[1+\exp\left(\frac{nF}{RT}(E^o_{O_X/R_d} - E\right)\right]^2} \tag{3.6}$$

Equation (3.6) is for the case of an ideal reversible redox reaction. However, due to the electrode matrix structure, the distribution of reaction redox centers may not be uniform and the interaction between the centers may cause a quasi-reversible behavior of the redox reaction. To take care of this quasi-reversible behavior, we may introduce a factor as did Conway and Gileadi [20]. This factor can be written as

$$-g\frac{C_{O_X}}{C^o_{O_X}}$$

which represents the lateral interaction energy. Thus, Equation (3.2a) can be modified to

$$\frac{C_{O_X}}{C^o_{O_X} - C_{O_X}} = \exp\left(\frac{nF}{RT}(E - E^o_{O_X/R_d}) - g\frac{C_{O_X}}{C^o_{O_X}}\right) \tag{3.7}$$

Then Equation (3.3) can be modified as

$$C_{O_X} = C^o_{O_X}\left(\frac{1}{1+\exp\left(\frac{nF}{RT}(E^o_{O_X/R_d}-E)+g\frac{C_{O_X}}{C^o_{O_X}}\right)}\right) \tag{3.8}$$

Equation (3.6) can then be alternatively expressed as

$$C_{pc}(E) = \frac{n^2F^2}{RT}dC^o_{O_X}$$

$$\frac{\exp\left(\frac{nF}{RT}(E^o_{O_X/R_d}-E)+g\frac{C_{O_X}}{C^o_{O_X}}\right)}{\left[1+\exp\left(\frac{nF}{RT}(E^o_{O_X/R_d}-E)+g\frac{C_{O_X}}{C^o_{O_X}}\right)\right]^2+g\exp\left(\frac{nF}{RT}(E^o_{O_X/R_d}-E)+g\frac{C_{O_X}}{C^o_{O_X}}\right)} \tag{3.9}$$

Equation (3.9) will be converted into Equation (3.6) when $g = 0$ and the situation corresponds to a completely reversible redox reaction. If $g > 0$, the redox reaction will be quasi-reversible. Figure 3.1 shows the calculated $C_{pc} - E$ curves at three different g values. We can see that with increasing g the peak height becomes low and the potential width at half peak ($\Delta E_{1/2}$) becomes wider, indicating the redox behavior has drifted from the ideal reversible situation.

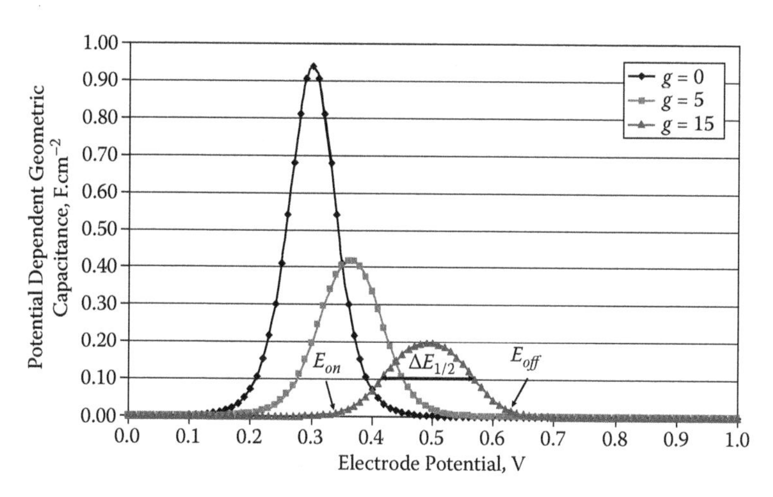

FIGURE 3.1
Calculated C_{pc}–E curves at three different g values. The following parameters are used for calculation: $n = 1$, $d = 1 \times 10^{-4}$ cm, $C^o_{O_x} = 1 \times 10^{-2}$ mol/cm^{-2}, $E^o_{O_x/R_d} = 0.3$ V, and $C^o_{O_x} / C_{O_x} = 0.5$. The g numbers are as marked in the legend box.

Actually, the pseudocapacitance expressed by Equation (3.9) is the momentary capacitance, which is a function of electrode potential. For the intrinsic pseudocapacitance (C_{sp}, F/g) of the redox reaction, an average value may make more sense and may be defined using the following equation by combining with Equation (3.9):

$$C_{sp} = \frac{AM_{mw}}{m} \sum_{j=1(E_{on})}^{j=n(E_{off})} C_{pc}(E_j)$$

$$= \frac{n^2F^2}{mRT} AdM_{mw}C^o_{Ox}$$

$$\sum_{j=1(E_{on})}^{j=n(E_{off})} \left(\frac{\exp\left(\frac{nF}{RT}(E^o_{Ox/R_d} - E_j) + g\frac{C_{Ox}}{C^o_{Ox}} \right)}{\left[1+\exp\left(\frac{nF}{RT}(E^o_{Ox/R_d} - E_j) + g\frac{C_{Ox}}{C^o_{Ox}} \right) \right]^2 + g\exp\left(\frac{nF}{RT}(E^o_{Ox/R_d} - E_j) + g\frac{C_{Ox}}{C^o_{Ox}} \right)} \right) \tag{3.10}$$

where A is the electrode geometric area (cm^2), d is the thickness of the electrode layer, M_{mw} is the molecule weight of the redox material (g), m is the weight of redox material inside the entire electrode layer (g), and E_{on} and E_{off} are the onset and offset potentials of the wave shown in Figure 3.1, respectively. Other parameters in Equation (3.10) have the same meanings as those in Equation (3.8).

Note that the difference between the onset and offset potentials ($E_{off} - E_{on}$) should be approximately equal to two times the potential width at half peak ($\Delta E_{1/2}$), that is, $E_{off} - E_{on} \approx 2\ \Delta E_{1/2}$. In capacitance measurement using linear scan cyclic voltammetry or cyclic voltammetry, the obtained voltammogram has the same shape as those in Figure 3.1 except the Y-axis is current density rather than capacitance. By integrating the current peak in the potential range of $E_{off} - E_{on}$ to obtain the charge quantity of the redox reaction (Q_{pc}, C), the intrinsic specific pseudocapacitance can be measured using:

$$C_{sp} = \frac{Q_{pc}}{m\left|E_{off} - E_{on}\right|} = \frac{Q_{pc}}{2m(\Delta E_{1/2})} \tag{3.11}$$

Note that because the value of $E_{off} - E_{on}$ is normally small (<0.3 V), the obtained intrinsic pesudocapacitance of the redox reaction is normally much higher than that of the double-layer. However, its contribution to total capacitance may not necessarily be that high because the operation voltage window of ES is much wider than $E_{off} - E_{on}$. The contribution of its charge quantity (Q_{pc}) to the overall electrode capacitance should be divided by a wider voltage window rather than the narrow window of $E_{off} - E_{on}$.

In addition, because $E_{off} - E_{on}$ is narrow, the charge and discharge can function only as a narrow voltage range. To achieve a wider operation window, multiple redox reactions in which each individual redox makes a contribution at a different potential range may be desired. Such systems include metal oxides and electroactive polymers listed in Table 3.1. They are discussed further in the following sections of this chapter.

If an ES electrode layer contains both double-layer and pseudocapacitive materials, the coupling of these two materials can boost total capacitance. However, the interaction between these two materials and their corresponding charge storage processes make the theoretical treatment complicated. We will discuss experiment measurements further in Section 3.2.4. For an ideal situation, the total capacitance (C^T) of this electrode should be the sum of double-layer capacitance (C_{dl}) and pseudocapacitance (C_{pc}):

$$\begin{aligned} C^T &= C_{dl} + C_{pc}(E) \\ &= C_{dl} + \frac{n^2F^2}{RT} dC^o_{Ox} \\ &\quad \frac{\exp\left(\frac{nF}{RT}(E^o_{Ox/R_d} - E) + g\frac{C_{Ox}}{C^o_{Ox}}\right)}{\left[1+\exp\left(\frac{nF}{RT}(E^o_{Ox/R_d} - E) + g\frac{C_{Ox}}{C^o_{Ox}}\right)\right]^2 + g\exp\left(\frac{nF}{RT}(E^o_{Ox/R_d} - E) + g\frac{C_{Ox}}{C^o_{Ox}}\right)} \end{aligned} \tag{3.12}$$

Figure 3.2 shows the calculated total capacitance as a function of electrode potential when the potential is scanned forward from low to high and backward from high to low (as in cyclic voltammetry). Note that the double-layer capacitance is treated as potentially independent with a value of 0.15 F/cm^2 for a carbon porous electrode layer [21].

Several types of redox reactions can produce pseudocapacitance, for example, surface underpotential deposition, lithium intercalation, bulk redox couple reactions, and electrically conducting polymers. In the following sections, we will discuss these reactions and their corresponding pseudocapacitances.

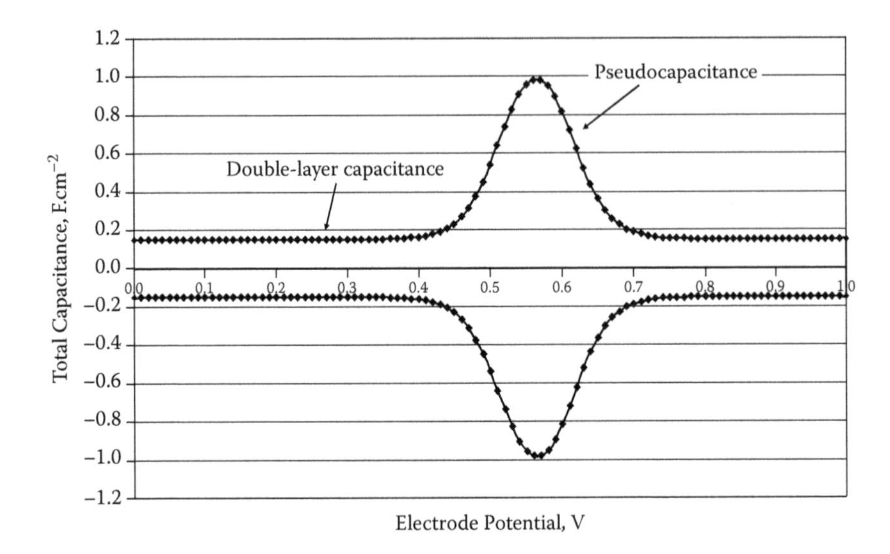

FIGURE 3.2
Calculated total capacitance as function of electrode potential. The following parameters are used for calculation: $n = 1$, $d = 1 \times 10^{-4}$ cm, $C^o_{O_x} = 1 \times 10\text{–}2$ mol/cm, $E^o_{O_x/R_d} = 0.3$ V, and $C^o_{O_x} / C_{O_x} = 0.5$.

3.2.2 Pseudocapacitance Induced by Underpotential Deposition

Underpotential deposition (UPD) of H^+ and other metals such as Pd and Cu on polycrystalline materials such as gold and platinum has been used as an example system in developing pseudocapacitance theory and practice [2]. For example, the reaction of H^+ on a Pt electrode surface, shown by the surface cyclic voltammogram in Figure 3.3, can be written as

$$H^+ + e^- + Pt \leftrightarrow Pt - H_{ads} \tag{3.II}$$

Reaction (3.II) is a reversible redox reaction with H^+ residing in the electrolyte solution and the H atom on the Pt surface. If the entire Pt surface is completely covered by a monolayer of H atoms, the saturated surface concentration of $Pt - H_{ads}$ can be expressed as Γ^o_{Pt} (mol/cm^2). If the Pt surface is not completely covered by $Pt - H_{ads}$, the surface concentration of $Pt - H_{ads}$ can be expressed as $\Gamma_{Pt\text{–}H}$ (mol/cm^2), and the Pt surface not occupied by adsorbed H atoms can be written as $\Gamma_{Pt} = \Gamma^o_{Pt} - \Gamma_{Pt\text{–}H}$ (mol/cm^2). Due to the reversibility of the electrochemical Reaction (3.II), a Nernst form similar to Equation (3.1) may apply:

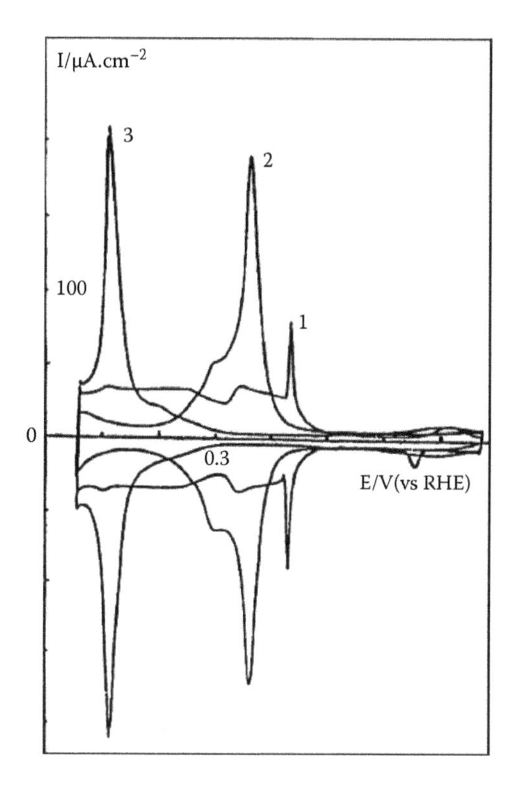

FIGURE 3.3
Current versus voltage of UPD. (1) Pt(111), (2) Pt(100), and (3) Pt(110) with hydrogen in H_2SO_4. *(Source:* Clavilier, J. et al. *Journal of Electroanalytical Chemistry and Interfacial Electrochemistry,* 295, 333. With permission.) Scan rate was 50 mV/s and temperature was 298 K for each case. (*Source:* Claviller, J. et al., *Journal of Electroanalytical Chemistry and Interfacial Electrochemistry,* 295, 333. With permission.)

$$E = E^{o}_{H^+/Pt-H} + \frac{RT}{nF}\ln\left(\frac{C_{H^+}\Gamma_{Pt}}{\Gamma_{Pt-H}}\right) = E^{o}_{H^+/Pt-H} + \frac{RT}{nF}\ln\left(\frac{C_{H^+}(\Gamma^{o}_{Pt-H} - \Gamma_{Pt-H})}{\Gamma_{Pt-H}}\right) \tag{3.13}$$

Or

$$\Gamma_{Pt-H} = \Gamma^{o}_{Pt-H}\left[\frac{1}{1 + \frac{1}{C_{H^+}}\exp\left(\frac{nF}{RT}(E - E^{o}_{H^+/Pt-H})\right)}\right] \tag{3.13a}$$

where C_{H^+} is the concentration of H^+ (mol/dm^3), n is the electron transfer number (here $n = 1$), and $E^o_{H^+/Pt-H}$ is the standard electrode potential of Reaction (3.II). As with Equations (3.4) through (3.6), the pseudocapacitance (F/cm^2) induced by H^+ underpotential deposition can be derived as

$$C_{pc}(E) = -\frac{idt}{dE} = -nF\frac{d\Gamma_{Pt-H}}{dt}\frac{dt}{dE}$$

$$= \frac{n^2F^2}{RT}\Gamma^o_{Pt-H}\frac{C_{H^+}\exp\left(\frac{nF}{RT}(E^o_{H^+/Pt-H} - E)\right)}{\left[1 + C_{H^+}\exp\left(\frac{nF}{RT}(E^o_{H^+/Pt-H} - E)\right)\right]^2} \quad (3.14)$$

Considering $n = 1$, $nF\Gamma^o_{Pt-H} = q^o_{Pt-H}$ (C/cm^2), which is the surface charge quantity when the entire surface is completely covered by a monolayer of H atoms, Equation (3.14) becomes

$$C_{pc}(E) = \frac{F}{RT}q^o_{Pt-H}\frac{C_{H^+}\exp\left(\frac{F}{RT}(E^o_{H^+/Pt-H} - E)\right)}{\left[1 + C_{H^+}\exp\left(\frac{F}{RT}(E^o_{H^+/Pt-H} - E)\right)\right]^2} \quad (3.15)$$

If differentiating Equation (3.15) with respect to electrode potential, the maximum pseudocapacitance $(C_{pc})_{max}$ induced by hydrogen adsorption and desorption on a Pt surface at

$$\frac{dC_{pc}(E)}{dE} = 0$$

can be found when

$$E = E^o_{H^+/Pt-H} - \frac{RT}{F}\ln\left(\frac{1}{C_{H^+}}\right)$$

$$(C_{pc})_{\max} = \frac{1}{4}\frac{F}{RT}q^o_{Pt-H} \quad (3.16)$$

For H on polycrystalline Pt, the well recognized value of q^o_{P-H} is 2.1 × 10^{-4} C/cm^2. Using this value and other parameters' values in Equation (3.16) such as $C_{H^+} = 1.0 \text{ mol/dm}^3$, and T = 298 K (25°C), the obtained pseudocapacitance can be as high as 2045 F/cm^2 at standard electrode potential ($E^o_{H^+/Pt-H}$), which is more than 100 times the double-layer capacitance.

In the literature [2], the coverage of the surface saturated, the surface occupied, and the surface unoccupied by H atoms are expressed as θ^o_{Pt}, θ_{Pt-H}, and θ_{Pt}, respectively:

$$\theta^o_{Pt} = \frac{\Gamma^o_{Pt}}{\Gamma^o_{Pt}} \tag{3.17}$$

$$\theta_{Pt-H} = \frac{\Gamma_{Pt-H}}{\Gamma^o_{Pt}} \tag{3.18}$$

$$\theta_{Pt} = \frac{\Gamma_{Pt}}{\Gamma^o_{Pt}} = \frac{\Gamma^o_{Pt} - \Gamma_{Pt-H}}{\Gamma^o_{Pt}} = 1 - \theta_{Pt-H} \tag{3.19}$$

A Nernst equation similar to Equation (3.12) can be written as

$$E = E^o_{H^+/Pt-H} + \frac{RT}{F}\ln\left(\frac{C_{H^+}(1-\theta_{Pt-H})}{\theta_{Pt-H}}\right) \tag{3.20}$$

or

$$\theta_{Pt-H} = \frac{C_{H^+}\exp\left(\frac{F}{RT}(E^o_{H^+/Pt-H} - E)\right)}{1 + C_{H^+}\exp\left(\frac{F}{RT}(E^o_{H^+/Pt-H} - E)\right)} \tag{3.21}$$

In this way, Equation (3.15) can be alternatively expressed as

$$C_{pc}(E) = \frac{F}{RT} q^o_{Pt-H}\theta_{Pt-H}(1-\theta_{Pt-H}) \tag{3.22}$$

When $\theta_{Pt-H} = 0.5$, the maximum pseudocapacitance will be induced; the result is the same as that when $E = E^o_{H^+/Pt-H}$ predicted by Equation (3.15). Note that the treatment above is based on the totally reversible electrochemical adsorption and desorption of H^+ (or an ideal Langmuir isotherm model with a monolayer adsorption) on a metal surface.

As coverage increases, repulsions occur between the orbitals of the adsorbate and metal. Surface electron redistribution will also occur locally along the surface. Depending on these interactions, the capacitance will decrease and spread over a larger potential window as shown in Figure 3.1. High levels of surface interaction during charging can also reduce the reversibility of the deposition.

The cyclic voltammograms shown in Figure 3.3 illustrate that, even for monocrystalline structures, multiple states are represented by many peaks in the curve. This means that there are multiple states below a monolayer with quasi-stable surface organization or by partial charge transfer from the anion [2,22].

As noted previously, the high maximum capacitance induced by these types of redox processes is not maintained over the available voltage range, which is normally limited to 0.6 V for most materials. Further the two-dimensional surface limitation, high costs of noble and rare metals, and low available surface area reduce usefulness for energy storage compared to other pseudocapacitive materials.

3.2.3 Pseudocapacitance Induced by Lithium Intercalation

Cathode materials that exhibit lithium ion intercalation represent an interesting transition technology between standard battery and ES behavior. In a battery, the lithium is driven thermodynamically into a graphitic anode material and upon discharge the lithium reacts and inserts itself into the intercalation host cathode where it is more thermodynamically stable. The host cathode is a crystalline transitional metal oxide or sulfide (e.g., MoS_2, TiS_2, and V_6O_{13}). The cathode intercalation mechanism requires a faradic charge or driving force for the lithium cation to absorb and deposit into the cathode material lattice structure. Therefore, electrode potential increases linearly with the occupancy fraction (state of charge) of the three-dimensional lattice sites. An example of this behavior (Figure 3.4) illustrates a near pseudocapacitive nature of the sorption isotherm.

The pseudocapacitance of the intercalation cathode can be utilized in ES through combination with another lithium intercalation electrode, a lithium source electrode, or a capacitive double-layer electrode for the anode. Capacitance for a TiS_2 electrode can reach as high as 500 $F.g^{-1}$ over a large potential range, providing high energy density [24]. Intercalation exhibits a clear state separation during absorption seen in Figure 3.5 for a MoO_2 host material [2].

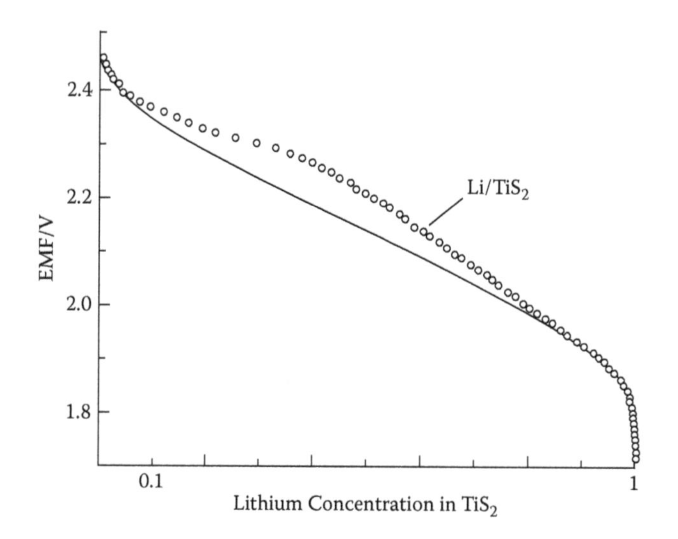

FIGURE 3.4
Discharge curve for Li^+ intercalation into TiS_2 during electrode discharge. (*Source:* Conway, B. E. and W. G. Pell. *Journal of Solid State Electrochemistry*, 7, 637–644. With permission.)

The intercalation occurs in the layer lattice sites of the cathode's crystal structure and it is subjected to diffusion limitations of lithium within the lattice [2]. The chemical diffusion coefficient (D) relates Li flux to the concentration gradient, while the behavior for a dilute Li species is represented as the jump diffusion coefficient (D_j) [25,26]. D_J describes the collective statistical mobility for lithium ions diffusing through lattice sites by hopping events. Monte Carlo simulation of the diffusion in TiS_2 crystal structure illustrates that as Li concentration increases, the z-lattice parameter contracts and makes it harder for continued lithium diffusion (see Figure 3.6) [25]. Therefore, the diffusion rate of Li inside the host material restricts charge rates so that the corresponding ES could not be used in high power demanding applications. The system may be more applicable for hybrid battery/ES devices.

3.2.4 Pseudocapacitance Induced by Redox Couples

3.2.4.1 Pseudocapacitance Induced by Dissolved Couples

Pseudocapacitance can also be induced by a redox reaction in which both the reductant and oxidant are dissolved in the electrolyte. For a reversible redox reaction, the Nernst equation like Equations (3.1) and Equations (3.2) to (3.6) are all applicable. In this case, Equation (3.6) can be alternatively rewritten as

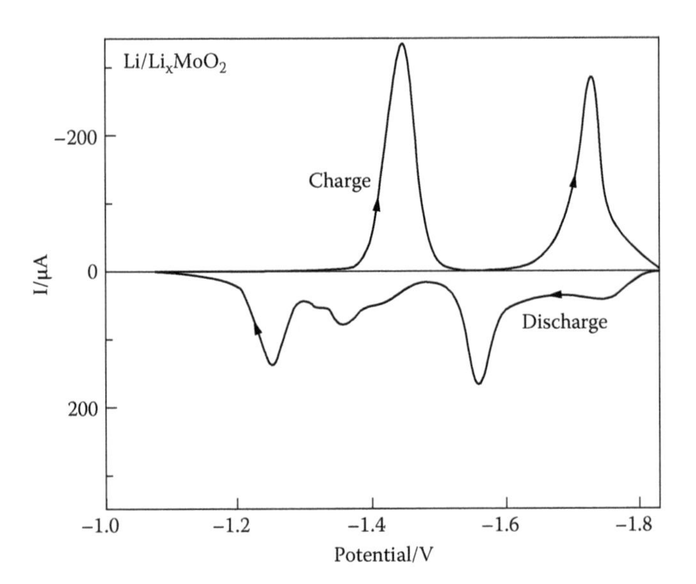

FIGURE 3.5
Cyclic voltammogram for pseudocapacitance induced by lithium intercalation in MoO_2. Note very slow scan rate of 9 uV/s. (*Source:* Conway, B. E. 1999. *Electrochemical Supercapacitors.* New York: Plenum. With permission.)

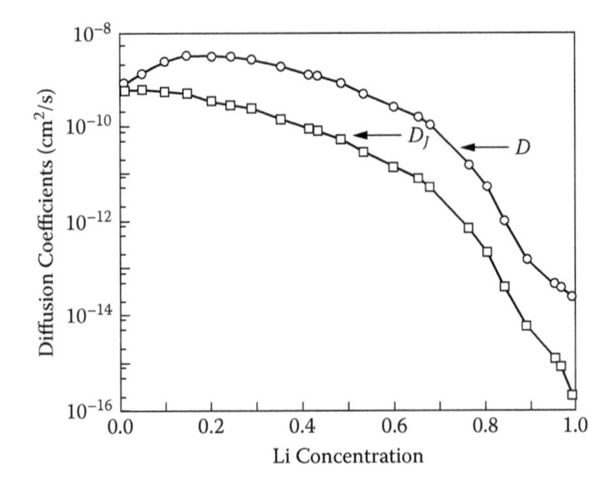

FIGURE 3.6
Effect of increasing concentration for lithium lattice occupation on chemical (D) and jump (D_J) diffusion coefficient rates. (*Source:* Bhattacharya, J. and A. Van der Ven. 2011. *Physical Reviews B*, 83,.1–9. With permission.)

$$C_{pc}(E) = \frac{n^2F^2}{RT} C^o_{Ox} \frac{\exp\left(\frac{nF}{RT}(E^o_{Ox/R_d} - E\right)}{\left[1+\exp\left(\frac{nF}{RT}(E^o_{Ox/R_d} - E\right)\right]^2} \tag{3.23}$$

where C^o_{Ox} is the dissolved total concentration of the redox material. Let us consider a one-electron redox system with C^o_{Ox} = 5.0 × 10^{-3} mol/cm^3 at a potential of $E = E^o_{Ox/Rd}$ (25°C and ambient pressure). Note that the unit of C_{pc} (E) is F/cm^3 rather than F/cm^2 as in Equation (3.6). The calculation according to Equation (3.23) gives a maximum pseudocapacitance of 4697 F/cm^3, which is much higher than that possible through double-layer storage induced by a carbon porous electrode (~10 to 20 F/cm^3) [21].

Equation (3.23) suggests that increased reactant concentration will lead to a much greater amount of charge storage, making the liquid form a strong storage medium. However, the diffusions of redox couples from the electrolyte solution to the electrode surface or from the electrode surface to the electrolyte solution are major concerns that limit high power applications. An example of the dissolved redox couple system can be seen in Figure 3.7 for a ferricyanide–ferrocyanide couple [1]. It is clear that the dissolved couple exhibits diffusion control.

This diffusion resistance compromises the use of the undissolved couple for ES devices. However, when bound by a polyvinylpyridine polymer to a gold collector, diffusion becomes much less significant. This highlights why application of undissolved redox couples is more useful in reversible ES designs. Again, the short voltage range available to the single transition limits the energy storage capability compared to transition metals with multiple redox transition states available within the stable charge region of the electrolyte–active material combination.

3.2.4.2 Pseudocapacitance Induced by Undissolved Redox Couples

We discussed the undissolved redox couples in Section 3.2.1. However, those discussions mainly focus on a single redox process. In practice, several important requirements for redox couples should be mentioned here: (1) the redox couple should exhibit multiple reversible redox states and remain stable over a large potential window; (2) the redox material should be very electrically conductive and have the accessible structure needed to support performance (normally, high conductivity allows rapid distribution of charge within the structure); and (3) a high diffusion rate of charge balancing ions (e.g., metal oxides) as protons within the material matrix.

Many transition metals have been considered as possible sources of reversible pseudocapacitance including Ru, Ir, W, Mo, and Co oxides [2]. RuO_2, shown in Figure 3.8, was identified early as an excellent candidate based

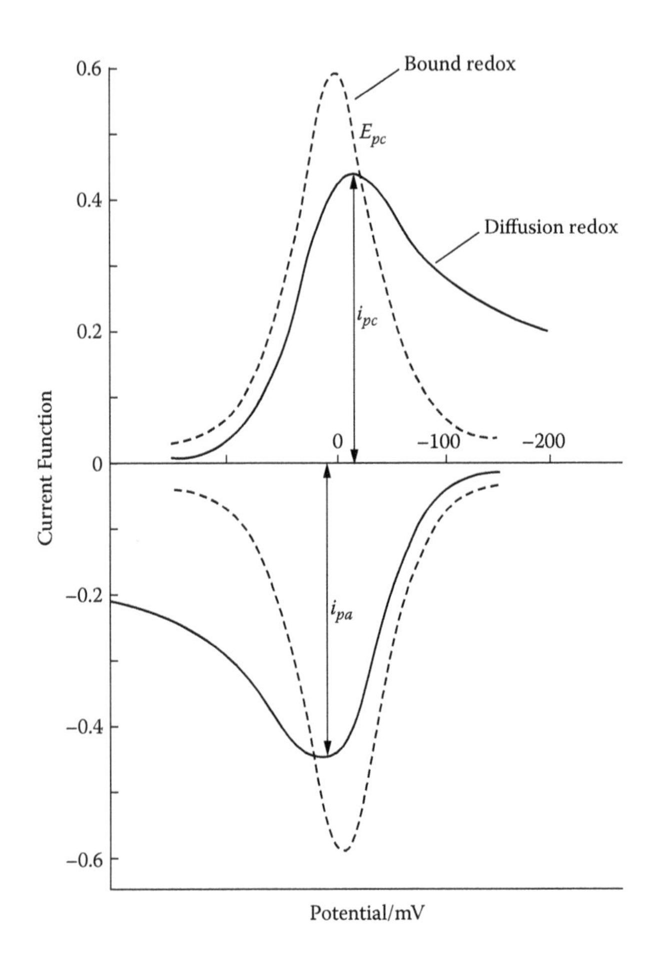

FIGURE 3.7
Cyclic voltammograms of pseudocapacitance for ferricyanide and ferrocyanide couple free in solution (solid line), compared to ferrocyanide bound to a collector with PVP (dashed line). (*Source:* Conway, B. E., V. Birss, and J. Wojtowicz. 1997. *Journal of Power Sources*, 66, 1–14. With permission.)

on its electrochemical properties, strong reversibility, very high conductivity, and wide potential range of 1.4 V and high capacitance. Wide potential is possible due to the three available redox reactions ($Ru^{2+}\leftrightarrow Ru^{3+}$, $Ru^{3+}\leftrightarrow Ru^{4+}$, and $Ru^{4+}\leftrightarrow Ru^{6+}$). Unfortunately, Ru is a noble metal and too expensive for ES applications.

The attractiveness of cheap nickel oxide materials led to studies of different crystal and nano morphologies that offer improved performance and reversibility via improvements to surface area. Table 3.2 illustrates the variability possible for this pseudocapacitive material, highlighting the importance of optimizing internal diffusion and conductivity by controlling morphology and choosing the correct crystal structure. This table also

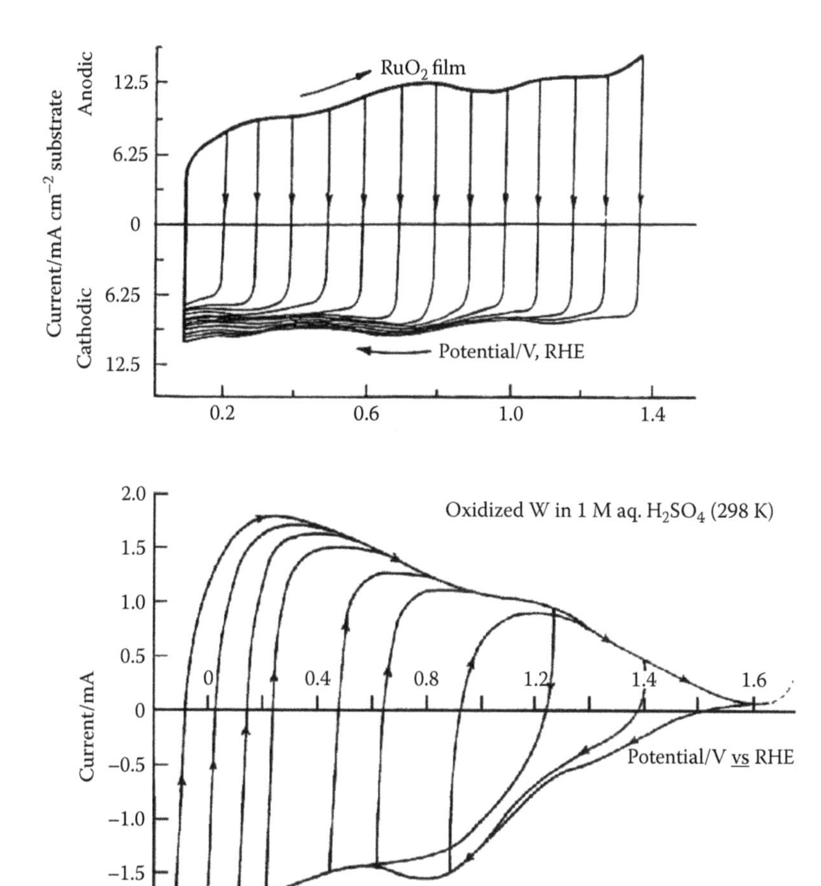

FIGURE 3.8
Cyclic voltammograms of reversible redox capacitance on RuO_2 (top) and W oxide (bottom) in acidic electrolyte. (*Source:* Conway, B. E., V. Birss, and J. Wojtowicz. 1997. *Journal of Power Sources*, 66, 1–14. With permission.)

indicates that the sweep rate and/or current density are also important in different power applications.

However, the cycle life is a major limitation because the material phase may be changed over time through faradic processes, leading to a much shorter cycle life than that possible with double-layer capacitance. This statement holds true for many other redox materials including transition metal candidates, providing variable results among studies of the same active material.

Regarding metal oxide, it is worthwhile to point out that the coupling mechanism between proton insertion and the redox electron transfer mechanism plays a considerable role in pseudocapacitance generation. For example, depending on electrode film structure and origin, the hydration mechanism

TABLE 3.2

Comparison of Experimental Results for NiO Redox Couple [3]

Material	Phase	Morphology	Capacitance (F/g)	Electrolyte
NiO	Cubic	Porous ball	124	3 wt% KOH
NiO	Cubic	Nanoflake	140	2 *M* KOH
NiO	Cubic	Thin film	590	3 wt% KOH
NiO	Cubic	Thin film	1110	1 *M* KOH
NiO	Monoclinic	Petal	710	6 *M* KOH
NiO	Monoclinic	Nanocolumn	390	1 *M* KOH
		Nanoslice	285	
		Nanoplate	176	
NiO	Cubic	Nanoflake	411	2 *M* KOH
NiO–MWNT	Cubic	Nanoflake	206	2 *M* KOH
NiO–carbon fabric			230	3 *M* KOH
$Ni(OH)_2$	Alpha $Ni(OH)_2$	Thin film	3152	3 wt% KOH
$Ni(OH)_2$ on activated carbon	Beta $Ni(OH)_2$	Irregular	194	6 *M* KOH
$Ni(OH)_2$ on activated carbon	Beta $Ni(OH)_2$	Nanoparticle	315	6 *M* KOH
$Ni(OH)_2$ on graphene	Beta $Ni(OH)_2$	Nanoparticle	1335	1 *M* KOH

Source: Sun, X. et al. 2011. *Journal of Materials Chemistry*, 21, 16581–16588. With permission.

will be limited to the surface layer accessible to the electrolyte—therefore of little use for charge storage. Nonetheless, proton insertion reactions allow a pathway for enhanced proton mobility:

$$O^{2-} + H^{+} \leftrightarrow OH^{-} \tag{3.III}$$

In the case of RuO_2, the reversible hydration mechanism allows protons to hop between RuO_2 lattice sites, opening the possibility to charge subsurface layers slowly. For oxides prepared electrochemically or by sol gels, the material produced is hydrous. The enhanced mobility effect becomes more prominent through the ionization of water trapped in the metal-hydrated matrix. The H^+ and OH^- from the ionized water molecule can insert into the transition metal lattice sites. The extra-hydrated sites within the material bulk can enhance mobility and optimize performance.

The hydration coupling and multistate electron transfer can create a great deal of overlap between potential regions and certainly contribute to current leveling within the potential window. In the case of RuO_2, the hydrated oxide (RuO_2-xH_2O) is capable of achieving 750 F/g as a film compared to 350 F/g for the nonhydrous type [27]. The high capacitance and 1.4 V potential window led to stacked cell capacitors with maximum performance of energy density of 8.3 Wh/kg and power density of 30 kW/kg, respectively [2].

Recent work combining carbon black and RuO_2 showed individual cells with 24 Wh/kg and 4 kW/kg [28]. Other recent work has shown even higher capacitance, with RuO_2 nanoparticles with carbon supports showing 1500 F/g, 40 Wh/kg, and 2 kW/kg in aqueous electrolytes [6]. This illustrates the high energy density and performance that are possible for pseudocapacitance induced by undissolved redox couples. The overall coupled form of the hydration-coupled redox mechanism is seen for any hydrous metal oxide (transition metal M of the form MO_x-nH_2O):

$$MO_x(OH)_y + aH^+ + ae^- \leftrightarrow MO_{x-a}(OH)_{y+a} \tag{3.IV}$$

As noted earlier, the high cost and toxicity of RuO_2 limit its global application to ES devices [2]. Ir is an even more expensive (and thus limited) alternative. Other highly reversible redox materials such as W and Co [2] illustrate high capacitance at low charge rates but have smaller potential windows (<1 V). It seems that MnO_2 is a safer, low cost alternative to RuO_2. It exhibits a nearly double-layer-like curve through the oxide transition from Mn^{3+} to Mn^{4+} which has a potential window of 1 V. However, its conductivity is much lower and the rate of proton coupling diffusion is slower. As a result only a small fraction of films are electroactive. Pure thick films and composite electrodes containing binder and conductive additives can only reach 150 to 250 F/g [8,29]. However, controlled thin films (smaller than a few micrometers) on conductive substrates can achieve more comparable pseudocapacitances of 900–1300 F/g because all the MnO_2 is active in the redox process [29].

Limitations in applications of pseudocapacitance on a practical scale continue to be tackled by attempting to combine high area and conductive carbon with thin film coatings or using controlled nanostructures to reduce charging restrictions that limit high rate and high energy pseudocapacitive devices. However, with increased structural importance, it is even more important to avoid irreversible degradation caused by phase transformations, physical relaxation, or dissolution seen during cycling.

Dissolution of the oxidant or reductant during a redox reaction can greatly reduce the reversibility of the system. The dissolution can alter an electrode material from an undissolved state to a dissolved one. In an undissolved state, it exhibits a controlled three-dimensional morphology that is closely linked to the conductive transport pathways of the current collector. When a redox reaction moves the material into a dissolved state, the charge is lost. Then the process becomes heavily diffusion limited and if the electrode was designed for charging based on its undissolved state, the diffusion will likely generate an overpotential on the material and the capacitance will be irreversible.

Further, the crystal phase or morphology within the electrode structure is irreversibly lost. An example of this situation is seen for dissolution of MnO_2 in its Mn^{2+} and Mn^{7+} states [30,29]. The $Mn^{4+} \leftrightarrow Mn^{2+}$ redox occurs at 0.47 V versus NHE, while the $Mn^{4+} \leftrightarrow Mn^{7+}$ reaction occurs at 1.19 V versus NHE under neutral pH (~6.5). In a symmetric design, this restricts the potential window

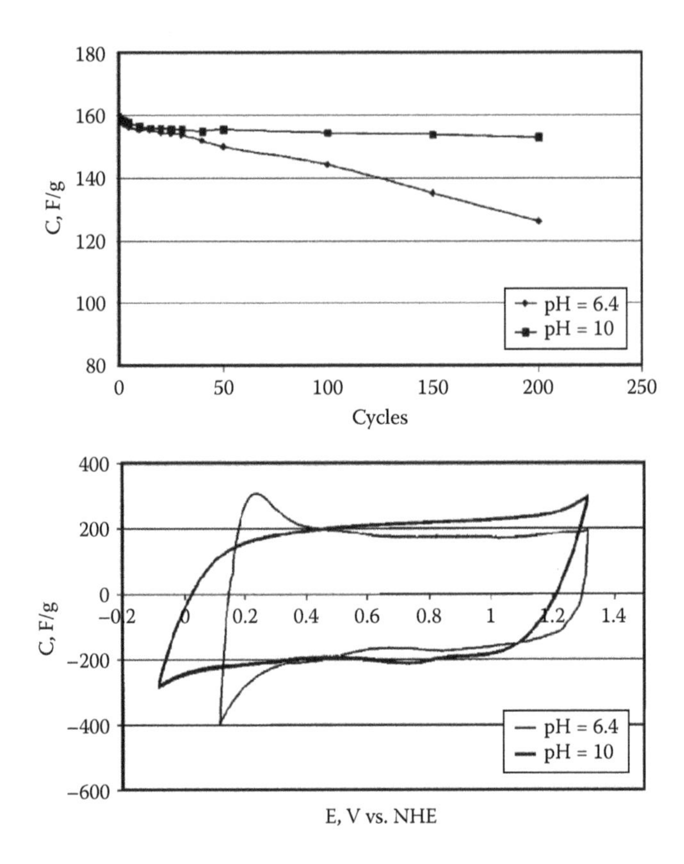

FIGURE 3.9
Top: Capacitive degradation of MnO_2 at two pH levels. Bottom: Cyclic voltammograms of MnO_2 illustrating effect of pH on degradation onset potential. (*Source:* Raymundo-Piñero, E. et al. 2005. *Journal of the Electrochemical Society*, 152, A229. With permission.)

to only 0.6 to 0.7 V to avoid irreversibility. As an example, Figure 3.9 shows the degradation within 300 cycles caused by a potential imbalance within the electrode. One solution is to utilize MnO_2 as an asymmetric positive electrode so the $Mn^{4+} \leftrightarrow Mn^{2+}$ dissolution reaction can be avoided. Another way to control the irreversibility caused by dissolution is to control electrolyte pH to shift reaction potentials. Figure 3.9 indicates that by moving to basic conditions (pH = 10) the $Mn^{4+} \leftrightarrow Mn^{2+}$ redox occurs at 0.05 V versus NHE while the $Mn^{4+} \leftrightarrow Mn^{7+}$ occurs at 1 V versus NHE. The shift enables symmetric designs to be reversibly cycled within the 0.6 V potential range available.

3.2.5 Pseudocapacitance Induced in Electrically Conducting Polymer (ECP)

A small suite of electronically conductive polymers including PEDOT, Ppy, and PANI exhibit behavior falling somewhere between a double-layer and redox mechanism. Depending on the material it is possible that charge is stored by electron transfer to the polymer with ion pairing to stabilize the charge. Alternatively, the nature of the polymer system may allow resonance throughout the ring structure along the polymer backbone. This creates a delocalization of charge allowing n or p doping within the polymer. The doped polymer can act as a double-layer system.

In both cases, the process must involve ion intercalation into three-dimensional chain entanglements of the polymers [2]. In general, polymers offer competitive capacitance values, advantages in cost, and a variety of deposition targets and techniques. Their potential window is normally around 0.7 to 0.8 V, putting them on par with most metal oxides. However, chemical degradation of the polymer and swelling from ion intercalation may reduce their cycle lives compared to that of reversible pseudocapacitance created by redox on metal oxides.

In addition, overcharging can create polymer degradation, and in the undoped state, polymer conductivity is much lower [32]. This low conductivity can create a charge isolation state that at practical charge rates can lead to irreversible degradation to the polymer structure. To reduce the effect of the isolation state and reduce damage from swelling, a polymer can be coupled with carbon supports in thin films and high-area nanostructures similar to high-capacitance metal oxides.

3.2.6 Coupling of Differential Double-Layer and Pseudocapacitance

In Section 3.2.1, the ideal coupling of double-layer and pseudocapacitance was discussed briefly. In practice, one important factor to consider when looking to optimize capacitance is the interaction between faradic charge transfer and double-layer charge storage. In general, it has been shown theoretically that coupling between the two capacitive types is possible [1,33,34].

In composite carbon or oxide electrodes, it can be difficult to quantify coupling effects because of overlapping potential regions. In addition to coupling between charge types, it is also important to recognize that some improvement can be attributed to increasing the reversibility of the pseudocapacitive component. High conductivity carbon support also contributes to boosting the fraction of active pseudocapacitive material. An example that allows us to observe this relationship can be seen in Figure 3.10 [35]. The pure PANI electrode could not effectively charge at the 10 mV/s scan rate applied. After introduction of the conductive CNT–graphene, the PANI is able to charge effectively but quantifying the contribution of the double-layer charge to the system remains difficult.

Coupling effects are much more pronounced for carbon electrodes that contain small amounts of surface functionality. These functionalities can induce a clear pseudocapacitive component that significantly boosts capacitance even for carbons with lower surface areas. Heteroatom functional groups induced by specific carbonization procedures or doping reactions

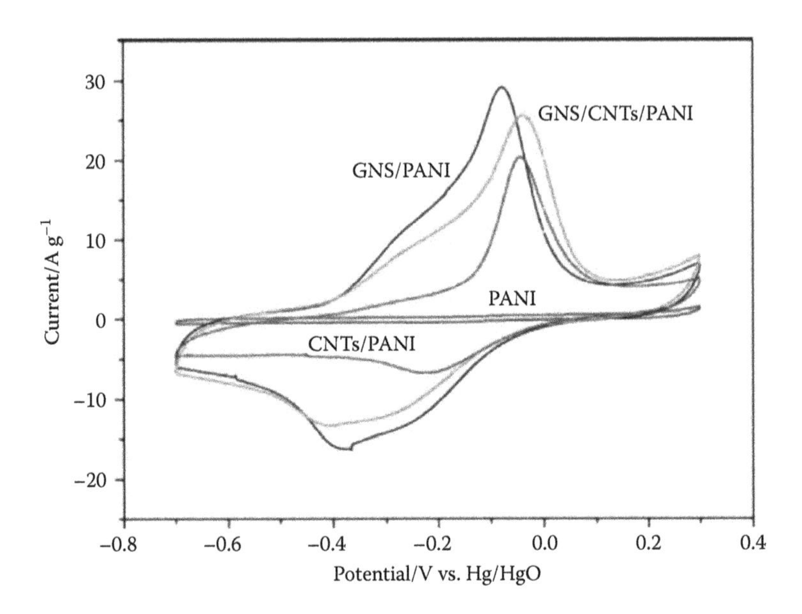

FIGURE 3.10
(See color insert.) Cyclic voltammograms of different PANI composites and their effects on performance. (*Source:* Zhang, K. et al. 2010. *Chemistry of Materials*, 22, 1392–1401. With permission.)

often include various oxygen and nitrogen groups. It is difficult to discern which groups contribute most heavily to performance or by what mechanism. In this regard, nitrogen groups are often infused into carbon systems to boost performance via wettability and pseudocapacitance [36]. As proof that electron transfer is occurring, it has been shown that pyridinic nitrogen groups within the carbon structure can also boost conductivity [37]. The C=N bond allows the pyridinic group to take part in the electronic resonance of the carbon ring structure. For example, with changing potential, the conductivity drops in acidic medium, indicating a redox is occurring that removes the nitrogen center from the system of carbon conjugation, as shown in Figure 3.11.

Redox reactions of different nitrogen groups including this pyridinic redox and other charge transfer processes can result in large pseudocapacitances [17]. Figure 3.12 illustrates the possible reaction pathways for different nitrogen functionalities. In each case, the carbon ring structure stabilizes the intermediate during charge transfer, resulting in a fast redox process [38]. Due to this coupling of redox nitrogen groups with carbon, even a relatively low surface area carbon can exhibit performance of 130 to 170 F/g with a surface area of only 400 m^2/g. This is comparable to commercial active carbon (Norit Super 50) that exhibits 120 F/g with 1400 m^2/g [39].

Analyses of some carbon materials show the clearest case of coupling with a pseudocapacitive component when exposed to an acidic or basic electrolyte. Oxygen groups known to contribute to capacitance include redox pairs

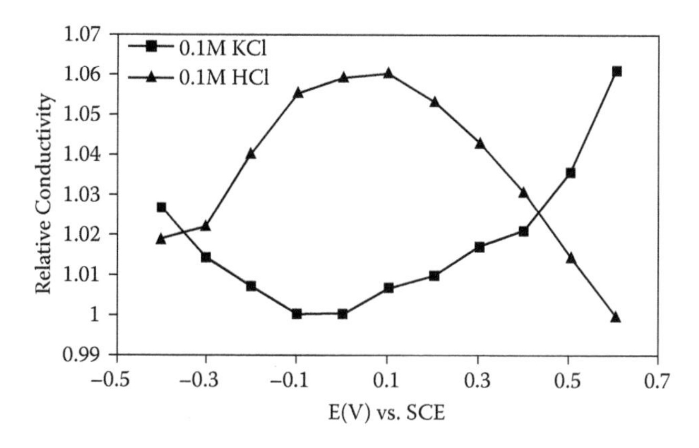

FIGURE 3.11
Dependence of conductivity on potential state for nitrogen-rich carbon material in acidic and neutral electrolyte media. (*Source:* Pollak, E., G. Salitra, and D. Aurbach. 2007. *Journal of Electroanalytical Chemistry*, 602, 195–202. With permission.)

involving quinone, ether, and pyrone-like surface functionalities along carbon edge planes [17,39,40]. The free protons in acidic solution clearly contribute to pseudocapacitive humps within the CV around 0 V versus Hg–$HgSO_4$ when a significant percent of oxygen is present on the material surface (Figure 3.13) [39].

The oxygen-rich precursor (14 to 15% oxygen after pyrolysis) was able to achieve 200 F/g for a surface area of only 270 m^2/g. Carbonization of oxygen and nitrogen-rich precursors has shown capacitance as high as 250 F/g in acid where the surface area was only 750 m^2/g (9.6% oxygen, 2.6% nitrogen) [41]. These oxygen redox interactions clearly illustrate the potential of coupling between pseudocapacitance and the ion pairing seen at the double-layer.

$$> C = NH + 2e^- + 2H^+ \leftrightarrow\, > CH - NH_2$$

$$> C - NHOH + 2e^- + 2H^+ \leftrightarrow\, > C - NH_2 + H_2O$$

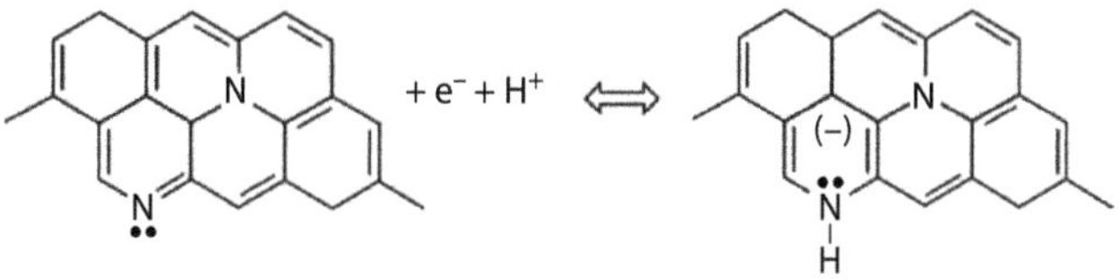

$$+ e^- + H^+ \Leftrightarrow$$

FIGURE 3.12
Electrosorption redox for nitrogen–carbon functionalities that contribute to pseudocapacitive peaks seen in acidic electrolyte. > = presence of carbon network. (Sources: Beguin, F., E. Raymundo-Piñero, and E. Frackowiak. 2010. Asymmetric systems. In *Electrical Double-Layer Capacitors and Pseudocapacitors*, 358–372; Frackowiak, E. et al. 2006. *Electrochimica Acta*, 51, 2209–2214. With permission.)

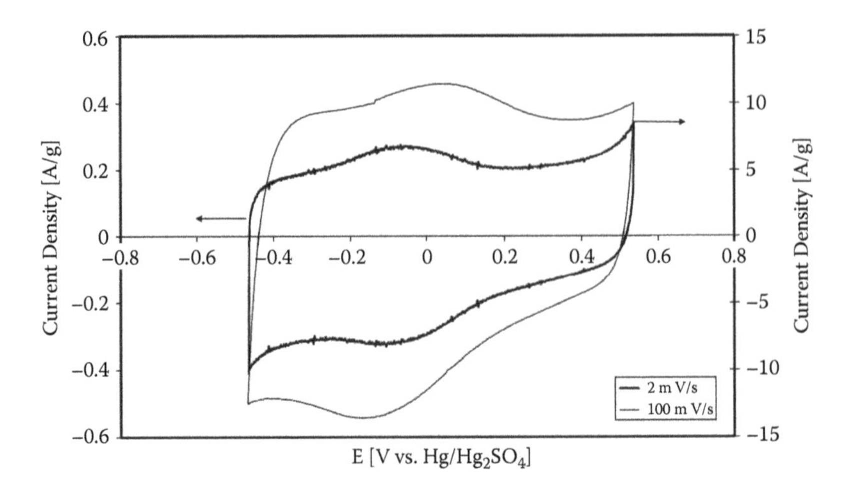

FIGURE 3.13
Cyclic voltammograms of carbon material rich with surface active oxygen functionalities (7.1% phenol, 3.5% quinone, 3.4% carboxylic) in acidic electrolyte. (*Source:* Raymundo-Piñero, E., F. Leroux, and F. Béguin. 2006. *Advanced Materials*, 18, 1877–1882. With permission.)

Performance is still dependent on overall porous structure and the positions of functionalities. Too much oxygen functionality significantly reduces the material conductivity, making it an insulator as seen with graphene versus graphene oxide [42].

3.3 Electrochemical Impedance Spectroscopy and Equivalent Circuits

Analysis based on electrochemical impedance spectroscopy (EIS; also called AC impedance spectroscopy) allows estimation of frequency behavior, quantification of resistance, and the ability to model equivalent circuits (ECs) of ES systems. The fundamental EC for a double-layer circuit, as discussed in Chapter 2, contains series resistance and double-layer capacitance. In addition, there is often a faradic parallel resistance from impurities in the carbon. In the pseudocapacitive case, the faradic resistance is a related reciprocal of the overpotential-dependent charge transfer [2,21].

$$R_f = \frac{dn}{di} \tag{3.24}$$

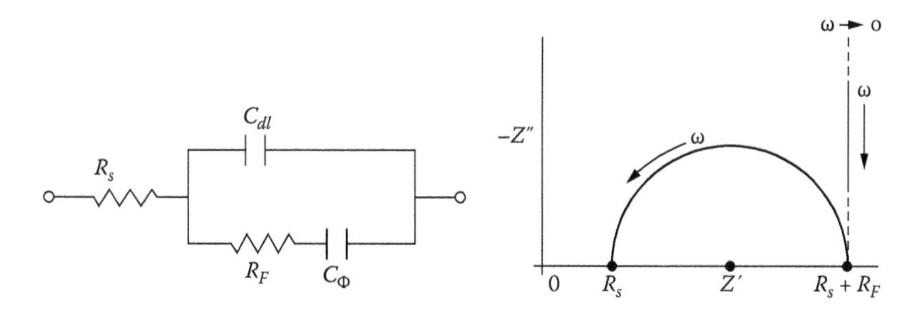

FIGURE 3.14
Left: Proposed equivalent circuit for simple nonporous pseudocapacitance. Right: Corresponding shape of EIS plot. (*Source*: Conway, B. E., V. Birss, and J. Wojtowicz. 1997. *Journal of Power Sources*, 66, 1–14. With permission.)

The pseudocapacitive capacitance is in series with this resistance. From these quantities, a simple circuit representation as shown in Figure 3.14 was compiled by Conway et al. [2]. The associated impedance profile allows for experimental determination of R_f and R_s based on the semicircular region (figure on right).

$$Z^{-1} = jw(C_{dl} + C_{pseudo}) \text{ when } w \rightarrow 0 \tag{3.25}$$

$$Z = jw\left(C_{dl}\right) \text{ when } jwC_{dl} >> \frac{1}{R_f} \tag{3.26}$$

Note that the equivalent circuit shown in Figure 3.14 is for a non-porous electrode. For porous electrodes, this EC definitely does not reflect the real situation [36]. For example, as shown in Figure 3.15, the EISs obtained using a graphene–MnO_2 porous electrode have a different shape (see Figure 3.14, right), suggesting that the model proposed in Figure 3.14 (left) is too simple to reflect the real situation.

Practical analysis of pseudocapacitors by EIS must also consider the effects of porous structures present in some pseudocapacitive materials. Morphologies will exhibit Warburg diffusion regions (45° phases) similar to those seen in porous double-layer devices. The region is induced by distribution of different resistance and capacitance between pores in the system at different sizes and distances from the collector. The spectra shown in Figure 3.15 illustrate both Warburg and double-layer charge skews with variation in layer thickness of MnO_2 [43]. Figure 3.16 shows the changing impedance region for PEDOT on steel after multiple cycles [16]. In order to simulate the cases shown in Figures 3.15 and 3.16, some modification to the EC are needed. For more detailed discussion, please see Chapter 7.

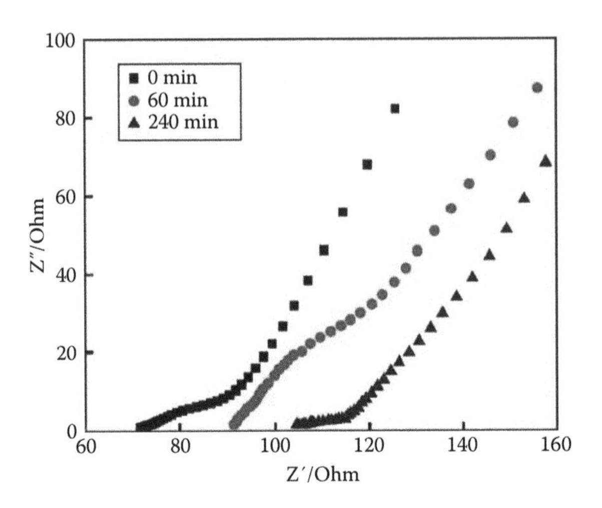

FIGURE 3.15
EIS of graphene–MnO_2 electrode with different MnO_2 electrodeposition times. (*Source:* Yu, G. et al. 2011. *Nanoletters*, 11, 2905–2911. With permission.)

3.4 Materials, Electrodes, and Cell Designs

3.4.1 Electrode Materials

As discussed previously, concerns of slow charging and irreversibility restrict practicality of pseudocapacitive electrodes deposited as thick films on current collectors. Therefore, both material selection and electrode design are very important in achieving high energy and power densities. A success story is an RuO_2-based pseudocapacitor used in military applications. However, the drawbacks are the expensive and toxic nature of the material. Many other oxides lack sufficient conductivity to perform well. Electric conductive polymers (ECPs) can be deposited and polymerized in situ, but can suffer from diffusion control. Further, ECP films are often well packed and brittle. This issue is exploited further by the volumetric changes in ECPs during ion insertion that causes consistent degradation.

As a result, more practical implementation includes the use of composites, particularly composites of pseudocapacitive materials with carbon nanotubes (CNTs). CNTs are common choices for support materials because they offer strong performance gains. They are mechanically strong so they serve as good supports by preventing damage from volumetric changes during cycling. Further, they act as conductive, strong anchoring points for thin films of oxide to be deposited.

Deposition of pseudocapacitive materials is done in situ on a carbon electrode formed by coating the collector through CVD growth, electrodeposition, or slurry. The CNT aspect ratio is important to promote porous

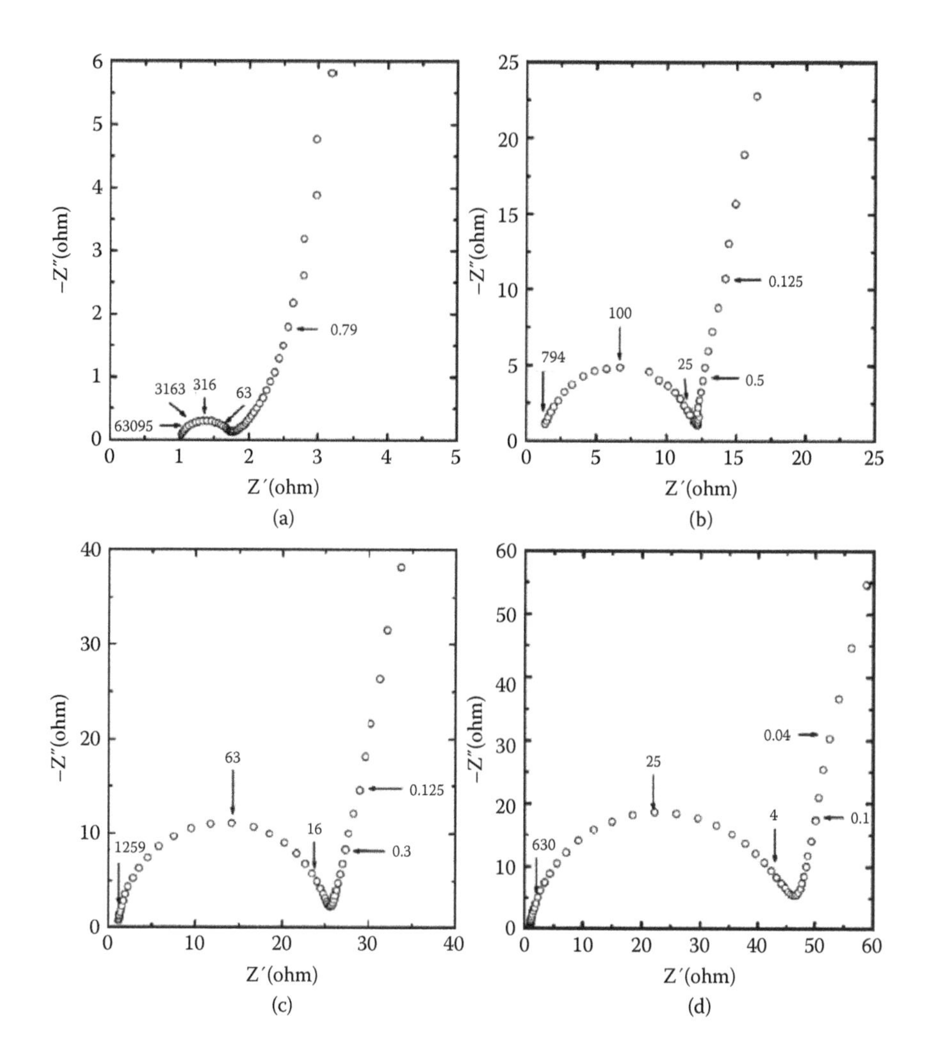

FIGURE 3.16
EIS of PEDOT–SS electrode in 1 *M* oxalic acid at 0.5 V (open circuit potential) after (a) 0, (b) 200, (c) 500, and (d) 1000 charge–discharge cycles All frequency values inside the figure are in Hz. (*Source:* Patra, S., and N. Munichandraiah. 2007. *Journal of Applied Polymer Science*, 106, 1160–1171. With permission.)

networks in the composite that enable rapid diffusion to the high capacitance pseudocapacitive component. As a result, the composite electrode can provide high capacitance and cycling ability at a practical scale for cell design [31]. Examples for PANI can be seen in Figure 3.17.

PANI provided the most uniform deposition and morphology, resulting in a capacitance of ~500 F/g at a 1 V window. It is of interest that in a symmetric two-electrode cell, the capacitance for the material dropped to 250 F/g and the scan window was reduced by the oxygen evolution and charge

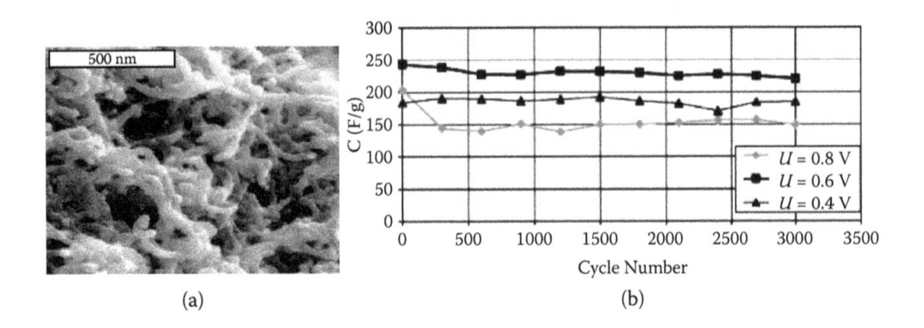

FIGURE 3.17
(a) SEM image of thin film ECP coating achieved by using CNT support. (b) Increased capacitance loss for PANI at increasing potential cycle range. (*Source:* Frackowiak, E. et al. 2006. *Journal of Power Sources,* 153, 413–418. With permission.)

isolation states to ~0.6 V. Figure 3.18 shows that the voltage window is important to the long-term retention of capacitance. Similar results were seen for PEDOT and Ppy; a 0.4 V window was required to avoid the onset of significant degradation. Symmetric cells made from CNT composites with MnO_2 showed acceptable charging behavior at practical scan rates on the low end of the spectrum [31]. Figure 3.18 illustrates that this is in opposition to conductive carbon black additives that exhibit resistance. This shows that CNTs are effective in boosting conductivity and porosity of a charge network to properly utilize available pseudocapacitance.

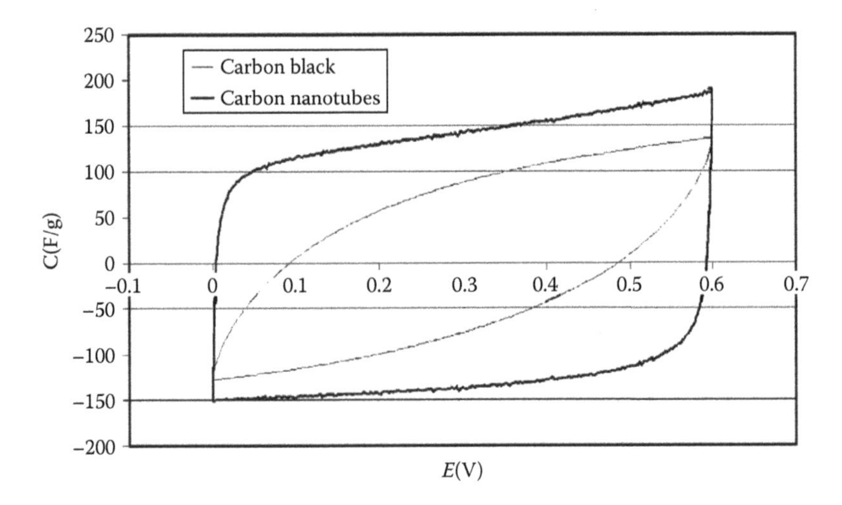

FIGURE 3.18
Cyclic voltammograms illustrating effects of CNTs within ECP composite, compared to using carbon black additives to improve pseudocapacitive efficiency and charge capability. (*Source:* Raymundo-Piñero, E. et al. 2005. *Journal of the Electrochemical Society,* 152, A229. With permission.)

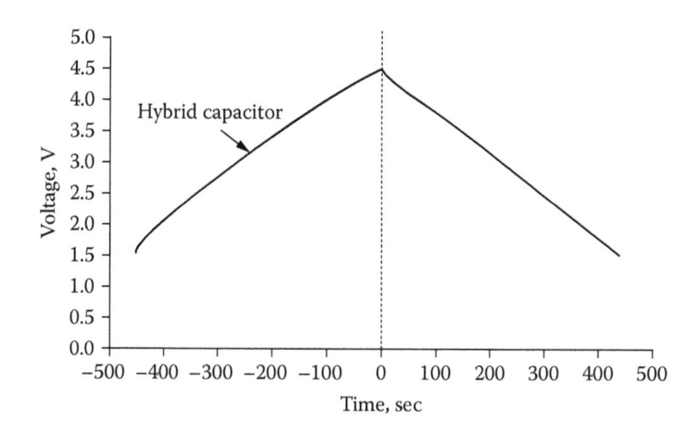

FIGURE 3.19
Charge–discharge performance of hybrid lithium ES. (*Source:* Khomenko, V., E. Raymundo-Piñero, and F. Béguin. 2008. *Journal of Power Sources*, 177, 643–651. With permission.)

3.4.2 Cell Designs (Symmetric versus Asymmetric)

In general, the low voltage window in symmetric designs remains a fundamental problem for maximizing energy density—a key reason for the application of pseudocapacitance. Increasing the low voltage windows of pseudocapacitive materials to harness increased energy density can be accomplished in a few strategic ways.

One method is to move away from pseudocapacitance and employ a lithium intercalation system to create a hybrid battery-supercapacitor. In this system, graphite (a common choice in battery systems) is a very low potential intercalation material. By appropriately matching the electrode mass, a stable half cell potential near 0.1 V can be maintained for changing lithium concentration in the electrode (Figure 3.19). The lithium stored acts as a supply for ion pairing at the cathode, which is a high performance carbon double-layer material in the form of active carbon [45] or graphene [46].

After slow initial charge and a few conditioning cycles, the graphite anode stabilizes and the active carbon, stable up to 5 V in lithium electrolyte, controls the potential window. As a result, the cell can be charged linearly between 1.5 and 4.5 V without disrupting the stability of the intercalation anode [45]. The cell design maintains 85% of its capacitance after 10,000 cycles with maximum energy density of 10 $kW.kg^{-1}$ and power density between 50 and 100 $Wh.kg^{-1}$ [45].

The second method involves choosing electrode materials for cell structures based on their use as positive or negative electrodes. Electrode fabrication still follows a CNT composite design with material deposited in situ by chemical or electrodepositing techniques. Analysis of voltage windows shows that ECPs and carbons in acidic media cycle primarily in the negative potential region, delaying the onset of hydrogen evolution [30]. Meanwhile,

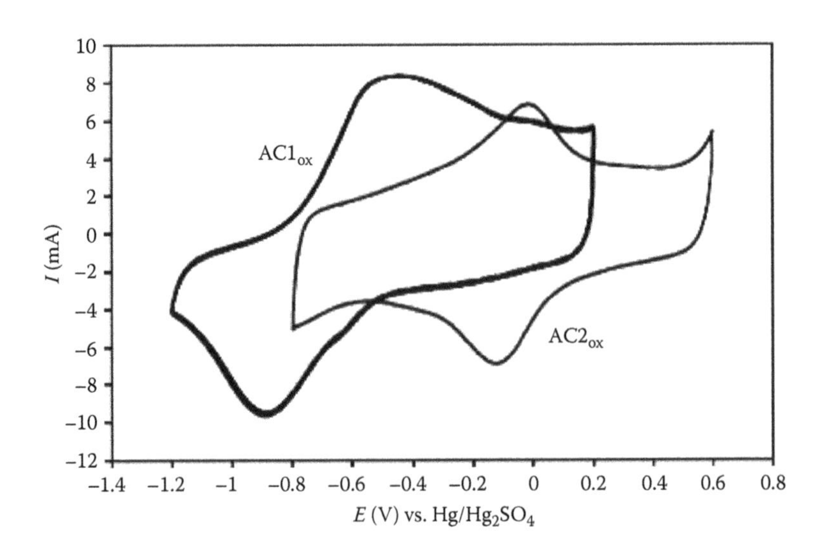

FIGURE 3.20
Cyclic voltammograms of differently treated active carbon (AC) materials illustrating overlap region and overall potential window. (*Source:* Khomenko, V., E. Raymundo-Piñero, and F. Béguin. 2010. *Journal of Power Sources*, 195, 4234–4241. With permission.)

oxides cycle within more positive voltage regions, delaying oxygen evolution. This creates an optimal pairing where, in a two-electrode cell, the anode and cathode will charge in opposing directions starting from some nominal redox equilibrium potential shared by the materials (Figure 3.20) [30].

Table 3.3 illustrates the significant energy and power density boost created by the larger potential windows for different structures. The MnO_2–carbon device in aqueous electrolyte shows stability for up to 185000 cycles with performance comparable to that seen for double-layer ES systems in organic electrolytes [47]. The system has also shown stable scalability with full size 380 F, 2 V modules consisting of a series of stacked cells. Asymmetric cells that succeed with an aqueous electrolyte significantly reduce packaging challenges, lower costs, and make the cells safer for users.

Another interesting fabrication technique is designing asymmetric carbon cells with controlled functionalities. Each electrode is based on a different carbon precursor and is subjected to different treatment to produce pseudocapacitance. The pseudocapacitance boosts performance but more importantly it extends the overpotential and prevents gas evolution (Figure 3.20) [48]. The results can achieve capacitance as high as 300 $F.g^{-1}$ at 1 $A.g^{-1}$. This capacitance is spread over a reversible potential window of 1.6 V, leading to energy densities of 10 to 30 $Wh.kg^{-1}$ and power levels of 10 to 20 $kW.kg^{-1}$ [48].

Based on the charge matching of the electrodes in a two-electrode configuration, a natural open circuit resting position of the device will exist between

TABLE 3.3

Electrochemical Characteristics of Symmetric and Asymmetric ESs Based on Different Active Materials

Electrode Materials		ES Characteristics			
Positive	**Negative**	V **(V)**	E **(Wh/kg)**	ESR **(Ω/cm²)**	P_{max} **(kW/kg)**
PANI	PANI	0.5	3.13	0.36	10.9
PPy	PPy	0.6	2.3	0.32	19.7
PEDOT	PEDOT	0.6	1.13	0.27	23.8
Carbon Maxsorb*	Carbon Maxsorb	0.7	3.74	0.44	22.4
PANI	Carbon Maxsorb	1	11.46	0.39	45.6
PPy	Carbon Maxsorb	1	7.64	0.37	48.3
PEDOT	Carbon Maxsorb	1	3.82	0.33	53.1
MnO_2	MnO_2	0.6	1.88	1.56	3.8
MnO_2	PANI	1.2	5.86	0.57	42.1
MnO_2	PPy	1.4	7.37	0.52	62.8
MnO_2	PEDOT	1.8	13.5	0.48	120.1

Source: Khomenko, V. et al. 2005. *Applied Physics A*, 82, 567–573. With permission.

* Carbon Maxsorb is a high-surface-area activated carbon.

the overlap regions. The overall potential window will be limited by the gas evolution onset that occurs on the less stable electrode. If this difference is large enough, it is possible to mismatch electrode mass to delay voltage rise on the shorter window electrode and optimize design efficiency.

3.5 Summary

In this chapter, several fundamental features of pseudocapacitance were discussed along with the fabrication challenges involved in harnessing them for increased energy. The key concepts covered include: (1) the fundamental theories and mechanisms behind pseudocapacitance; (2) how pseudocapacitance differs from double-layer storage and how they can be combined; (3) the importance of maintaining capacitive efficiency at practical charge rates, and how low-voltage windows prevent improvements in energy density; (4) challenges of irreversibility for different material types; (5) the fundamental principles of EIS analysis for pseudocapacitance; (6) the issues of symmetric pseudocapacitors; and (7) present design strategies for fabricating asymmetric cells to achieve high energy density at practical power performance.

References

1. Conway, B. E., V. Birss, and J. Wojtowicz. 1997. The role and utilization of pseudocapacitance for energy storage by supercapacitors. *Journal of Power Sources*, 66, 1–14.
2. Conway, B. E. 1999. *Electrochemical Supercapacitors*. New York: Plenum.
3. Sun, X. et al. 2011. Porous nickel oxide nano-sheets for high performance pseudocapacitance materials. *Journal of Materials Chemistry*, 21, 16581–16588.
4. Raistrick, I. D. 1992. Electrochemical capacitors. In *Electrochemistry of Semiconductors and Electronics: Processes and Devices*, New York: Noyes, 297–365.
5. Zheng, J. P. 1995. A new charge storage mechanism for electrochemical capacitors. *Journal of the Electrochemical Society*, 142, L6–L8.
6. Hu, C. 2004. Effects of substrates on the capacitive performance of $RuO_x \cdot nH_2O$ and activated carbon–RuO_x electrodes for supercapacitors. *Electrochimica Acta*, 49, 3469–3477.
7. Cottineau, T. et al. 2005. Nanostructured transition metal oxides for aqueous hybrid electrochemical supercapacitors. *Applied Physics A*, 82, 599–606.
8. Simon, P. and Y. Gogotsi. 2008. Materials for electrochemical capacitors. *Nature: Materials*, 7, 845-854.
9. Fan, Z. et al. 2006. Preparation and characterization of manganese oxide/CNT composites as supercapacitive materials. *Diamond and Related Materials*, 15, 1478-1483.
10. Yang, Y. and C. Huang. 2009. Effect of synthetic conditions, morphology, and crystallographic structure of MnO_2 on its electrochemical behavior. *Journal of Solid State Electrochemistry*, 14, 1293–1301.
11. Zhang, J. et al. 2011. Synthesis, characterization and capacitive performance of hydrous manganese dioxide nanostructures. *Nanotechnology*, 22, 125703.
12. Yan, J. et al. 2010. Preparation of graphene nanosheet/carbon nanotube/polyaniline composite as electrode material for supercapacitors. *Journal of Power Sources*, 195, 3041–3045.
13. Wang, Y. G., H. Q. Li, and Y. Y. Xia/ 2006. Ordered whisker-like polyaniline grown on the surface of mesoporous carbon and its electrochemical capacitance performance. *Advanced Materials*, 18, 2619–2623.
14. Zhang, J. et al. 2010. Synthesis of polypyrrole film by pulse galvanostatic method and its application as supercapacitor electrode materials. *Journal of Materials Science*, 45, 1947–1954.
15. Liu, R., and S. B. Lee. 2008. MnO_2/poly(3,4-ethylenedioxythiophene) coaxial nanowires by one-step coelectrodeposition for electrochemical energy storage. *Journal of the American Chemical Society*, 130, 2942–2943.
16. Patra, S., and N. Munichandraiah. 2007. Supercapacitor studies of electrochemi¬cally deposited PEDOT on stainless steel substrate. *Journal of Applied Polymer Science*, 106, 1160–1171.
17. Beguin, F., E. Raymundo-Pinero, and E. Frackowiak. 2009. Carbons for *Electrochemical Energy Storage and Conversion Systems*, 358–372.
18. Bard, A. J. and L. R. Faulkner. 1980. *Electrochemical Methods, Fundamentals, and Applications*, New York: John Wiley & Sons.

19. Hubbard, A. T. and F. C. Anson. 1966. Linear potential sweep voltammetry in thin layers of solution. *Analytical Chemistry*, 38, 58–61.
20. Gileadi, E. and B. E. Conway. 1965. *Modern Aspects of Electrochemistry*, Vol. 3, London: Butterworth.
21. Tsay, K. C., L. Zhang, and J. Zhang. 2012. Effects of electrode layer composition and thickness and electrolyte concentration on both specific capacitance and energy density of supercapacitor. *Electrochimica Acta*, 60, 428–436.
22. Conway, B. E. and E. Gileadi. 1962. Kinetic theory of pseudocapacitance and electrode reactions at appreciable surface coverage. *Transactions of the Faraday Society*, 58, 2493.
23. Conway, B. E. and W. G. Pell, Double-layer and pseudocapacitance types of electrochemical capacitors and their applications to the development of hybrid devices. *Journal of Solid State Electrochemistry*, 7, 637–644.
24. Grigorchak, I. I. 2003. Redox processes and pseudocapacitance of capacitors in light of intercalation. *Nanotechnologies*, 39, 770–773.
25. Bhattacharya, J. and A. Van der Ven. 2011. First principles study of competing mechanisms of non-dilute Li diffusion in spinel Li_x-TiS_2. *Physical Reviews B*, 83,.1–9.
26. Garcia-Belmonte, G. et al. 2006. Jump diffusion coefficient of different cations intercalated into amorphous WO_3. *Solid State Ionics*, 177, 1635–1637.
27. Zheng, J. P. and T. R. Jow. 1995. A new charge storage mechanism for electrochemical capacitors. *Journal of the Electrochemical Society*, 142, 50–52.
28. Wang, X. et al. 2008. Pseudocapacitance of ruthenium oxide/carbon black composites for electrochemical capacitors. *Journal of University of Science and Technology Beijing*, 15, 816–821.
29. Toupin, M., T. Brousse, and D. Belanger. 2004. Charge storage mechanism of MnO_2 electrode used in aqueous electrochemical capacitor. *Journal of the American Chemical Society*, 16, 3184–3190.
30. Khomenko, V. et al. 2005. High-voltage asymmetric supercapacitors operating in aqueous electrolyte. *Applied Physics A*, 82, 567–573.
31. Raymundo-Piñero, E. et al. 2005. Performance of manganese oxide–CNT composites as electrode materials for electrochemical capacitors. *Journal of the Electrochemical Society*, 152, A229.
32. Khomenko, V., E. Frackowiak, and F. Beguin. 2005. Determination of the specific capacitance of conducting polymer/nanotubes composite electrodes using different cell configurations. *Electrochimica Acta*, 50, 2499–2506.
33. Delahay, P. 1966. Electrode processes without separation of double-layer charging. *Journal of Physical Chemistry*, 70.
34. Delahay, P. and K. Holub. 1968. The admittance for an infinite exchange current. *Journal of Electroanalytical Chemistry*, 16, 131–136.
35. Zhang, K. et al. 2010. Graphene–polyaniline nanofiber composites as supercapacitor electrodes. *Chemistry of Materials*, 22, 1392–1401.
36. Lota, G. and E. Frackowiak. 2010. Pseudocapacitance effects for enhancement of capacitor performance. *Fuel Cells*, 10, 848–855.
37. Pollak, E., G. Salitra, and D. Aurbach. 2007. Can conductivity measurements serve as a tool for assessing pseudocapacitance processes occurring on carbon electrodes? *Journal of Electroanalytical Chemistry*, 602, 195–202.
38. Frackowiak, E. et al. 2006. Optimisation of supercapacitors using carbons with controlled nanotexture and nitrogen content. *Electrochimica Acta*, 51, 2209–2214.

39. Raymundo-Piñero, E., F. Leroux, and F. Béguin. 2006. A high-performance carbon for supercapacitors obtained by carbonization of a seaweed biopolymer. *Advanced Materials*, 18, 1877–1882.
40. Montes-Morán, M. A. et al. 2004. On the nature of basic sites on carbon surfaces: an overview. *Carbon*, 42, 1219–1225.
41. Raymundo-Piñero, E., M. Cadek, and F. Béguin. 2009. Tuning carbon materials for supercapacitors by direct pyrolysis of seaweeds. *Advanced Functional Materials*, 19, 1032–1039.
42. Shin, H. J. et al. 2009. Efficient reduction of graphite oxide by sodium borohydride and its effect on electrical conductance. *Advanced Functional Materials*, 19, 1987–1992.
43. Yu, G. et al. 2011. Solution-processed graphene–MnO_2 nanostructured textiles for high-performance electrochemical capacitors. *Nanoletters*, 11, 2905–2911.
44. Frackowiak, E. et al. 2006. Supercapacitors based on conducting polymer–nanotube composites. *Journal of Power Sources*, 153, 413–418.
45. Khomenko, V., E. Raymundo-Piñero, and F. Béguin. 2008. High-energy density graphite–AC capacitor in organic electrolyte. *Journal of Power Sources*, 177, 643–651.
46. Stoller, M. D. et al. 2012. Activated graphene as a cathode material for Li ion hybrid supercapacitors. *Physical Chemistry/Chemical Physics*, 14, 3388–3391.
47. Brousse, T. et al. 2007. Long-term cycling behavior of asymmetric activated carbon/MnO_2 aqueous electrochemical supercapacitor. *Journal of Power Sources*, 173, 633–641.
48. Khomenko, V., E. Raymundo-Piñero, and F. Béguin. 2010. A new type of high energy asymmetric capacitor with nanoporous carbon electrodes in aqueous electrolyte. *Journal of Power Sources*, 195, 4234–4241.

4

Components and Materials for Electrochemical Supercapacitors

4.1 Introduction

4.1.1 Traditional Capacitors

As discussed in Chapter 1, capacitors are known for their extremely high power density and cycle lifetimes, lending to their significance in the world of analog and digital electronics. An ideal capacitor is a passive device best visualized as two separated metal plates composed of conductive films such as stainless steel, nickel, or aluminum. The two plates allow a static charge to build up and be released quickly, resulting in high power density. Between the plates is a dielectric, a thin insulating material that has the potential to polarize in an electric field.

Charges in the dielectric move to generate an internal electric field and form equilibrium with the field present on the two plates. By using higher permittivity dielectrics, it is possible to store higher levels of charge in the internal field. Alternatively, capacitance is increased by reducing the separation between the plates. Further, a capacitor's performance can be altered by the breakdown potential of the dielectric. When breakdown occurs, all the stored energy between the plates is lost. To avoid dielectric breakdown, a minimum separation distance limits capacitance at higher operating voltages.

Dielectric materials vary by cost and the capacitance needed for a specific application. Glass, ceramic, and mica papers are high quality, low capacitance dielectrics with extremely high breakdown resistance. Conversely, metalized polymer foils such as polystyrene, polyethylene terephthalate (PET), and Teflon (PTFE) are single-piece dielectric films that offer better capacitor performance. In recent years, polymer foils have come to dominate the static capacitor market because they have better stability at high temperatures, can be manufactured at lower cost, and age better than dielectric papers.

4.1.2 Electrochemical Supercapacitors

Electrochemical supercapacitors (ESs) differ from traditional capacitors because they have lower power densities, greater charge storage densities, and different material requirements. Electrical double-layer capacitors (EDLCs) incorporate an electrolyte that allows charged ions to assemble on porous electrode surfaces with much higher areas than traditional capacitors. The charge is separated by a solvent cage layer at the interface of only 5 to 10 angstroms [1,2]. A combination of high surface area and small charge separation enables the generation of high energy density compared to traditional static capacitors.

Optimizing ES design involves the selection of appropriate electrodes, electrolytes, separators, and sealants. Electrode materials must be conductive and highly porous to increase charge storage ability. Non-metal porous electrodes are used in ECs rather than the metal plates of traditional capacitors; this creates a need for current collector plates to enhance the conduction of electrons in the capacitor. Collectors must be in strong contact with the electrode layer, remain stable during charge and discharge, and be highly conductive to enhance electron transport.

Electrolyte materials require high ion mobility to provide ions to the double-layer quickly. Improved electrolyte performance also requires optimization of operating voltages, toxicity, corrosion, and safety. Separators must be electronically insulating to prevent short circuits between the two electrode layers, and allow high ionic mobility from the electrolyte to the electrode surface. The separator choices in ES design include microporous or nonwoven polymers, glass, and cellulose derivatives. The most common ones are polymer separators that include polyolefins such as polypropylene (PP), polyethylene (PE), Teflon, PVdF, and PVC.

The chosen polymer and production method vary based on the electrolyte used.

Sealants must be non-conductive, prevent ion leakage between stacked cells, and resist corrosion and degradation. Sealant use depends on cell type; sealants are commonly made of low melting temperature PE material or viscous fast setting polymers such as epoxies. Sealants do not directly impact performance; they help control safety and moisture. A seal that fails can lead to short circuits within assembled cells.

Overall, careful selection and matching of ES materials can minimize resistances, avoid short circuits, reduce safety issues, support high ion mobility, increase operating voltage, and enhance the charge storage capacity of future ES devices [3].

Electrode materials in ESs fall into two categories based on storage mechanism: EDLC and pseudocapacitive materials. Each material in these categories presents its own benefits and challenges as this technology moves toward the next generation of ES devices. EDLC materials are carbons that provide physical charge storage at the interface between electrode and electrolyte.

Activated carbon (AC) materials made from coal pitch precursors are the industrial standards for ESs, although newer materials and process variation can alter the characteristics of AC materials.

Improvements to AC are under investigation in attempts to improve pore structures and increase surface areas through templating AC carbon, functionalization, and generating AC using different precursor materials. More advanced carbons, such as carbon nanotubes (CNTs), carbon nano-onions, and graphene, may offer a variety of improved properties because of their graphitic plane structures. These advanced carbons also exhibit very small size scales in at least one dimension, enabling the possibility of large area and improved power density through stronger electrolyte access.

Pseudocapacitive materials are used to increase energy density and undergo redox reactions to store more than one charge per reactive surface site and also allow storage deeper than the surface layer. Pseudocapacitance has been investigated in transition metal oxide materials that can shift between oxidation states quickly and in electrically conductive polymers that store charges along reactive groups in their polymer backbones.

Commercialized electrolyte materials in ES devices are dominated by organic electrolytes containing quaternary salts (tetraethylammonium tetrafluoroborate, $TEA^+ BF_4^-$) because of their moderate ionic conductivities and voltage windows [4]. The major solvents include acetonitrile and polypropylene carbonate. Aqueous electrolytes have very high ion conductivity rates and are easier to handle than organic electrolytes, but they are unstable at high potential (potential window limited to approximately 1 V). Common aqueous electrolyte solvents used in the development of electrode materials include sulfuric acid (H_2SO_4), potassium hydroxide (KOH), potassium chloride (KCl), and sulfates (K_2SO_4, Na_2SO_4). Attempts to design new and improved electrolyte performance have led to ionic liquids that are capable of high voltage windows (3 to 5 V). Also, attempts to reduce packaging complexity and minimize corrosion and safety problems have led to investigations of gel and solid-state polymers that combine both the separator and the electrolyte into a single component within an ES.

4.2 Anode and Cathode Structures and Materials

4.2.1 Overview of Battery Functions and Materials

Batteries are operated by converting the energy stored in chemical bonds to electrical energy. A constant potential is produced when a stable resistive load is applied and is equivalent to the difference between the half-cell potentials of the electrode materials. Figure 4.1 illustrates an example of the ion double-layer and nominal voltage generated by the half-cell reactions of

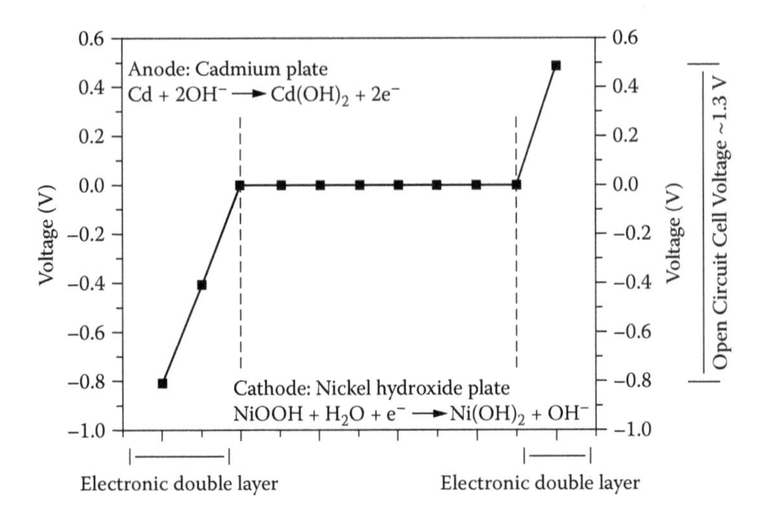

FIGURE 4.1
Voltage separation and characteristic ion diffusion layer from bulk caused by half-cell reactions in NiCd battery system.

the device [3,5]. The anode is more active than the cathode and spontaneously oxidizes, losing electrons and cations to the electrolyte solution. The electrons move through the circuitry as current to the cathode material; this reduces and creates an excess of anions in solution. The reaction proceeds at a constant rate until only a small amount of energy remains present in the cell, at which time the voltage produced by the cell will quickly drop off.

Depending on material properties and the nature of the reactions, the performance, safety, packaging, design, and application characteristics of batteries will vary. Table 4.1 showcases a number of commercial battery materials that dominate the energy storage market [5].

Figure 4.2 illustrates the lead acid battery that, until the invention of lithium storage technology, dominated the rechargeable battery industry ($18.4 billion of the $28 billion market in 2003) [5]. Lead acid batteries contain alloyed lead anodes and opposing lead oxide cathodes that share a sulfuric acid electrolyte. They are the oldest rechargeable devices and have lower volumetric (80 $Wh.L^{-1}$) and gravimetric energy density (40 $Wh.kg^{-1}$) than some of the more modern battery chemistries.

The main advantage of lead acid batteries is their ability to handle large surge currents and their low cost [3]. In the discharged state, the electrode plates become lead sulfates and most of the sulfuric acid in the system is absorbed, leaving the electrolyte as water. The water-heavy electrolyte lacks the abundance of ions that block gas evolution, making the cells susceptible to water loss during operation. The moisture loss dries out the battery and reduces its capacity over time, thus requiring maintenance. An important

TABLE 4.1

Commercial Battery Systems and Their Structures

Common Name	Nominal Voltage	Anode	Cathode	Electrolyte
Primary				
Leclanche	1.5	Zinc foil	MnO_2	Aq $ZnCl_2$_NH_4Cl
Zinc chloride	1.5	Zinc foil	Electrolytic MnO_2	Aq $ZnCl_2$
Alkaline	1.5	Zinc Powder	Electrolytic MnO_2	Aq KOH
Zinc–air	1.2	Zinc powder	Carbon (air)	Aq KOH
Silver–zinc	1.6	Zinc powder	Ag_2O	Aq KOH
Lithium–manganese dioxide	3	Lithium foil	Treated MnO_2	$LiCF_3SO_2$ or $LiClO_4$
Lithium–carbon fluoride	3	Lithium foil	CFx	$LiCF_3SO_2$ or $LiClO_4$
Lithium–iron sulfide	1.6	Lithium foil	FeS_2	$LiCF_3SO_2$ or $LiClO_4$
Rechargeable (Secondary)				
Lead–acid	2	Lead	PbO_2	Aq H_2SO_4
Nickel–cadmium	1.2	Cadmium	NiOOH	Aq KOH
Nickel–metal hydride	1.2	MH	NiOOH	Aq KOH
Lithium ion	4	Li(C)	$LiCoO_2$	$LiPF_4$ in organic solvent
Specialty				
Nickel–hydrogen	1.2	H_2(Pt)	NiOOH	Aq KOH
Lithium–iodine	2.7	Li	I_2	LiI
Lithium–sulfur dioxide	2.8	Li	SO_2 (C)	SO_2_LiBr
Magnesium–silver chloride	1.6	Mg	AgCl	Seawater

Source: Winter, M. and R. J. Brodd. 2004. *Chemical Reviews*, 104, 4245–4269. With permission.

development in module design is the valve-regulated lead acid (VRLA) design that can reduce gas emissions by over 95%. The valve remains closed during gas evolution, forcing oxygen recombination, and opens only if a set pressure level is reached [3].

Two of the other most common rechargeable battery systems are nickel–cadmium (NiCd) and nickel–metal hydride (NiMH) that utilize KOH electrolytes. The use of alkaline conditions means the charge discharge mechanism involves no change to the electrolyte composition and exhibits a very flat discharge profile [3]. The chemistry of NiCd provides longer cycle life, very low maintenance, and better charge retention than lead acid

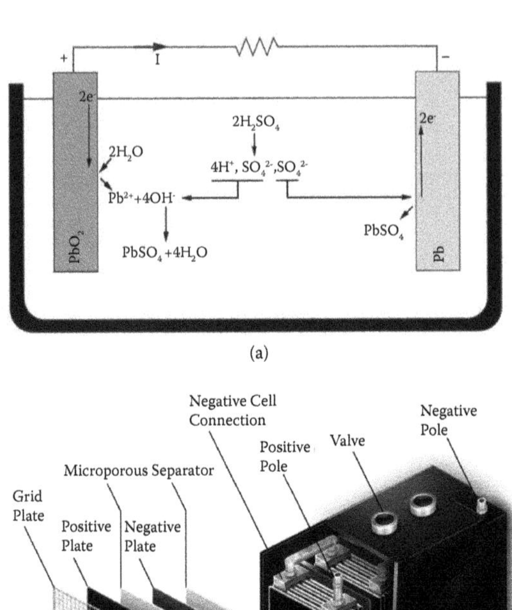

FIGURE 4.2
(See color insert.) (a) Lead acid battery showing anode, cathode, and sulfuric acid electrolyte. (b) Cross section of lead acid battery pack. Separation of plates is created by a nonconductive separator; cells are stacked within battery module. (*Sources: Worlds of David Darling Encyclopedia* (online). Lead–acid battery. http,//www.daviddarling.info/encyclopedia/L/AE_lead–acid_battery.html [accessed April 4, 2012]; Georgia State University. 2012. Lead–acid battery: hyperphysics (online). http,//hyperphysics.phy–astr.gsu.edu/hbase/electric/leadacid.html [accessed April 9, 2012]. With permission.)

systems. NiMH batteries have high energy density (100 $Wh.kg^{-1}$ and 430 $Wh.L^{-1}$) and slightly better cycle lives than lead acid systems. NiMH is limited by poor low temperature operation and memory effects that require battery management [3].

The lithium ion battery has emerged to capture 75% of the rechargeable battery market because lithium exhibits high redox potential, has long shelf life, and a high voltage window leading to high gravimetric energy (203 $Wh.kg^{-1}$) and volumetric energy (570 $Wh.L^{-1}$) densities [3]. The anode is graphite intercalated with lithium ions and the opposing plate consists of a lithium cobalt oxide source. The main issues with lithium technology are high cost, dendrite growth, safety problems, and the management required to prevent overcharge or discharge that can cripple performance.

Alternative cathode materials are utilized, depending on application, to mitigate the disadvantages of the lithium chemistry. These alternatives include lithium manganese oxide (lower cost, better safety, poor temperature stability), lithium iron phosphate (costly production, very good safety, low energy), and improved electrode designs that allow higher power density and solid polymer electrolytes to help reduce dendrite growth [3].

The appropriate electrode chemistry for a given application can be determined by considering an assortment of important parameters. The ideal battery electrode chemistry includes high storage ability (energy and power), long term stability, flat discharge shape, large operating temperature range, low cost, reduced safety issues, recharge capability, long cycle life, and short charge time [5]. Flat discharge shape allows for easy integration into electronic circuitry as a power supply and increased utilization of the energy stored in the battery. Storage ability is also dependent upon reaction efficiency; side reactions that emit stored energy as heat reduce usable energy.

After manufacture, batteries are held at a high chemical potential and their chemistry must remain stable during storage to extend shelf life before and during use. Stability issues include chemical side reactions, electrode phase changes, and corrosion reactions. All of these stability issues can reduce chemical storage capacity over the lifetime of a device. Ideally, rechargeable electrodes exhibit a return to the same material composition as before discharge. In real operation, electrode phase changes occur over time and can reduce cycle life and storage ability of a battery.

An example is the lithium device in which recharging can lead to long dendrite formation [3]. Eventually a shift in phase could lead to penetration of the separator, causing a short circuit. Side reactions lead to gas build-up or corrosion of casing materials. These can be concerns because electrode materials must operate safely in the designed application without causing thermal runaway, explosion, release of toxic gas, or corrosion of dangerous chemicals.

Issues arise when batteries are required to operate at high power, charge quickly, or act outside their designed temperature ranges. High loads tax a battery and the efficiency begins to drop due to side reactions. Limitations stem from an insufficient ability for ions to diffuse from bulk and for reaction rates to meet the power requirements put on the electrode materials. Thermal side reactions waste energy as heat, increasing the changes of by-product phase shift reactions. Stress caused from operating a battery beyond its rated power level can significantly decrease cycle lifetime through reduced storage ability and material corrosion.

4.2.2 Introducing Electrode Requirements for Electrochemical Supercapacitors

Static storage mechanisms in EDLCs efficiently store charges upon the electrodes at high rates. Unlike rechargeable batteries, ESs involve no chemical breakdown or redeposition of electrode materials during operation. This lowers the risk of electrode phase changes during operation and enables long electrode cycle lives. The static charge storage in EDLCs enables the anode and cathode to be interchangeable. Unlike specific electrode requirements seen in many batteries, the anode and cathode of an EDLC are composed of the same material. If the positive and negative current collector terminals and cell casing are also made of similar materials, theoretically the EDLC has no true polarity [8]. The important requirements for the optimization of electrode materials include:

- Long and stable cycle life ($>10^5$)
- Minimal irreversible redox processes
- High specific surface area
- Thermodynamic stability for a large potential window of operation
- Ability to control morphology, pore size, particle size, and material distribution
- Surface wettability
- High electrical conductivity
- Sufficient thermal conductivity to reduce heat build-up within cell
- Strong mechanical properties

Expanding upon the EDLC mechanism, pseudocapacitive materials used in ESs can be used to store more charge and energy. Pseudocapacitive materials undergo reversible redox reactions that are faster than those in batteries. Multiple redox states are available and each reaction will occur at a specific voltage as the device charges and discharges. The fast redox rate allows an increase in energy density while maintaining a large portion of the power density available in EDLC electrodes.

The main drawback is that pseudocapacitive materials suffer from poor cycle lives compared to EDLC devices. This knowledge can be utilized to design composite electrodes where pseudocapacitive materials can be deposited onto an EDLC support to provide highly porous structure. Further, in asymmetric (also known as hybrid) cells, pseudocapacitive material can be used at one electrode while EDLC materials can be used on the other. Hybrid designs are meant to optimize energy density while maintaining the power density available to EDLC-based capacitors. Both pseudocapacitive and EDLC contributions can be significantly altered by key factors such as surface area, structure, conductivity, and interaction with functional groups on the material surface.

4.2.3 Electrode Conductivity

Good electrical conductivity is important in enabling ES devices to operate at high power. During discharge, charges stored within the material must effectively maneuver through the thick electrode layers and out to the circuit. Electron transport is dependent upon the quality of conductive pathways within the material and the conductivity of the material used in the electrode. If the conduction through the electrode is not strong enough or the path is too tortuous, performance will suffer.

High power will not be achievable, as charges will not be able to respond quickly enough to match the rated load of the circuit. This is because the high capacitance provided by a poorly conductive material at very low power will be much lower at a desirable power output. Charges cannot organize efficiently to match the load and some charges will dissipate as heat to meet the current demanded by the circuit load. Eventually the device will no longer be able to meet the power demands and the ES will no longer store any charge. For poorly conducting electrodes, this will happen at much lower current density.

Metals used in traditional capacitors utilize highly conductive metals as electrode materials that can achieve very high power; however, ESs commonly use carbons as the active electrode components. Not all carbon is sufficiently conductive to support high power operation. Graphitic planes are highly conductive but temperatures used in the activation processes are limited to prevent complete restructuring into nonporous graphite.

A balance is achieved between conductivity of the active carbon and porosity, which in effect translates to better capacitive performance. The result is that many active carbon materials lack sufficient conductivity to support long range or short range conduction of current within the electrode layer. To restrict the range over which carbon conduction must occur, metal collectors are used as supports for the carbon layer. Further, when the active carbon is unable to effectively conduct over short distances, specifically designed small carbon additives (e.g., carbon black, Super P) are used to increase the conduction of the electrode.

4.2.4 Surface Area for EDLC Design

Increased electrode surface area plays an important role in performance. Enhanced area allows more electrolyte ions to organize at the electrode surface. Larger pores and channels in the electrode layer increase the accessibility and speed at which ions can organize onto the electrode pores from bulk electrolyte. Area plays a significant role for deciding on the materials for the electrode layers and the electrolyte. The pore structures of materials vary (macroporous >50 nm, mesoporous <50 nm, microporous <2 nm) and the ability to control the type of porous area available will lead to increased optimization between accessible power and maximum charge storage.

Maximizing the amount of surface area provides the greatest number of active sites in the material and improves the performance of the ES. However, surface area analysis is conducted with gas that can penetrate pores that are too small for ions so blocked areas may exist. Blocked pores or tightly bound planes prevent ions from organizing at the surface and can be due to poor long range order or collapsed channels within the electrode. To prevent blockages, materials are often designed with periodic structures or large macroporous channels. Chemical activation with KOH is another way to increase surface area by unblocking pores and creating new ones in the carbon materials. However, if the macropores are too big, they become macroscopic voids that detract from otherwise usable surface area.

4.2.5 Pore Structure for EDLC Design

In the past, desirable surface area encompassed macroporous structures for the increased flow of ions and mesoporous structures to maximize surface area. Ions were viewed as restricted from entering micropores. The comparably large solvent cage around the ions blocked entrance into the smaller pores [9,10]. Recent discoveries by Chmiolo et al. and Huang et al. [11–15] have shown that this is not entirely true. Figure 4.3 shows that as pore size decreases into the sub-5 nm range, capacitance also decreases.

The standard parallel plate model (applicable to region IV in Figure 4.3) of capacitance begins to fail below pore sizes of 5 to 10 nm (region III):

$$\frac{C}{A} = \frac{\varepsilon_r \varepsilon_o}{d} \tag{4.1}$$

Adjusting to account for pore curvature enables accurate modeling of region III as capacitance continues to decrease [1]:

$$\frac{C}{A} = \frac{\varepsilon_r \varepsilon_o}{b \ln\left(\frac{b}{b-d}\right)} \tag{4.2}$$

where b is pore radius, A is surface area of the electrode, d is distance between pore surface and the ion, ε_0 is the permittivity in a vacuum, and ε_r is the relative permittivity of the electrolyte. However, below 1 nm, a large increase in capacitance is observed. Further research reveals that around 1 nm the capacitance exhibits a sharp increase reaching a maximum at the ion size [13]. The increase in capacitance below 1 nm is characterized as the ions shedding their solvent cage, entering the micropore, and contributing to the storage mechanism of the double-layer. A pore structure can then be modeled as an electric wire in a cylinder [1]:

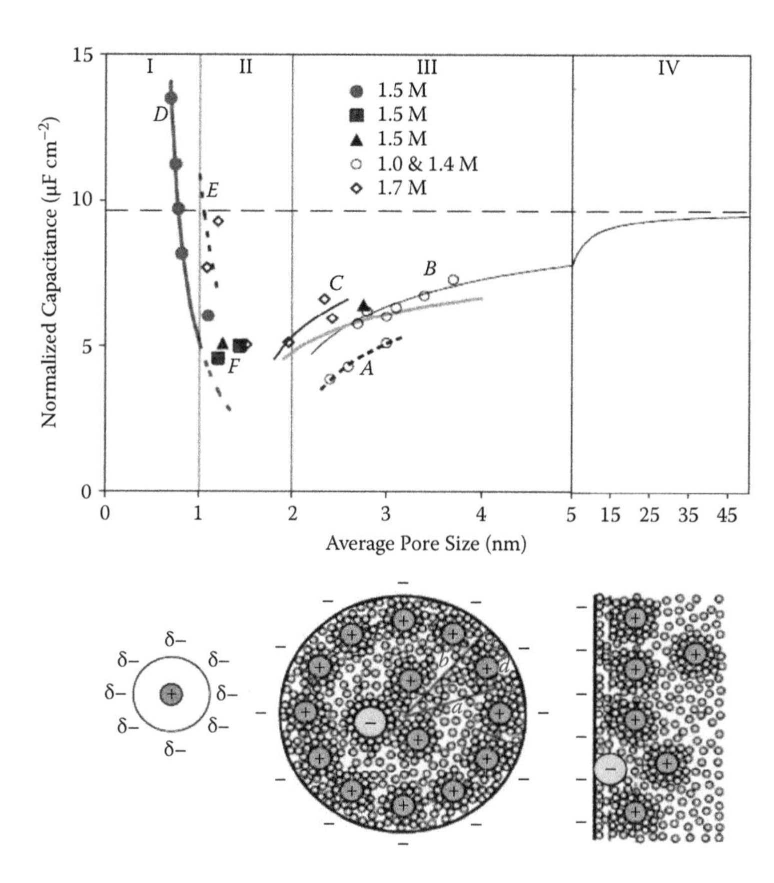

FIGURE 4.3
(See color insert.) Capacitance tested with various ionics in acetonitrile (TEAMS: 1.7M, TEABF4: 1,1.4,1.5M) for various carbon structures. Templated mesoporous carbon (A, B), activated carbon (C), microporous carbide derived carbon (D, F), and microporous activated carbon (E). The bottom images from right to left illustrate model of planar EDLC with negligible curvature, EDLC with pores of non-negligible curvature, and model single ion wire within cylindrical pore. The models can accurately estimate capacitance in their pore regions. (*Source:* Simon, P. and Y. Gogotsi. 2008. *Nature: Materials*, 7, 845–854. With permission.)

$$\frac{C}{A} = \frac{\varepsilon_r \varepsilon_o}{b \ln\left(\frac{b}{a_o}\right)} \tag{4.3}$$

where a_o is the effective size of the ion without its solvent cage. The solvent cage is shed when it becomes thermodynamically favorable to lose the ion solvent cage layer and enter the micropore system. Simulations have shown the energy penalty for entering a micropore to be small beyond a 1 nm range due to intermolecular force combinations [16]. This low energy penalty can be overcome by the driving potential of the capacitor during charging. This discovery illustrates potential for improvement with subnanometer- and nanometer-sized pore micropore structures to increase the capacitances of electrode materials. It also highlights the complex tunable pore design required to optimize electrode structures for EDLC application. At the same time, complex pore structure and material choices for electrolytes can significantly alter performance characteristics, advantages, and drawbacks.

Other important properties to consider in EDLCs are pore regularity and distance. In classical electrochemistry, ion transport time has a quadratic link to transport length [17]. Regular pore structures provide reliable short diffusion pathways between storage sites, but pore defects disrupt pore regularity, which increases the pore interspacing and causes ion scattering. Irregular pore spacing and ion scattering detract from ion transport speed and throughput, resulting in fewer accessible ions and lower power. Scattering can also be caused by poor interfacial wetting of the electrode. Depending on the character of the electrolyte solvent, hydrophobic or hydrophilic dopants or functional groups can be applied to the material.

4.2.6 Functionalization Effects on EDLCs

Small functional groups on the electrode layer materials, such as carbon particles, improve wetting with the electrolyte and allow increased solvent penetration within small pores of the electrode materials. For example, nitrogen dopants and oxygen functional groups can improve wettability in aqueous systems by altering surface effects [18]. Further, active heteroatoms, such as oxygen and nitrogen, give carbon materials a very stable covalently bound acidic (oxygen, electron acceptor) or basic character (nitrogen, electron donor) that introduces a pseudocapacitive component on top of the EDLC capacitance [19].

Pseudocapacitive metal oxide and metal nitride coatings deposited and adsorbed on carbon supports are also effective in increasing the maximum energy density possible for an ES and are discussed in detail later in this chapter. Oxygen dopants such as carboxyl, carbonyl, and hydroxyl groups can improve wetting and open surface area for increased capacitance and higher power [20]. Activated carbons can incorporate oxygen groups through steam activation during carbonization or through a secondary KOH activation reaction. An overabundance of oxygen functionalities disrupts the graphitic π-bonding network of the material and reduces conductivity of the material too much. In active carbons, this can occur when activation temperatures and activation reagent concentrations are too high.

Graphene oxide (GO), the precursor material of graphene, contains a very large number of oxygen groups introduced by a harsh oxidation reaction from bulk graphite. The precursor material is insulating until it is reduced with a strong reduction technique (sodium borohydride, high temperature, and hydrazine) that aims to remove all the oxygen functionalities and restore the graphitic plane to create conductive few-layer graphene for use in ESs.

Lin et al. [20] illustrate that a complete reduction is not necessary to achieve acceptable graphene capacitance. By using a low temperature reduction (150°C) of GO in DMF under a nitrogen environment for time varying from 1 to 6 hr, they were able to optimize the electrolyte-wetting benefit of oxygen functionality against the capacitance developed by restoring graphitic plane conductivity (3 hours was optimal). After 2000 cycles, the capacitance even increased because of the increased oxygen content adsorbed from the aqueous electrolyte over time (230 $F.g^{-1}$ at 1 $A.g^{-1}$ and 0 cycles; 250 $F.g^{-1}$ at 1 $A.g^{-1}$ and 2000 cycles) [20]. This achieved a higher level than hydrazine-reduced graphene and thermally or surfactant exfoliated graphene (100 to 200 $F.g^{-1}$). This further illustrates the importance of balanced oxygen functionalities.

Nitrogen functionalities can be incorporated into carbon materials by treating them with ammonia or using precursor materials that contain nitrogen compounds, such as melamine, polyacrylonitriles, or polyvinylpyridine [18]. Nitrogen can form strong covalent bonds with carbon that can regulate electrical and chemical properties of carbon because of its comparable size and five valence electrons. Dopants can manifest in the form of pyridinic, pyrrolic, and quaternary bonds that contribute a large number of electrons to the delocalized graphitic π network within the carbon structure, and can alter the surface and catalytic properties of the carbon structures [21].

Theory further suggests that nitrogen dopants increase the positive charge present on adjacent carbon atoms [18]. The effects of nitrogen on ES performance were illustrated by Jeong et al. [21] when investigating nitrogen effects on few-layer graphenes by achieving high capacitance (250 $F.g^{-1}$ at 1 $A.g^{-1}$), increased hydrophilic character, continued performance at high rates (30 $A.g^{-1}$ and 175 $F.g^{-1}$), and stability of the increased capacitance for more than 10,000 cycles.

Belanger et al. [88] propose enhanced performance through pseudocapacitance using diazonium chemistry to covalently bond specific functional groups to carbon surfaces. Diazonium salts have a general form seen in Figure 4.4. The end group R can be selected to provide surface functionalities such as hydrophilicity (e.g., COOH), hydrophobicity (e.g., CF_3), or electroactivity (e.g., NO_2).

Anthroquinone (AQ) is a diazonium cation that offers a large pseudocapacitive peak (Figure 4.5) within the range utilized by ESs (–0.05 to 0.35 V) [22,23]. The cation is formed by the reaction of aminoanthroquinone (AAQ) dissolved in acetonitrile with injections of tertiary butylnitrite for varying times to create AQ cations. In situ grafting occurs when nitrogen

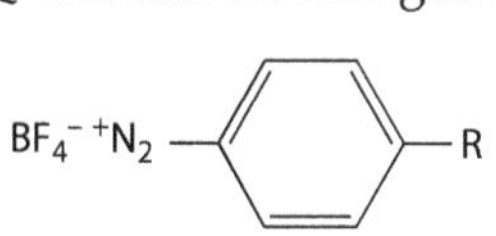

$BF_4^- \ {}^+N_2$ — (benzene ring) — R

FIGURE 4.4
General form of diazonium salt.

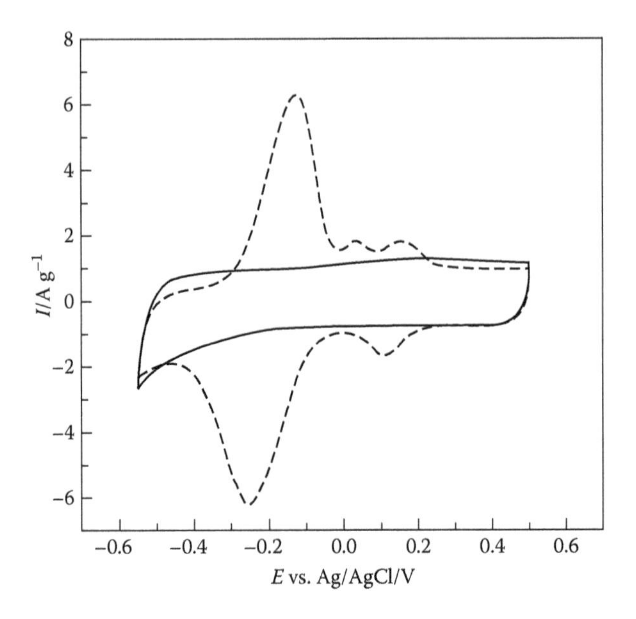

FIGURE 4.5
Cyclic voltammogram comparison of unmodified AC fabric and pseudocapacitive peak generated by addition of AQ functionality on carbon surface. Testing carried out in 1 M H_2SO4 at 0.1 $A.g^{-1}$. (*Source:* Pognon, G. et al. 2011. *Journal of Power Sources*, 196, 4117–4122. With permission.)

leaves the cation, creating a radical that binds with carbon (Figure 4.6) [23]. Alternatively, a stable azo bond (C-N=N-C) can form with the carbon prior to radical formation.

The presence of AQ functionalities on carbon provides the opportunity for oxidation reactions on either of the oxygen sites present on each AQ group. By altering the reaction time, functionalization is shown to reach a practical maximum of 11 to 14% before the pseudocapacitive peak created by AQ begins overlapping with the hydrogen evolution potential of the aqueous electrolyte. The large AQ functional groups block the pore systems present in ACs and can significantly lower surface area that can also reduce EDLC capacitance [24]. However, the EDLC component of capacitance still

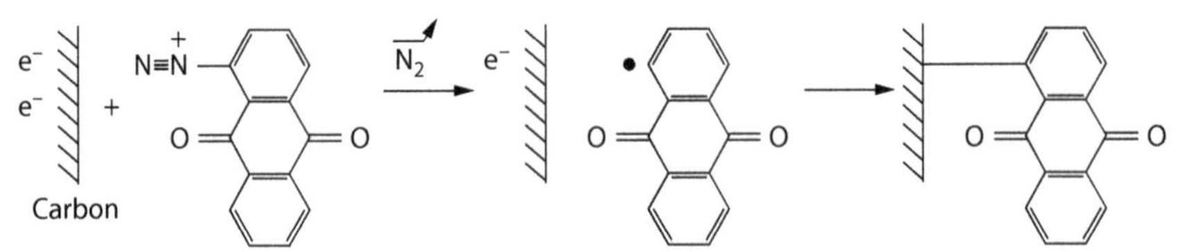

FIGURE 4.6
Reaction mechanism for chemical modification of carbon by spontaneous reduction of anthroquinone diazonium cations. (*Source:* Pognon, G. et al. 2011. *Journal of Power Sources*, 196, 4117–4122. With permission.)

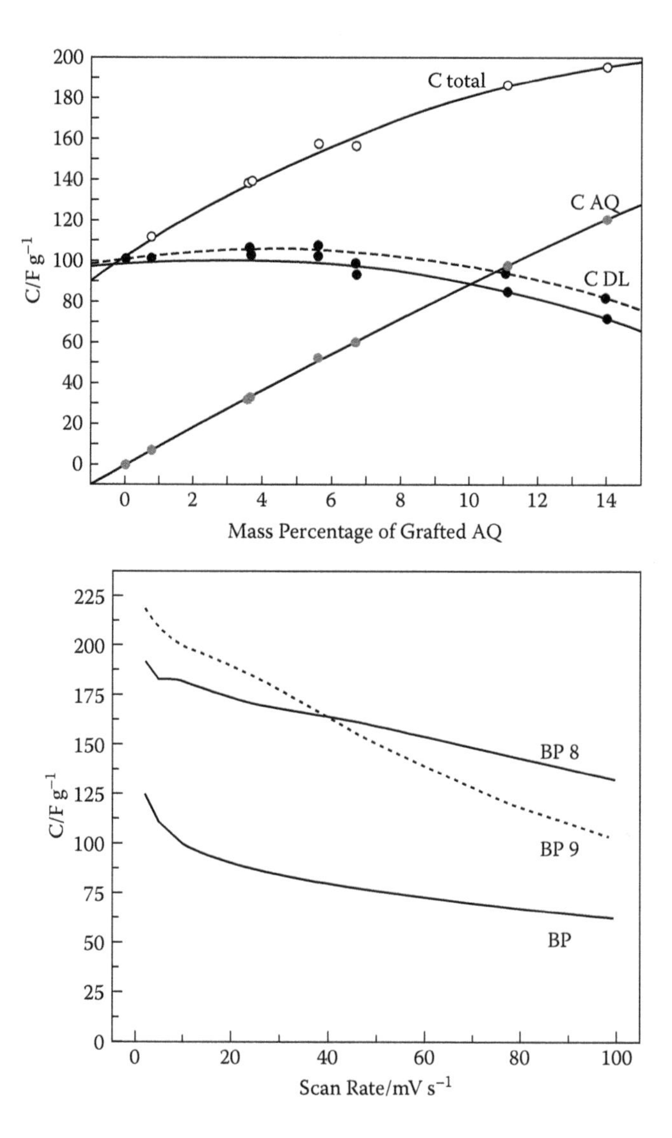

FIGURE 4.7
Top: Plot of total capacitance, AQ capacitance, and EDLC capacitance of modified AC cloth as function of AQ loading. Bottom: Persistence of total capacitance for CV at increasing scan rates where BP, BP 8, and BP9 represent 0 wt%, 11 wt%, and 14 wt%, respectively. (*Source:* Pognon, G. et al. 2011. *Journal of Power Sources*, 196, 4117–4122. With permission.)

maintains around 75% of its capacitance with 11 wt% AQ loading [24]; see Figure 4.7, top.

In addition to the large remaining EDLC component, a pseudocapacitive component is introduced by the functionality that almost doubles the overall capacitance at both low scan rate (10 mV.s^{-1}) and at faster scan rates (100 mV/s); see Figure 4.7, bottom. The double oxidation creates a large capacitance within the peak range that averages 370 $F.g^{-1}$ for 11 wt% AQ—more than three times the 120 $F.g^{-1}$ produced by the unmodified AC carbon fabric (1500 $m^2.g^{-1}$) [22]. This high capacitance in the narrow voltage region highlights AQ-modified materials for positive electrodes in asymmetric cell designs [22].

4.2.7 Series Resistance in EDLC Design

The series resistance (ESR) affects power loss and stems from the internal resistances within capacitor materials. Traditionally, the equivalent series resistance represents deviation from the pure ideal capacitor (90° phase, no resistance) due to the presence of an ohmic resistance generated in the dielectric. At very high frequencies, a phase delay occurs because the molecular relaxation of the dielectric is kinetically limited and polarization cannot occur quickly enough [25]. In supercapacitors, the dielectric losses within the double-layer start to occur only when frequencies greater than hundreds of megahertz are used—the ohmic losses do not dominate. Larger ESR resistances are generated due to the nature of the porous structure and multi-component cell design:

- External contact resistances
- Electrolyte solution resistance
- Separator transport resistance due to scattering and insufficient ion conduction
- Internal electrode resistance due to insufficient electron conduction
- Internal interparticle contact resistance
- Electrode–collector contact resistance

The reported effective ESR is taken from a Nyquist plot of real versus imaginary impedance. At high frequency approaching infinity, the imaginary impedance tends to zero as the capacitor becomes an AC short circuit. The ESR can be taken from the phase delay or intercept of the real impedance at high frequency. Alternatively, the ESR can be determined from the voltage drop seen during constant current discharge. These techniques are discussed in more detail in Chapter 7. As previously discussed, the ESR present in ESs is a very important factor in device performance.

Power loss due to resistance creates a significant amount of heat that can limit operating power output and performance characteristics. As devices age, degradation sets in, causing an increase in series resistance. Continued device operation at extreme temperatures or high power, lack of proper cooling, or contamination through gas leakage or permeation will increase degradation rates and the ESR will increase, leading to reduced performance and shorter cycle life.

4.2.8 EDLC Electrode Materials

4.2.8.1 Activated Carbons

Carbons act as excellent conductors, are chemically stable, and have high surface areas, making them the preferred materials for double-layer electrodes in modern ESs. Carbon, however, comes in many varieties and not all are applicable to electrode materials. The industrial standard and most basic high surface area carbon material is activated carbon (AC). It is widely used because of moderate cost and easy preparation.

AC materials can be generated from a number of precursor materials through carbonization and high temperature annealing in an inert atmosphere. Materials are pitches or resins derived from coal and petroleum that exhibit liquid phase shifts, allowing alignment during graphitization, creating heavily microporous areas that require activation for successful use in ESs.

In general, high surface area alone is not easily correlated to capacitance; pore structure seems to be critical. Mesopores are important to allow sufficient ion diffusion kinetics to support high power and area (20 to 50% suggested) [26]. Surface level micropores allow large area for storage but limit ion mobility to loss of their solvation shells. Macropores are important to provide high throughput ion channels.

Alternative carbon precursors are derived from more structured compounds such as wood, polymers, and hard shells that do not exhibit liquid phases during carbonization and thus maintain alignment and rigidity. Carbons derived from natural biomass exhibit large voids inherent to the natural structure of the material, reducing the high volumetric area needed for ESs. Table 4.2 lists available activated carbon powders along with their activation pathways and properties [27]. Capacitance testing of carbon materials on the list showed that KOH activation resulted in the highest capacitance, 154 $F.g^{-1}$ in organic electrolyte.

Activated carbons can often achieve capacitances as high as 100 to 200 $F.g^{-1}$ in aqueous electrolyte systems and 50 to 150 $F.g^{-1}$ in organic media [28]. An AC material prepared by Gryglewicz et al. [26] showed 160 $F.g^{-1}$ in aqueous electrolyte for a material made from coal and activated by steam (SA = 1270 $m^2.g^{-1}$). Wu et al. [29] prepared AC's from firewood using steam activation resulting in a maximum capacitance of 120 $F.g^{-1}$ in acidic electrolyte.

KOH activation was used on petroleum coke by Wang et al. [30] to produce material with 1180 $m^2.g^{-1}$ and maximum capacitance of 160 $F.g^{-1}$ in aqueous electrolyte. Kierzek et al. [31] illustrated that high surface area carbon (3000 $m^2.g^{-1}$) produced with KOH activation showed improved capacitance of 300 $F.g^{-1}$ in aqueous electrolyte. Table 4.2 shows performances in organic electrolytes and illustrates that (1) high surface area AC does not directly correlate to higher performance and (2) precursor materials and activation mechanisms are important to electrochemical performance.

TABLE 4.2

Various Activated Carbon Powders Generated via Different Methodologies and Their Performances in Organic Solvent

Carbon	Supplier	Precursor	Activation Process	Country	BET SA (m^2/g)	Average Pore Size (nm)	Capacitance (F/g)
Grade 1 Cloth	MarkeTech International	Resorcinol Formaldehyde	None	U.S.	402	4.14	28
RP-15	Kuaray	Phenolic resin	Steam	Japan	1318	1.94	90
NK-260	Kuaray	Mesophase pitch	KOH	Japan	2040	1.92	154
Nuchar RGC	MeadWestvaco	Mixed hardwoods	Chemical	U.S.	1622	2.99	82
Supra 50	Norit	Coconut shell	Steam	Netherlands	1989	2.05	81
TDA-1	TDA	Sucrose	Carbon dioxide	U.S.	2053	2.04	86
Generation 1	University of Kentucky	Filtration carbon	Carbon dioxide	U.S.	1907	2.46	88

Source: Smith, P. and T. Jiang. 2009. *High Energy Density Ultracapacitors*. http://www1.eere.energy.gov/vehiclesandfuels/pdfs/merit_review_2009/energy_storage/esp_22_smith.pdf With permission.

Note: Electrodes were made with polymer binders and conductive carbon black additives. Electrolytes were made with acetonitrile and 2 M $LiBF_4$.

The Maxsorb commercial carbon made in Kansai, Japan is produced by KOH activation of petroleum coke at 700°C and achieves over 3100 $m^2.g^{-1}$ [32]. Wan et al. [33] demonstrated that Maxsorb carbon can produce capacitance in aqueous electrolyte of 225 $F.g^{-1}$ at 1 $A.g^{-1}$ current density. In the same test, they showed that polymeric-activated carbon derived from KOH activation of resorcinol–formaldehyde exhibited only 1673 $m^2.g^{-1}$ surface area but was able to reach 325 $F.g^{-1}$ in aqueous electrolyte [33].

4.2.8.2 Templated Active Carbons

As carbon electrode technology moves forward, researchers have room to play with advanced carbon structures that have higher pore orders and increased conductivities. Closed pores produced by the activated carbon process reduce charge rate capabilities and overall storage capacities. Template systems offer to improve this deficiency by creating long range orders within carbon structures. The process commonly involves mesoporous silica or zeolite templates as seen in Figure 4.8 [34,35]. Results of studies of template systems have shown performance exceeding 300 $F.g^{-1}$. The costs of templates currently limit their use for building a theoretical understanding of the effects of pore size on capacitance and ion kinetics [36]. With improved

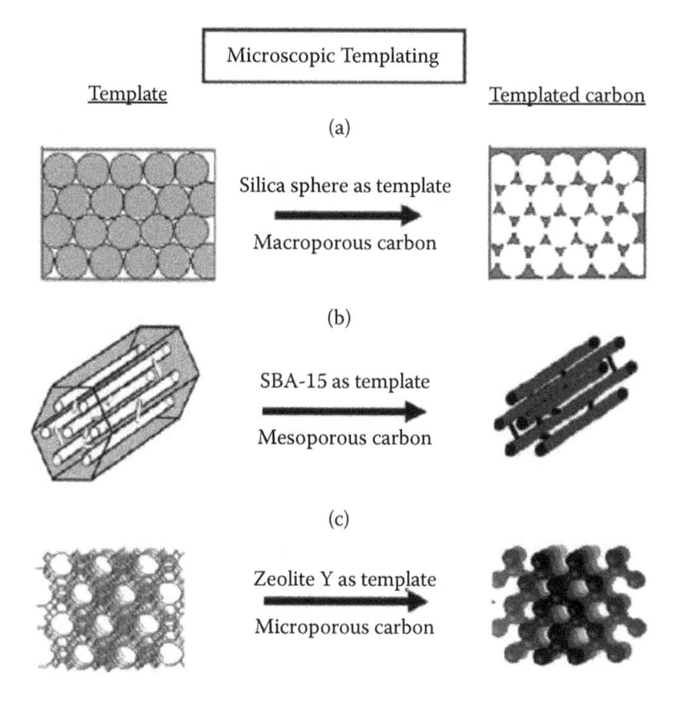

FIGURE 4.8
Templating for pore types. (*Source*: Zhang, L. L. and X. S. Zhao. 2009. *Chemical Society Reviews*, 38, 2520–2531. With permission.)

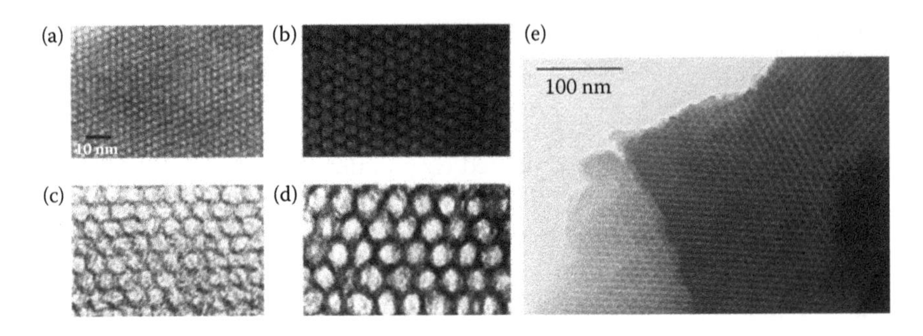

FIGURE 4.9
(a)–(d) TEM images of MCM-41 with pore sizes ranging from 2 nm (a) to 10 nm (d). (*Source*: Beck, J. S. et al. 1992. *Journal of the American Chemical Society*, 114, 10834–10843. With permission.) (e) TEM image of carbon prepared from MCM-48 using pitch as carbon source. Both images illustrate highly ordered structures compared to activated carbon materials. (*Source:* Vix-Guterl, C. et al. 2005. *Carbon*, 43, 1293–1302. With permission.)

production methods and lower costs, these methods may become accessible to commercial energy storage applications over time.

Zeolite templated carbons exhibit high performance without extensive activation due to the high surface areas and long range orders of their porous structures. Wang et al. [37] used zeolite X (670 $m^2.g^{-1}$, 1.4 nm pore size) and 8 hr of chemical vapor deposition (CVD) to introduce carbon into the template. The resulting template CNX-2 (2700 $m^2.g^{-1}$) had capacitance of 158 $F.g^{-1}$ (at 0.25 $A.g^{-1}$) and energy density of 25 $Wh.g^{-1}$, respectively, in aqueous electrolyte. Due to the ordered nature of the pore structure, over 97% of the capacitance was retained at rates of 2 $A.g^{-1}$. The zeolite could produce dense carbon pore structure (1.07 $cm^3.g^{-1}$) and improved volumetric capacitance versus most activated carbon materials [37].

Portet et al. [38] synthesized templated carbon (Y850) from zeolite Y using acetonitrile with nitrogen doping. Y850 had a surface area of only 1800 $m^2.g^{-1}$, but still attained performance of 146 $F.g^{-1}$ in organic electrolyte [38]. Nishihara et al. [39] obtained 168 $F.g^{-1}$ in organic electrolyte using zeolite X with pore CVD-coated acetylene as a carbon precursor. Ania et al. [34] illustrated that templated carbons with lower surface areas (1680 $m2.g^{-1}$) could still show good performance of 300 F.g–1 in aqueous electrolyte due to the accessibility of the ordered micropores for storage.

Mesoporous carbon structures produced by templating with MCM-48 and SBA-15 could produce larger ion conducting channels (Figure 4.9) [28]. The ion channels allow electrolyte to access the microporous area of the precursor carbon and provide enhanced performance versus activated carbons of equal surface area. Templated carbon derived from MCM-48 was used by Vix-Guterl et al. [40] to produce carbon capable of 115 $F.g^{-1}$ in organic electrolyte. Further, Kim et al. [41] developed nitrogen doped carbon capable of 182 $F.g^{-1}$ in aqueous electrolyte by templating on SBA-15. In an organic electrolyte, Fuertes et al. [35] demonstrated carbon derived from SBA-15 (3 nm) with

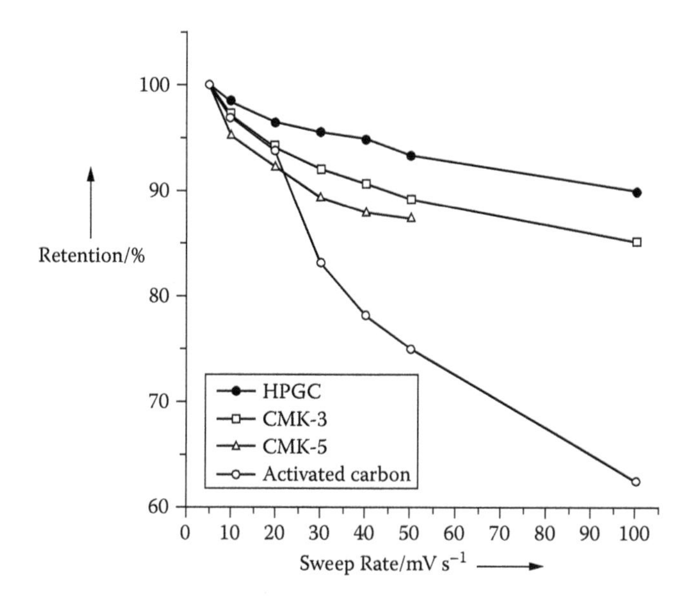

FIGURE 4.10
Retention of capacitive performance with increasing sweep rate and power for various carbon materials. (*Source:* Wang, D. W. et al. 2008. *Angewandte Chemie*, 47, 373–376. With permission.)

98 F.g^{-1} in organic electrolyte at scan rates of 10 mV.sec. Also, Lufrano et al. [42] used CMK-3 carbon generated by SBA-15 (8 nm pore size) to demonstrate 132 F.g^{-1} using a flexible Nafion® electrolyte.

Silica nanospheres can also be used to template carbon with interesting results. Lei et al. [45] produced nanospheres of 2.7 nm, resulting in mesoporous carbon area of 2400 m2.g^{-1}. The carbon sphere material showed capacitances of 225 F.g^{-1} and 180 F.g^{-1} in aqueous and organic electrolytes, respectively. The capacitance in organic electrolyte showed only a small drop compared to the aqueous electrolyte, leading to a high energy density of 62.8 Wh.kg^{-1} at low power density under 1 kW.kg^{-1}. However, the high energy density faded quickly, leading to moderate energy of 9 Wh.kg^{-1} at high power density of 30k W.kg^{-1} [45].

Based on the results from various microporous and mesoporous templates, it is clear that templated carbons are capable of improving upon most AC materials, but they still lack the macropores needed to support higher energy density at high power output. Improving upon this, Wang et al. [46] developed the templated carbon HPGC that showed a better capacitance retention (Figure 4.10) and improved electrode performance at high power.

The HPGC material is a composite of ordered micropores, mesopores, and macropores as illustrated in Figure 4.11. The composite structure can create short diffusion pathways for the electrolyte to enter the microporous walls and generate a high energy density of 22.9 Wh.kg^{-1} in an organic electrolyte. Furthermore, the accessibility of the pores allowed the energy density

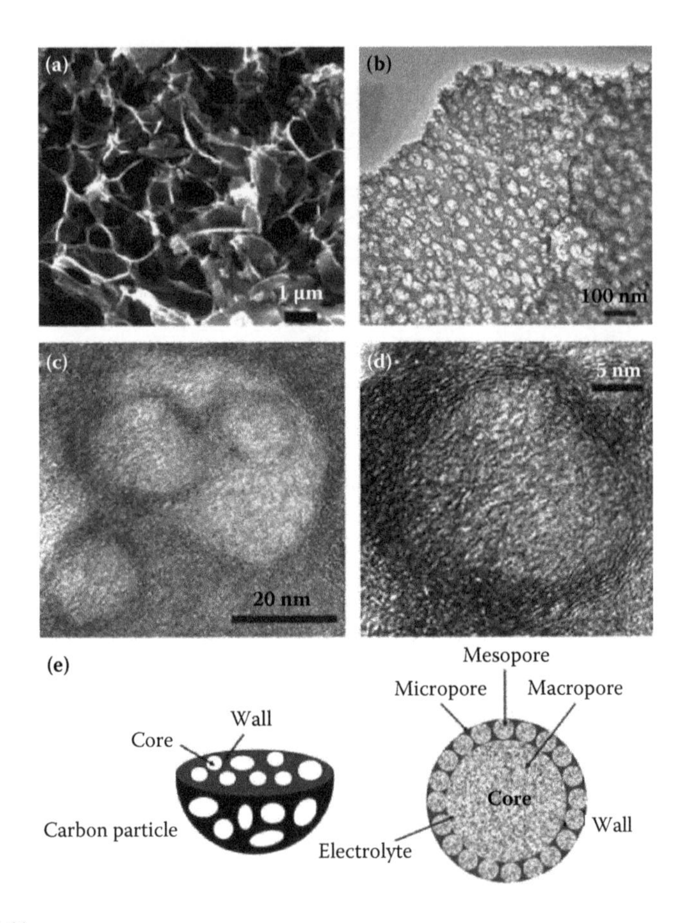

FIGURE 4.11
(a) SEM image of macroporous cores of HPGC. (b) TEM image of mesoporous walls. (c) TEM image showing micropores. (d) TEM image of localized graphitic mesopore walls. (e) Three-dimensional hierarchical structure. (*Source:* Wang, D. W. et al. 2008. *Angewandte Chemie*, 47, 373–376. With permission.)

to remain above 20 Wh.kg^{-1}, even for high power rates of 23 kW.kg^{-1} [46]. This example demonstrates the promise of EDLC electrodes with refined pore structures for obtaining both high energy and power density.

4.2.8.3 Carbon Nanotubes

More advanced carbons are being heavily studied for their potential in controlling structures and enhanced properties compared to activated carbon. Carbon nanotubes (CNTs; Figure 4.12) are cylindrical nanostructures that exhibit a near-1-dimensional structures. The cylinders are composed of graphitic carbon walls in the forms: single wall (SWNTs), double wall (DWNTs), and multiwall nanotubes (MWNTs). Their cylindrical radii are on the order

FIGURE 4.12
Comparison of (a) aligned CNT electrode array and (b) entangled CNT matrix.

of nanometers and length can reach millimeter scales. The graphitic planes in a CNT produce electrically conductive carbon tubes (conductivities as high as 1000 $S.cm^{-1}$) depending on CNT type [28]. An ordered array or a loosely entangled layer of CNTs has high surface accessibility, leading to high ionic conductivity [47]. The combination of rapid ion diffusion and high electrical conductivity produces devices of higher power than AC electrodes.

Hu et al. [47] demonstrated the power density of paper-based CNT electrodes. CNTs were brushed onto paper and through strong Van der Waals forces adsorbed to the cellulose fibers in the paper, forming a strong, porous, and conductive electrode. Considering only the weight of active CNT material, the device showed an exceptional maximum power density of 200 $kW.kg^{-1}$ and a maximum energy density of 30 $Wh.kg^{-1}$. Yoon et al. [48] shows that aligned CNT films can maintain a near rectangular CV curve at high scan rates of 1 $V.s^{-1}$. At the same rate, the AC material (1800 $m^2.g^{-1}$) tested showed high levels of resistance.

MWNT array electrodes produced by Honda et al. [49] illustrate the potential of CNTs for high power capacitors. At low current density, the MWNTs exhibited a low capacitance of only 15 $F.g^{-1}$ due to poor surface area. The TEM image in Figure 4.13 illustrates a single MWNT with highly conductive graphitic walls and tight interspacing that restricts surface area [50]. However, the combination of very low electrode equivalent series resistance (ESR, 1.9 Ωcm^2) and high ionic conductivity could enable the MWNT electrode to retain 12 $F.g^{-1}$ or 2.2 $Wh.kg^{-1}$ for high current density of 200 $A.g^{-1}$, corresponding to power density of 125 $kW.kg^{-1}$. When the AC cloth was charged at rates above 10 $A.g^{-1}$, large resistance prevented the development of any energy or capacitance on the electrode. Maximum power of the device is calculated to be 3.2 $MW.kg^{-1}$ [49].

CNT's are fabricated by a number of different methods including arc discharge, CVD, high pressure carbon monoxide (HiPco), and laser ablation. Laser ablation mainly produces SWNTs but is more expensive than the CVD and arc discharge techniques [51]. Alternatively, CVD methodology allows

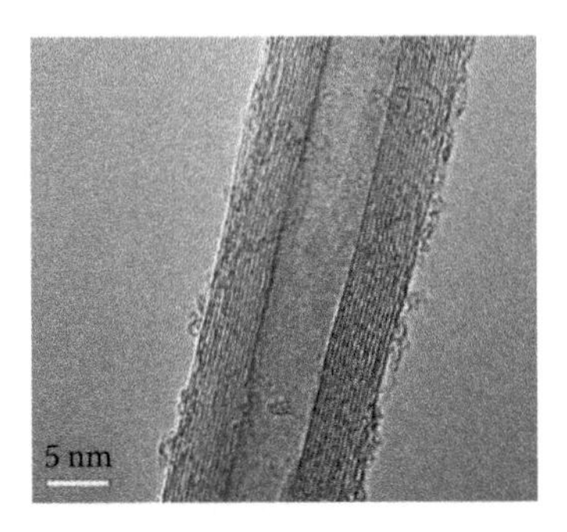

FIGURE 4.13
TEM image of bare MWNT used for high power supercapacitor. Low surface area is caused by tight interspacing between graphitic carbon walls. (*Source:* Zhou, R. et al. 2010. *Nanotechnology,* 21, 345701. With permission.)

lower operating temperatures and increased potential for moving to the higher scales required to decrease the high cost of CNTs.

All CNT production strategies are expensive due to the high energy processes, extensive purification, scalability issues, and size control required for commercialization. The high cost, especially for SWNTs, is a limiting factor in their adoption into ESs. Another major limitation to the adoption of CNTs in ESs is the low surface area, less than 500 $m^2.g^{-1}$ compared to areas of AC devices. Commercial HiPco production generates exclusively SWNTs that attain BET surface areas as high as 800 $m^2.g^{-1}$ [52]. This surface area is still far below those of AC materials.

The high theoretical surface areas of graphite planes are not observed for a number of reasons. First, MWNTs and DWNTs exhibit greatly reduced areas compared to SWNT's; the tight interlayer spacing limits the surface areas of multi-wall structures to the outer wall of the tube. Also, individual CNTs are known to bundle together due to the strong Van der Waals forces between their basal planes. Further entanglement of the CNTs can also occur during the cycle life of the electrode, reducing capacitance significantly [31].

Bundled CNT's prevent electrolyte access to all but the outermost tubes, significantly reducing the active surface areas of the materials [52]. To optimize accessible area, steps must be taken to separate bundles by sonication and by additions of stabilizers or dispersants [53]. Similar to ACs, electrochemical oxidation with KOH can be performed to increase the surface areas and the capacitances of CNTs [52]. The oxidation increases area by uncapping the nanotubes and exposing more internal surface area. Too much oxidation or large concentrations of dispersants can negatively alter CNT performance, so care must be taken to optimize the properties.

Capacitive performance of SWNTs was shown to reach 180 $F.g^{-1}$ in KOH by Lee et al. [54]. Another example by Nui el al. [55] shows SWNT capacitance of 102 $F.g^{-1}$ in acidic medium. CNT electrodes built with MWNT active material exhibit a lower range of capacitance between 5 and 135 $F.g^{1}$ depending on surface area, CNT quality, and type [31]. This shows that even SWNTs that exhibit moderate surface area underperform AC materials when gravimetric capacitance is considered.

However, the low gravimetric capacitance of CNTs is offset by their higher packing density that creates volumetric capacitances exceeding those of many AC materials [52]. ACs also suffer degradation of capacitance more rapidly than CNTs at high rates of operation because they exhibit less ionic and electrical conductivity. As a result, CNTs offer superior performance in high power electrodes. The higher conductivity along with the large aspect ratio enables a lower percolation threshold (0 to 4% for CNTs) compared to conductive carbon black additives (3 to 15%) [56].

An example of this is shown by Liu et al. [57] through the addition of a small percentage of CNTs to AC, resulting in a lower resistance and a higher capacitance of 180 $F.g^{-1}$, compared to 130 $F.g^{-1}$ when carbon black binder was used. The low percolation means that no carbon additives are needed in CNT electrodes to lower internal electrode resistance. Further, sheet resistances were observed to be as low as 1 to 10 $\Omega.cm^2$ for CNT-based films [47,49] and this suggested that heavy metal current collectors are not needed to provide conduction along the length of an electrode.

As a result, complete devices can be designed with less dead weight and improved gravimetric performance. The strong ionic conduction throughout entangled CNT electrodes could create conductive, lightweight current collectors for pseudocapacitive electrodes [36]. Zhou et al. [50] created a super aligned CNT film with low surface area (100 $m^2.g^{-1}$) and baseline EDLC capacitance of only 5 $F.g^{-1}$. However, the strong binding of MnO_x nanoparticles to the CNTs and conductive structures enabled performance of 245 $F.g^{-1}$ (including CNT collector mass) at a very high current density (155 $A.g^{-1}$) and reliable stability over 2500 cycles [50].

Vertically aligned CNT "forests" created by CVD can have highly ordered structures, resulting in even higher ionic conductivity compared to film-based CNT electrodes. The increased packing density improves volumetric capacitance. Recently, an improved ("super growth") CVD process was developed by Hate et al. [58]. This process involved the addition of water into the CVD chamber during deposition, which could increase the activity and lifetime of the metal catalysts for growth. The super growth generated ordered SWNT forests with heights of millimeters. The method is highly efficient, exhibiting short growth times of only 10 min and yielding SWNT with 99.9% purity without further purification steps [58].

Similar to standard CVD, vertical alignment of the CNTs is performed through the application of a high electric field during growth. When CNT growth is complete, further alignment occurs through a "zipping effect" caused by immersing a CNT electrode in solvent. The effect is the by-product of the solvent surface tension and the Van der Waals interactions of the CNTs as the electrode dries. The super growth process enables a surface area of

1000 $m^2.g^{-1}$ and upon electrochemical oxidation to remove the CNT tips, the surface area can reach 2200 $m^2.g^{-1}$ [52].

Hiraoka et al. [59] illustrate that high levels of oxidation caused by temperatures beyond 500°C reduced performance by burning the CNTs and enabling heavy bundling. An optimized process was utilized to create a CNT device with intimate contact to the metal collector and surface area beyond 2000 $m^2.g^{-1}$. Minimization of bundling could lead to 80% availability of the area to electrolyte ions. The resulting CNT electrode showed high capacitance of 115 $F.g^{-1}$ (24.7 $Wh.kg^{-1}$) in organic electrolyte at a current density of 1 $A.g^{-1}$. The maximum power density was calculated to be 98.9 $kW.kg^{-1}$ and due to the dense CNT packing, maximum volumetric power density was determined to be 60.1 $kW.L^{-1}$ [59].

4.2.8.4 Carbon Onions

Onion-like carbon (OLC) is composed of quasi-spherical concentric graphitic shells. Pech et al. investigated them for use in EC devices [60]. OLCs have only moderate specific areas of 500 $m^2.g^{-1}$ and are produced by the detonation of diamond powders at temperatures of 1800°C. This is a low valued compared to AC results, but the small ordered spherical particles in OLC can avoid the porous network seen in high area AC materials. Furthermore, the continuous curvature ensures mesoporous channels for electrolyte access and prevents bundling problems common in CNTs. This means that surface area is fully accessible to electrolyte. Figure 4.14 shows that charge transport distances (<10 nm) through the carbon walls are short and regular [60].

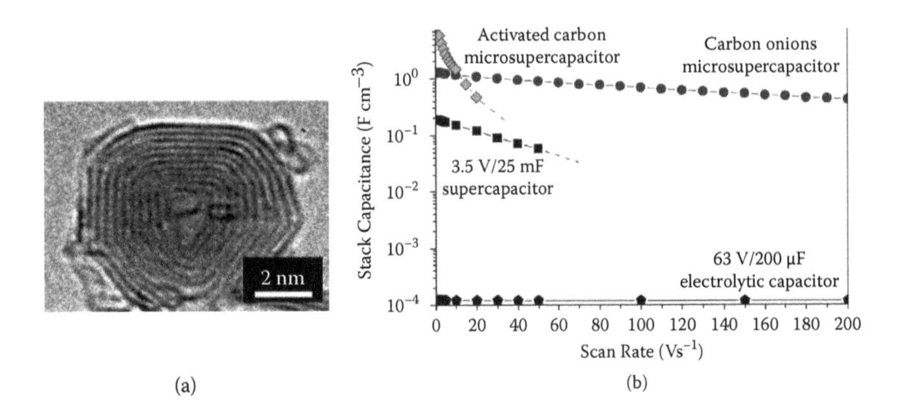

FIGURE 4.14
(a) TEM image of cross section of carbon onion (2 to 10 nm diameter). Volumetric capacitance of ordered OLC spheres compared to AC device built by same process. (b) Comparison of devices in (a) to other AC supercapacitive devices and electrolytic capacitors that exhibit the highest energies among traditional capacitor designs. (*Source:* Pech, D. et al. 2010. *Nature: Nanotechnology*, 5, 651–654. With permission.)

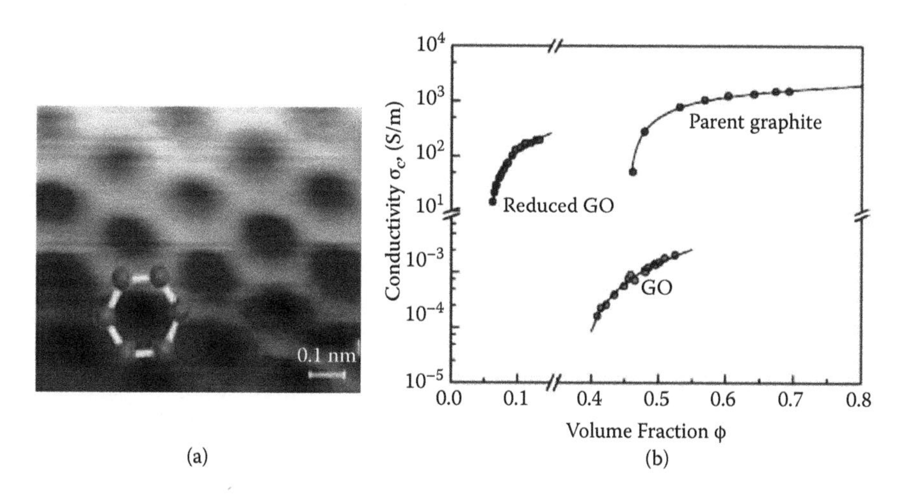

FIGURE 4.15
(a) STM image of atomic lattice structure of graphene sheet showing long range crystal order and hexagonal ring structure. (*Source:* Rao, C. N. R. et al. 2009. *Angewandte Chemie*, 48, 7752–7777. With permission.) (b) Comparison of conductivity of chemically oxidized and reduced graphite materials with its conductive graphite parent. (*Source:* Stankovich, S. et al. 2007. Carbon, 45, 1558–1565. With permission.)

To test device performance, interdigitated electrodes were printed by electrophoretic deposition onto a gold-coated glass wafer to compare AC against OLC [60]. Electrophoresis occurred when charged particles moved predictably in solution based on an applied electrical force (12 mV zeta potential for OLC). To stabilize the carbon materials, Mg^+ ions were added to solution and acted as a binder. The resulting electrodes were tested by cyclic voltammetry (Figure 4.15b). At high scan rates, the OLC-based electrode outperformed AC because the ionic delay in the unordered carbon could reduce AC performance as average power was increased [60]. Further, the electrophoretic deposition technique allows the patterned application of EDLC materials for integrated circuitry.

4.2.8.5 Graphene

Graphene is a new advanced carbon material with unique morphology that distinguishes it from other materials in the EDLC market. Graphite is a highly ordered carbon structure consisting of many tightly stacked sheets that exhibit angstrom level interspacing due to strong π-π bonding between the basal lattice planes of the graphene. A graphene sheet consists of many carbon atoms arranged into a large two-dimensional crystal lattice (Figure 4.15a). When separated into fewer layers through the application of some physical or chemical energy (single- or few-layer graphene), the material takes on very different properties from its bulk graphite state [61].

Graphene is mechanically robust, exhibiting a quantum Hall effect at room temperature and undergoing ballistic conduction of charge carriers along the basal planes, resulting in good conductivity in the material [61]. Stankovich et al. [62] illustrate in Figure 4.15b that the conductivity of chemically reduced graphene along its basal plane, on the order of 200 $S.m^{-1}$, was close to that of bulk graphite. The high conductivity helps reduce internal resistance in the materials and provides power gains similar to those exhibited by CNTs.

The most important properties for the ES market include the ability of graphene to reach the high theoretical surface area of 2630 $m^2.g^{-1}$ of its parent graphite and the highly regular pore spacing across the crystal lattice [63]. Due to the atomic thickness, negligible diffusion transport distances result in low ionic resistance. The high theoretical surface area of the fully exposed graphene sheets could provide a maximum theoretical capacitance of 550 $F.g^{-1}$ [64].

The combination of good electrical conductivity, low internal resistance, and high surface area makes graphene a competitive material for EDLCs. However, as with CNTs, aggregation and poor macroscale controls still limit the full potential of graphene materials. Restacking of graphene sheets can occur in solution over time during annealing and during drying procedures, leading to reduced surface area. Stoller et al. [63] were one of the first groups to utilize chemically reduced graphene (CMG) for ES applications. They produced a CMG graphene that exhibited an area of 705 $m^2.g^{-1}$ and capacitances reaching to 107 $F.g^{-1}$ in KOH and 100 $F.g^{-1}$ in acetonitrile using cyclic voltammetry with a 20 $mV.s^{-1}$ scan rate.

Many graphene-based materials report 100 to 200 $F.g^{-1}$, outperforming CNT devices in an aqueous electrolyte [65]. Most EDLC electrodes used in ESs use thick material layers that are opaque and fairly brittle. For example, Yu et al. [66] produced flexible, uniform, 25 nm thick layers of graphene that offered up to 70% optical transparency (Figure 4.16) and capacitance of 135 $F.g^{-1}$.

Other techniques have been applied to help boost the performance of graphene in ESs by altering the graphene structure. Zhu et al. [67] investigated the application of thermal KOH activation from AC materials to graphene materials. Similar to its activity in AC devices, KOH was shown to boost the surface area of TEGO and MEGO graphenes up to 3100 $m^2.g^{-1}$ by restructuring the carbon, exposing hidden graphene sheets and producing extra pores.

MEGO powder was weighed and mixed with KOH to undergo a secondary activation reaction (800°C, 400 torr, 1 hr) that increased the surface area significantly and generated a well defined pore distribution (Figure 4.17) of micropores (~0.8 nm) and mesopores (~4 nm). As a result, the device was able to achieve 165 $F.g^{-1}$ (stable up to 6 $A.g^{-1}$) in organic electrolyte [67].

Jeong et al. [21] studied the effects of doping graphene planes with nitrogen to enhance local electronic interactions and improve binding with ions in solution. Plasma treatment was used to impregnate nitrogen functionalities into defects along the basal planes of graphene sheets. The presence of nitrogen functionalities could generate cross linking along the graphene

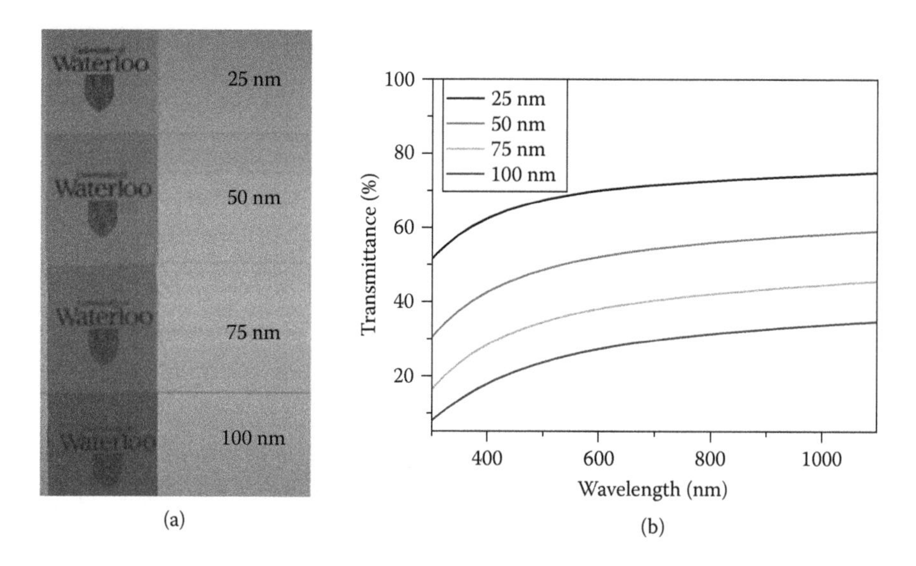

FIGURE 4.16
(See color insert.) (a) Visible demonstration of thin film transparency for prepared graphene thin films. (b) UV transmission spectra of graphene thin films with varying thicknesses. (*Source:* Yu, A. et al. 2010. *Applied Physics Letters*, 96, 253105. With permission.)

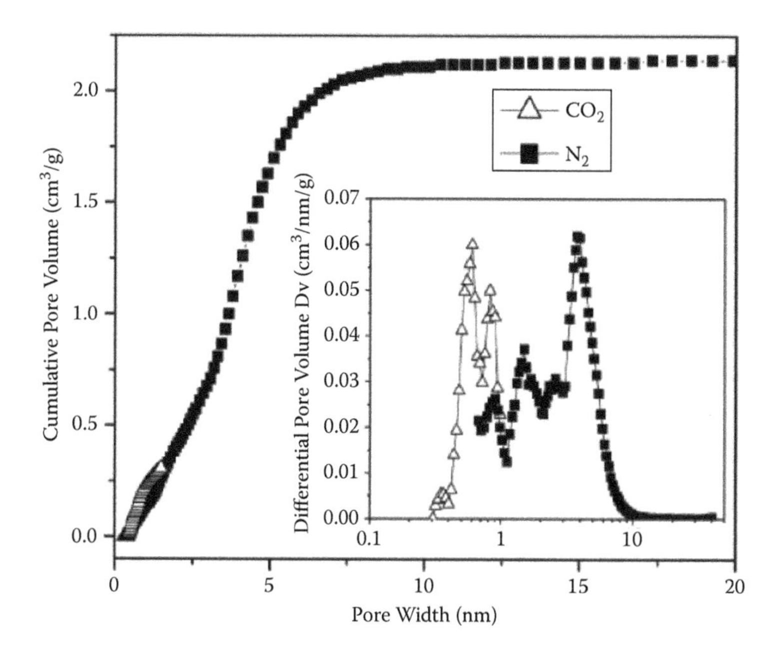

FIGURE 4.17
Differential pore size distribution curve created by KOH activation of MEGO and determined from BET surface area analyzer. (*Source:* Zhu, Y. et al. 2011. *Science*, 1537. With permission.)

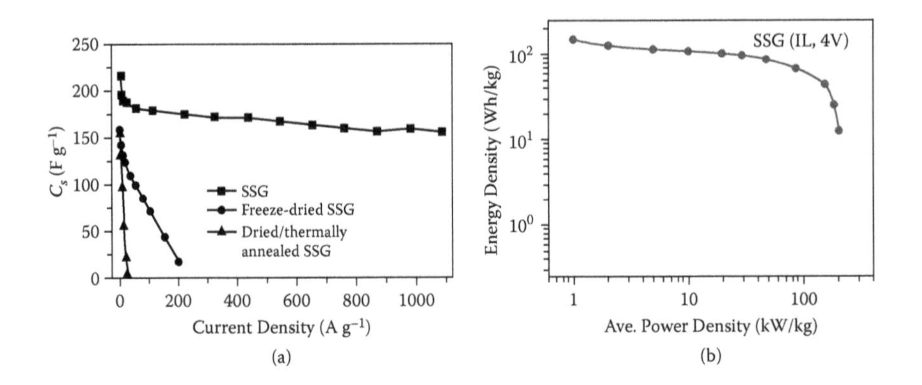

FIGURE 4.18
(a) Capacitive performance of chemically reduced graphene over large range of current density for different handling methodologies. (b) Ragone plot illustrating strong energy performance and average power density in ionic liquid electrolyte. (*Source:* Yang, X. et al. 2011. *Advanced Materials*, 23, 2833–2838. With permission.)

plane and create wrinkled graphene sheets with high levels of curvature. An optimal plasma exposure time of 1 min could lead to a high performance of 250 F.g^{-1} (at 1 A.g^{-1}) in aqueous electrolyte. Even at high current density of 30 A.g^{-1}, the performance still held 175 F.g^{-1}.

Yang et al. [68] reinvestigated the source of the restacking phenomena in graphene materials that limited device performance. Instead of improving performance by altering the graphene, they drew inspiration from the irreversible cell damage and collapse that occurred when moisture levels dropped to critical levels. They synthesized a solvated graphene film by keeping the active material wet after the reactions and during storage. Performance testing (Figure 4.18) illustrated that the film capacitance can retain 156 F.g^{-1} at a current discharge rate of 1080 A.g^{-1} in an aqueous H_2SO_4 electrolyte [68].

At 1 A.g^{-1} the capacitance of the wet film reached ~190 F.g^{-1}. Thermally dried graphene shows a complete voltage drop due to resistance that prevents the development of capacitance beyond a current density of 10 A.g^{-1}. Freeze-dried graphene fares better by restricting the pathways for pore collapse during the drying phase, but neither sample exhibited the same high power, high energy performance of the wet graphene electrode.

Due to the high ionic conductivity of the graphene and short transport distances between solvated sheets, the electrode was also tested in an ionic liquid electrolyte that had a stable operation window of 4 V [68]. By utilizing ionic liquid, the maximum energy and peak power densities were calculated at 150 Wh.kg^{-1} and 770 kW.kg^{-1} respectively. Figure 4.18 shows that energy density remains above 100 Wh.kg^{-1} for average power as high as 50 kW.kg^{-1} and with energy density approaching that of a modern battery system [68]. The high power and energy stability indicate the importance of keeping graphene samples wet between synthesis steps to maintain porosity and performance.

4.2.8.6 Carbon Nanofibers

Carbon nanofibers (CNFs) are long hollow fibers composed of ordered arrangements of graphene sheets. CNFs are of two types: highly graphitic and lowly graphitic. The highly ordered graphitic CNFs are produced by catalytic CVD using a metal catalyst such as nickel, iron, or cobalt particles in solution to promote CNF growth [69]. Alternatively, fixed catalysts on a support can be used to create a highly graphitic CNF array. Highly graphitic CNFs exhibit strong conductivity of 1000 $mS.cm^{-1}$, but suffer from stacked graphite planes in the walls of the tubules.

The stacking leads to low surface area of only 10 to 50 $m^2.g^{-1}$ and restricts capacitance to only 1 to 10 $F.g^{-1}$ [69]. Further, the tight stacking and stability of the graphitic walls prevent efficient activation of the CNF and reduce the usefulness of highly graphitic production methods in ES applications. Lowly graphitic CNFs are mostly amorphous and are created by the carbonization of polymer precursors. One production method is to use polymer blends and remove one polymer during carbonization to produce a CNF web template with 100 to 500 $m^2.g^{-1}$. However, the reduction in graphitic character restricts conductivity and capacitance to 100 $mS.cm^{-1}$ and below 100 $F.g^{-1}$ in aqueous electrolyte. An effective alternative is to use electrospinning. Subsequent carbonization of the polymer nanofiber webs could be created [69].

Electrospun polymers exhibit low levels of molecular defects, thus optimizing strength and creating order that leads to higher conductivity of 700 to 900 $mS.cm^{-1}$ after carbonization at temperatures between 700 and 800°C [69]. The amorphous character of the CNFs allows effective functionalization and activation of the lowly graphitic form. Lowly graphitic PAN-based CNFs synthesized by Kim et al. with steam activation (1100 $m^2.g^{-1}$) at 750°C showed 120 $F.g^{-1}$ at 1 $A.g^{-1}$ in KOH electrolyte [70]. They also showed that polyamic acid (PAA) fibers steam activated at 750°C (1400 $m^2.g^{-1}$), produced 160 $F.g^{-1}$ at 1 $A.g^{-1}$ in KOH electrolyte [71].

Barranco et al. [69] electrospun CNF fibers by using a polymer blend containing a phenolic resin and a high density polyethylene (PE). The blend was carbonized at 800°C producing CNFs of 450 $m^2.g^{-1}$ (700 $mS.cm^{-1}$) and then KOH activated at 750°C to increase surface area to 1500 $m^2.g^{-1}$. The KOH activation reduced conductivity to 400 $mS.cm^{-1}$ but boosted capacitance to 180 $F.g^{-1}$ (from 50 $F.g^{-1}$ not activated) at 1 $A.g^{-1}$ (20 $mA.cm^{-2}$) in KOH electrolyte [69].

Yan et al. [72] utilized the accessible webs created by the long fibers to derive a CNF composite with PANI (Figure 4.19). The base capacitance of the CNFs inactivated managed 310 $F.g^{-1}$ in H_2SO_4 at 2 $A.g^{-1}$ and electrical conductivity was 950 $mS.cm^{-1}$. The improved performance is likely due to improved layer-by-layer methodology for collection of the nanofibers and illustrates the macroscopic mechanical control over optimizing performance of CNFs produced by electrospinning. The CNF paper (10 to 15 μm) produced by layer-by-layer electrospinning required no conductive carbon black additives and was very flexible. After a simple rapid mixture polymerization,

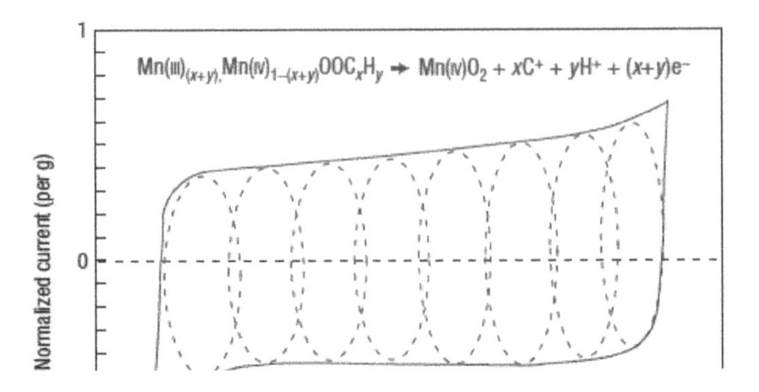

FIGURE 4.19
SEM images of: (a) PANI CNF composite web and (b) CNF web support (50,000× magnification). (*Source:* Yan, X. et al. 2011. *Nanoscale*, 3, 212–216. With permission.)

PANI nanoparticles uniformly coated the electrode and increased capacitance to 638 $F.g^{-1}$ and maintained 90% capacitance after 1000 cycles [72].

The constant deposition rate that creates uniform layers and simple processing of economical polymer solutions make CNFs good options for industrial applications. The ability to mechanically control the deposition quality and pattern is also a strong advantage for post-production improvements to quality and optimizing throughput. Electrospun deposition of fibers makes CNF production compatible with in-line, roll-to-roll manufacturing techniques of ESs. The nozzles can also easily be multiplexed to increase throughput and improve scalability, which increases the value of CNFs for industrial applications.

4.2.9 Pseudocapacitive Materials

4.2.9.1 Storage Overview

Pseudocapacitor electrodes follow faradic reactions to store charge via redox reactions at specific potential windows during charge and discharge. Pseudocapacitance is of interest in ES systems because of the potential for increased energy storage in comparison with carbon EDLC materials. These redox reactions occur at the surface layer of an electrode where thin layers are important to the successful application of pseudocapacitive materials. These materials can be used in electrodes to greatly improve energy density by chemical storage, unlike physical ion gradients that store charges in EDLC-based electrodes.

The higher energy available in chemical storage is balanced with challenges in maintaining power and controlling the durability of a pseudocapacitive material. Similar to cycling issues in batteries, ESs utilizing pseudocapacitance are more likely to irreversibly reconfigure over time because of constant chemical changes within the material. Reconfiguration of electrode materials leads to a loss of carefully selected morphology, reducing capacitance and leading to a greatly reduced cycle life.

EDLC-based electrodes can handle over a half million cycles, while high energy batteries may manage a few hundred to a few thousand charge cycles. One of the design challenges for pseudocapacitive electrodes is to avoid compromising cycle life. Electrolyte choice can also affect operating voltage range, reactivity, and power output of a total system and further alter the cycle life of a pseudocapacitive material and its performance. Inherently the presence of redox reactions and high energy storage will prevent use over half a million cycles as seen with EDLC devices. However, it is expected that 100,000 cycles would be sufficient for many ES applications and that should be achievable with proper electrode design [73].

Pseudocapacitive reactions involving underpotential deposition and partial electron transfer from anion chemisorption are present in all ESs and account for a small percentage of charge storage [74,75]. However, in pseudocapacitors, these processes can account for a much larger percentage of overall capacitance. More importantly, interfacial redox reactions involving transition oxide thin films store large amounts of charge through easily accessible surface oxidation states. Some of the more desirable transition metal oxide materials exhibit multiple oxidation or adsorption states within the range of EC operation, further boosting pseudocapacitive energy storage.

In conducting polymers, the ion uptake during doping allows ion intercalation throughout the electrode and exposes the electrolyte ions to more than just the surface interface of the material. The result is capacitance that can reach as high as 10 times that of carbon materials [74]. Unfortunately, ion uptake into the electrode material during oxidation causes swelling and strain on the material. This internal stress can result in material shifts, cracks, reduced contact with the substrate, and dislocation of crystal structures to more energetically favorable positions.

These changes can reduce the surface area and ion accessibility for future charge cycles. To deal with the challenges of pseudocapacitance, thin layers of material are frequently used on carbon supports that provide both high area for performance reasons and space for the materials to swell without generating damage. Stable crystal morphology is also important to resisting degradation over time.

4.2.9.2 Transition Metal Oxides

Transition metal oxides derived from ruthenium (RuO_2), iron (Fe_3O_4), vanadium (V_2O_5), tin (SnO_2), and manganese (MnO_2) are widely used in the research and application of pseudocapacitance [76–78]. Oxide materials are known to exhibit multiple oxidation states at specific potentials, and selection of materials with multiple stable states within an electrolyte's potential window allows maximum capacitance to develop.

Another benefit of crystalline metal oxides is that their high conductivities allow charge propagation along the lattice structures of thin surface layers. However, conductivity is lower in the commonly used hydrous metal oxides that are amorphous. It is possible to increase conductivity through crystallization but this significantly reduces active surface area via the removal of water and elimination of pore spacing within the material structure [79].

Ruthenium dioxide (RuO_2) is widely used because it is highly reversible, exhibits very high capacitance, and presents good cycle life. The fast redox reaction of RuO_2 follows the mechanism in the following reaction and contains three oxidation states ($0 \leq b \leq 2$) within a 1.2 V potential window:

$$RuO_x(OH)_v + bH^+ + be^- \leftrightarrow RUO_{x-b}(OH)_{y+b} \tag{4.4}$$

The ability of the oxidizing materials to transition from state to state allows overlapping of charge windows and can provide a constant current over operating potential. RuO_2-based devices show strong performance of 350 $F.g^{-1}$ at 100 $mV.s^{-1}$ scan rates [80]. Zheng et al. [81] illustrated that anhydrous salt forms (RuO_2-xH_2O) of the metal oxide could achieve optimum performance as high as 750 $F.g^{-1}$ when annealed at temperatures just below the crystallization point of the material. Chen et al. [82] showed results reaching 1500 $F.g^{-1}$ using a porous carbon support.

However, ruthenium is a rare earth mineral that is highly toxic and its high cost prevents its market use. Due to the high cost of RuO_2, deposition techniques must be optimized to deposit small quantities and enhance material utilization within a device. To further optimize use, composites with high-surface-area activated carbon supports can be used. Despite the strong performance characteristics of RuO_2, inevitable supply and demand issues push the market toward other pseudocapacitive materials.

A study of alternative transition oxides by Cottineau et al. yields important information about the performances of oxides of vanadium (V_2O_5), iron (Fe_3O_4), and manganese (MnO_2) in aqueous electrolytes [83]. The synthesis of powders showed MnO_2 and V_2O_5 had amorphous structures, while iron had a propensity to crystallize and form magnetite crystals. Composite electrodes were made on AC and tested with cyclic voltammetry (Figure 4.20).

The potential windows show that the oxide materials coupled with the neutral electrolyte kinetically limit water decomposition (unrestricted water decomposition occurs at $V < -0.57$ for hydrogen and $V > 0.67$ for oxygen versus Ag–AgCl reference at neutral pH) [83]. Fe_3O_4 pushes the negative potential limit while MnO_2 and V_2O_5 show stability up to +1 V. AC materials show the onset of oxygen evolution beyond 0.6 V, but the kinetic limitation of hydrogen evolution on AC in K_2SO_4 allows operation down to –1 V (Figure 4.21) [83]. This highlights the potential for an asymmetric device with increased potential window of 2 V using positive MnO_2 and negative AC–Fe_3O_4 electrodes. The curve shapes of the oxides can be seen in Figure 4.22.

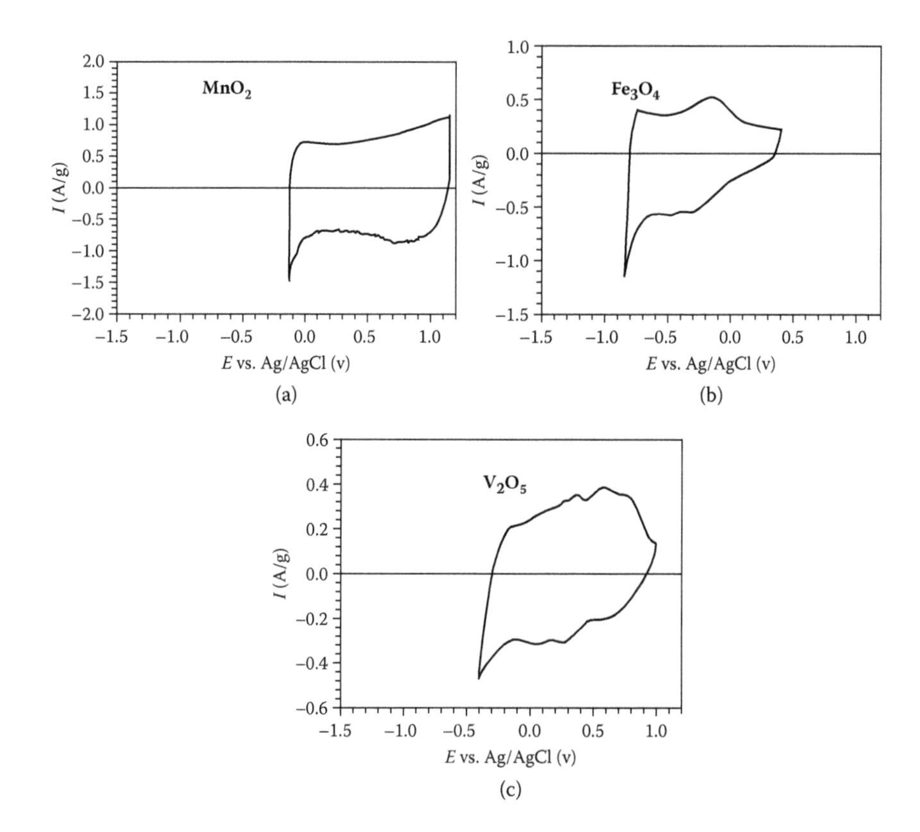

FIGURE 4.20
CV of composite electrodes illustrating shapes, with various transition metal oxide layers in K_2SO_4. (*Source:* Cottineau, T. et al. 2005. *Applied Physics A*, 82, 599–606. With permission.)

The figure shows the near rectangular shape of MnO_2 compared to the broad redox peaks seen in V_2O_5 and Fe_3O_4.

Performance testing of the electrodes can be seen in Table 4.3 [83]. V_2O_5 shows the highest specific capacitance but exhibits poor cycle life; capacitance fades considerably after only a few hundred cycles. The high capacitance, better cycle stability, and potential as a positive electrode material for composites make manganese the emerging alternative to RuO_2.

MnO_2 is safer and costs less than ruthenium. Synthesis is mostly performed by electrochemical deposition to produce optimum capacitive behavior [84] and avoid the particle aggregation common in sol–gel synthesis. A study of different electrochemical deposition types suggests that potentiodynamic deposition produced MnO_2 with the highest performance [84]. A series of oxidation states create a quasi-rectangular curve (Figure 4.21) similar to EDLC.

Restrictions to the use of MnO_2 include a lack of oxidation states available at negative voltages and the irreversibility of the reduction from Mn(IV) to

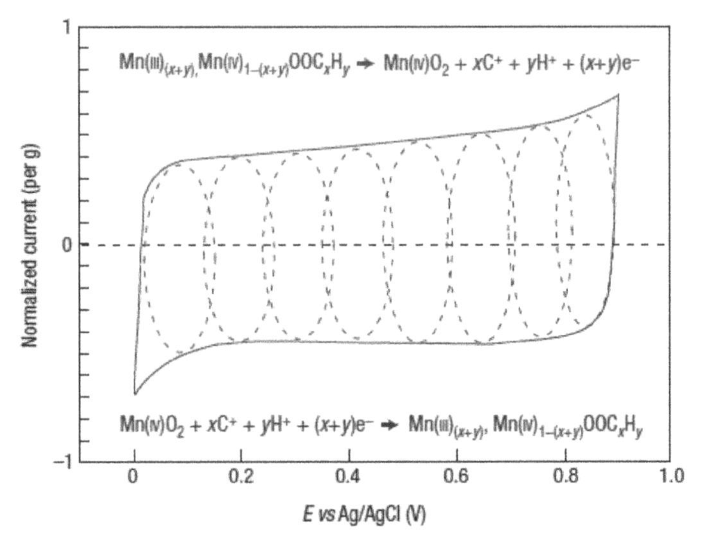

FIGURE 4.21
MnO_2 undergoing cyclic voltammetry analysis in 0.1 *M* K_2SO_4 and consecutive redox reactions that lead to pseudocapacitive charge. Positive current drives reactions between Mn(III) and Mn(IV) states. (*Source:* Simon, P. and Y. Gogotsi. 2008. *Nature: Materials*, 7, 845–854. With permission.)

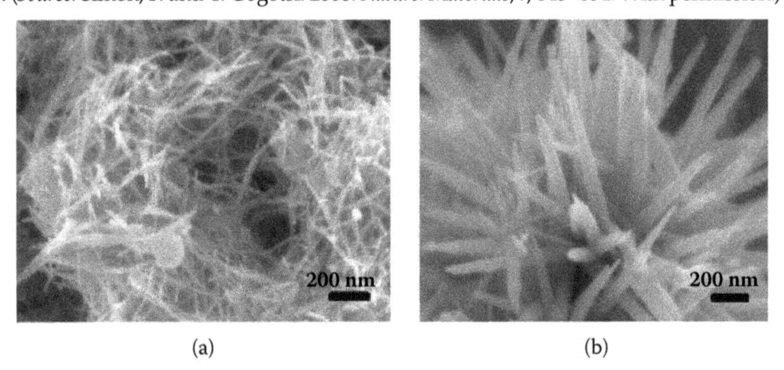

FIGURE 4.22
SEM images of crystalline MnO_2 materials with interesting (a) belt and (b) urchin morphologies. (*Source:* Zhang, J. et al. 2011. *Nanotechnology*, 22, 125703. With permission.)

TABLE 4.3
Performances of Composite Transition Metal Oxide Electrodes Using K_2SO_4 Electrolyte Evaluated by CV

Compound	BET (m_2/g)	Crystal Structure	ΔU versus Ag/AgCl (V)	Specific Capacitance (F/g)
MnO_2	200	Amorphous	0 to 1	150
Fe_3O_4	200	Magnetite	−0.7 to 0.2	75
V_2O_5		Amorphous	−0.2 to 1	170
AC	2800		−1.3 to 0.8	130

Source: Cottineau, T. et al. 2005. *Applied Physics A*, 82, 599–606. With permission.

Mn(II) that results in dissolution of negative electrode material at a cell voltage of 0.5 V (corresponding to approximately –0.25 V on the negative MnO_2 electrode) during the charging cycle [85]. Due to slight variations in electrode mechanisms, the positive Mn electrode experiences a set of reduction reactions from Mn(III) to Mn(IV) that are stable and reversible [83,86]. The combination indicates the practical use of MnO_2 is in hybrid systems that employ pseudocapacitive positive electrodes with different negative electrodes.

Layer shape and thickness are also important parameters in developing MnO_2 electrodes. Pure films of MnO_2 exhibit lower pseudocapacitance (150 $F.g^{-1}$) than RuO_2 (600 $F.g^{-1}$) in aqueous electrolyte [1]. The capacitance of MnO_2 is restricted by poor electronic conductivity. Further, evaluation of the oxidation stress on thicker films shows that only the top few layers of material are available for use because of limited ion accessibility [87]. This means that the high capacitance of MnO_2 is not easily achieved at practical mass loading or higher power.

The use of conducting metal or carbon substrates with controlled thin film deposition (<100 nm) of MnO_2 demonstrates that a much higher maximum performance of 1200 $F.g^{-1}$ was possible for low mass loading [1]. Belanger et al. showed that MnO_2 capacitance could reach 900 to 1380 $F.g^{-1}$ when mass loading was low (5 to 30 $\mu g.cm^{-2}$) [88]. Zhitomirsky et al. studied the improved performance retention possible by doping MnO_2 with conductive additives such as silver [89] and CNTs [90] at material loading as high as 150 to 300 $\mu g.cm^{-2}$. Using CNTs to generate an open surface network created higher power electrodes containing amorphous MnO_2 with capacitance up to 568 $F.g^{-1}$ [91].

Controlling hydrothermal synthesis of MnO_2 could create thin layers of crystalline material with a variety of nanoarchitectures such as nanowires, nanobelts, nanorods, and hollow nanospheres and urchins. Morphology and crystal type (α-MnO_2, β-MnO_2, γ-MnO_2, δ-MnO_2, and ξ-MnO_2) are controlled by temperature, pH, mole ratios, and reaction time. Yang et al. [92] demonstrated thin 17 nm layers of δ-MnO_2 nanoflowers that exhibited a maximum capacitance of 260 $F.g^{-1}$ at low current of 70 $mA.g^{-1}$. Zhang et al. [93] showed α-MnO_2 nanourchins (basic pH) and belts (acidic pH) with maximum capacitances of 161 $F.g^{-1}$ and 262 $F.g^{-1}$, respectively, at 250 $mA.g^{-1}$. However, the performance of these materials (Figure 4.22) was only 199 $F.g^{-1}$ and 121 $F.g^{-1}$ for current discharge at 1 $A.g^{-1}$ in aqueous electrolyte compared to EDLC electrodes [93]. Despite the interesting morphology, performance of amorphous MnO_2 remains superior.

4.2.9.3 Transition Metal Nitrides

Metal nitrides are transition metal materials (titanium, vanadium, and molybdenum) that are receiving attention for pseudocapacitive study because they exhibit high electronic conductivities compared to transition oxides. Liu et al. [94] studied molybdenum nitride (Mo_xN) and showed that

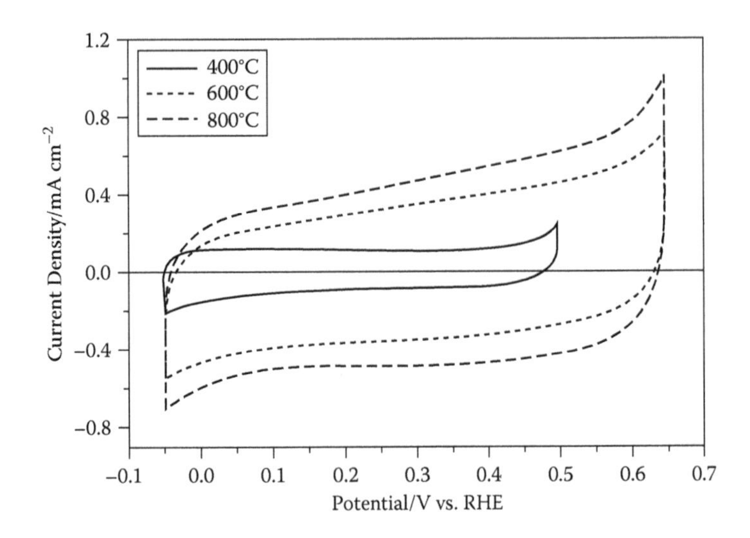

FIGURE 4.23
CV profiles for MoxN film electrode deposited at different temperatures. (*Source:* Liu, T. C. 1998. *Journal of the Electrochemical Society,* 145, 1882. With permission.)

the material had a capacitive behavior comparable to ruthenium dioxide. Figure 4.23 shows that performance was optimized by depositing the material at 800°C [94]. However, the Mo_xN film suffers from decomposition in electrolyte at only 0.7 V, which limits practical application.

Choi et al. [95,96] tested titanium (TiNxCly) and vanadium nitride (VNxOyClz) nanocrystals synthesized through a two-step process inside a glove box to limit the high levels of air sensitivity exhibited by precursors. The metal chloride precursors were dissolved in chloroform and then treated with ammonia to create a nitride powder before crystallizing at high temperatures between 400 and 1000°C [95]. Films were prepared onto a nickel collector using a paste containing 85% nitride, 5% conductive carbon, and 10% PVdF binder in an N-methylpyrrolidone (NMP) solvent.

Both nitride materials showed smaller crystallite sizes, higher nitrogen content, and higher surface area at the lower temperature of 400°C. The conductivity was higher at 1000°C but capacitance was significantly lowered at temperatures above 400°C (Figure 4.24) [96]. Titanium nitride achieved moderate capacitance values of 150 F.g–1 in aqueous electrolyte. However, the high capacitance exhibited by 400°C TiN decreased by 72% after 400 cycles, whereas the TiN synthesized at 500°C and above was stable [96].

Vanadium nitride has high electrical conductivity (1.67×10^6 $S.m^{-1}$) compared to the limiting conductivity of vanadium oxide powders (V_2O_5-nH_2O, 1×10^{-4} $S.m^{-1}$) [95]. Unlike titanium nitride, which does not exhibit any oxygen functionalities, vanadium nitride was confirmed by FTIR spectroscopy to have an oxygen monolayer (V_2O_5, 0.5 nm) on the surface of the spherical nitride crystal (400°C sample: 6.33 nm crystal size, 38 $m^2.g^{-1}$ BET area). The

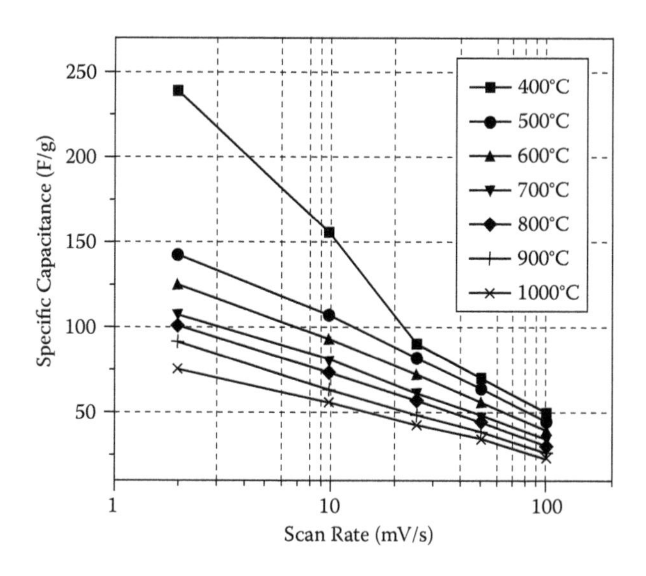

FIGURE 4.24
Specific capacitance for TiN nanocrystals synthesized at different deposition temperatures and tested at varying scan rates in 1 *M* KOH. (*Source:* Choi, D. and P. N. Kumta. 2006. *Journal of the Electrochemical Society*, 153, A2298–A2303. With permission.)

oxide-coated nitride powders exhibit high surface conductivity (8×10^3 S.m^{-1}), much closer to ruthenium dioxide (2.8×10^6 S.m^{-1}) than pure vanadium oxide powder. The oxide layer is suspected to enhance the capacitive performance of the vanadium nitride as seen in Figure 4.25 [95].

Maximum capacitance of over 1300 F.g^{-1} is observed at low scan rates, and at higher scan rates of 100 mV.sec^{-1} performance still exhibits 550 F.g^{-1}. This is much better than other reported values (350 F.g^{-1} at 5 mV.sec^{-1}) for vanadium oxide [95]. High scan rates of 2 V.sec^{-1} for the vanadium nitride electrode still yield 190 F.g^{-1}, indicating that vanadium nitride is capable of high power density. Stability issues due to dissolution of the oxide layer caused a large decrease in original capacitance after 1000 cycles. However, by controlling the pH of the electrolyte, stable capacitance of 400 F.g^{-1} was observed for over 1000 cycles on the vanadium nitride electrodes [95].

4.2.9.4 Conducting Polymers

Conducting polymers constitute another category of promising pseudocapacitive materials. The most common ones include polypyrrole (PPy), polyaniline (PANI), and poly-(3,4)-ethylenedioxythiophene (PEDOT). This group is of particular interest due to low cost and ease of synthesis. These compounds can be polymerized directly onto a collector material via EPD. Alternatively, the polymerization can be done within surfactant emulsions

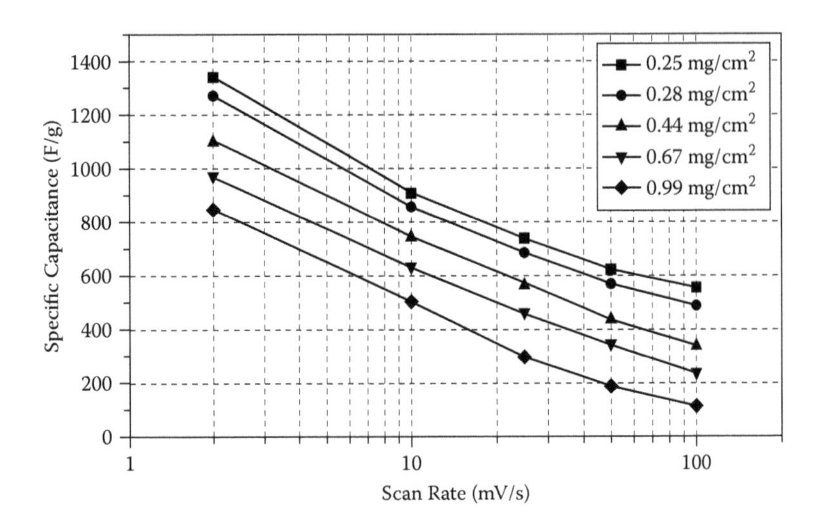

FIGURE 4.25
Specific capacitance for vanadium nitride nanocrystals synthesized at 400°C and tested at varying scan rates and mass loadings onto collector in 1 *M* KOH. (*Source:* Choi, D. G. E. Blomgren, and P. N. Kumta. 2006. *Advanced Materials*, 18, 1178–1182. With permission.)

to generate particles of specific size and shape [97]. The conducting polymer storage mechanism involves accumulation of charge via proton doping interactions throughout the materials backbone (Figure 4.26).

However, as a result of direct charge uptake and fast discharge inherent to an ES, conductive polymers are subject to swelling and cracking over time. The relaxation of the polymer matrix can result in a large amount of irreversible reduction over the first few hundred cycles. To reduce this relaxation effect, slower charging must be used. The upside is that high specific capacitance values of 400 $F.g^{-1}$ can be achieved throughout the material and not just at the material surface [99].

Conducting polymers, like most other pseudocapacitive materials, limit the potential range of the electrode to regions where reversible redox reactions occur for that particular material. In the case of polymers, a lack of oxidation states limits the potential window and also a physical polymer breakdown caused by over-oxidation occurs when charges cannot redistribute quickly enough within the polymer matrix [100]. The poor stability during oxidation and the irreversible capacitance challenges highlight conductive polymers as interesting materials for composite cathodes.

Work by Yan et al. on PANI and graphene composites showed an electrode capacitance as high as 1000 $F.g^{-1}$ in an aqueous electrolyte [101]. The same research also showed that incorporation of CNT additives helped enhance percolation and mechanical strength during the doping process, enabling the electrode to retain 94% of the original capacitance after 1000 cycles compared to retention below 50% without the CNT additive [101].

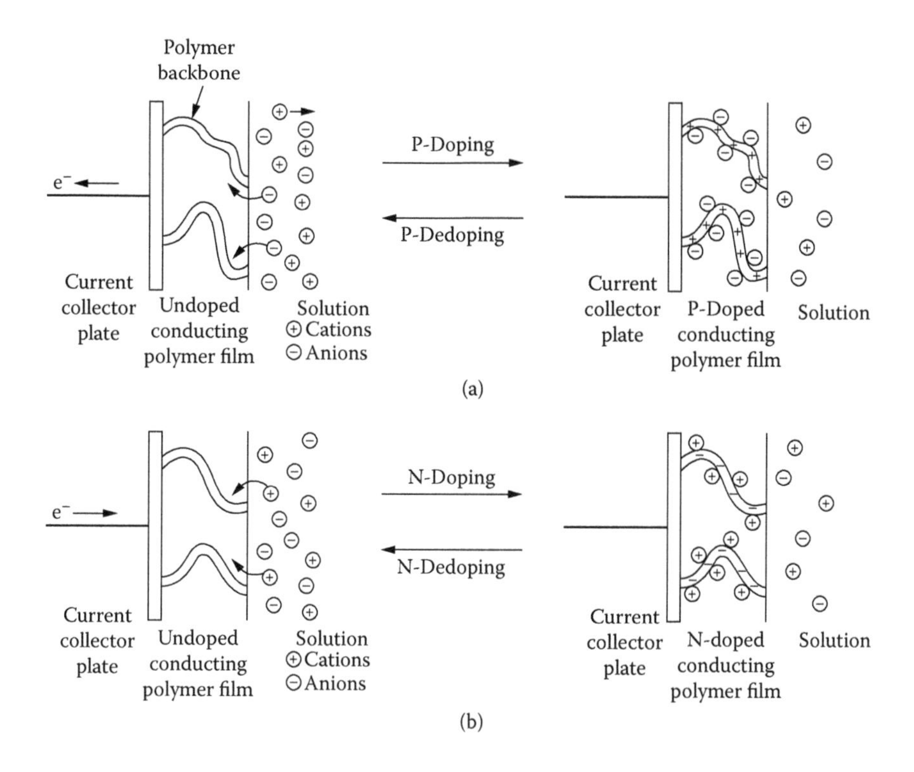

FIGURE 4.26
The *p*-doping (a) and *n*-doping (b) of polymers as they undergo charging and discharging. (*Source:* Rudge, A. et al. 1994. Conducting polymers as active materials in electrochemical capacitors, *Journal of power sources*, 47, 89–107.)

Controlling the nanoscale features of conducting polymer electrodes allows for a number of important performance improvements [102]: (1) higher surface area creates increased contact with electrolyte, improving charge rates through better charge uptake into the polymer; (2) short path lengths provide faster transport of ions into the polymer backbone; and (3) more space and less material reduce strain created from operation and improves cycle life.

Liu et al. [103] designed a composite electrode made of MnO_2 and PEDOT. An anodized alumina template was used along with a one-step electrochemical co-deposition process to create coaxial nanowires with PEDOT shells and MnO_2 cores (Figure 4.27). Material structure could be controlled by varying the deposition voltage and the result showed the mechanical stability of the polymer. However, capacitance of only 185 $F.g^{-1}$ was achieved at 25 $mA.cm^{-2}$ charge rates. Conduction across the thin polymer layer was quick but the performance limitation arose from the low conductivity in the MnO_2 core [103].

Highly porous carbon templates provide a stronger conduction pathway and composite polymer carbon electrodes can handle charge rates more

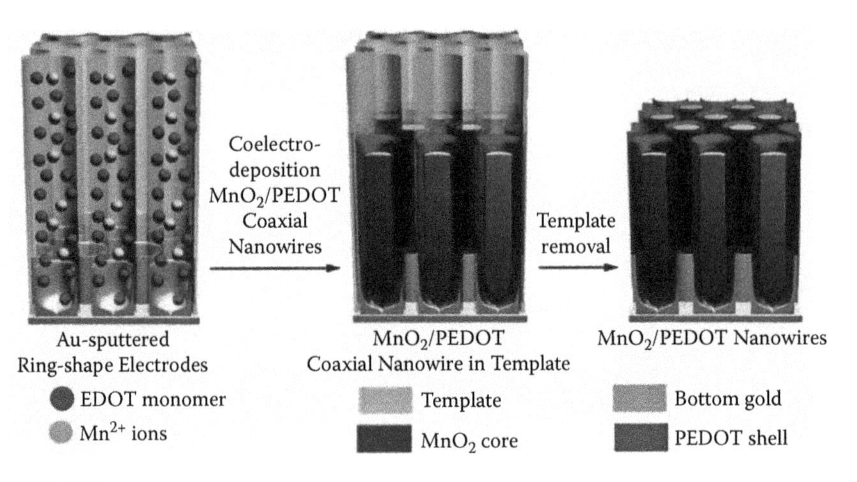

FIGURE 4.27
(See color insert.) Steps of electrochemical co-deposition of high energy density MnO_2–PEDOT nanowires. (*Source:* Liu, R. and S. B. Lee. 2008. *Journal of the American Chemical Society*, 130, 2942–2943. With permission.)

comparable to EDLC devices. An example of this can be seen in the work of Xia et al. [104], who reported the growth of loosely packed PANI thorns on a mesoporous carbon template (Figure 4.28). At 0.5 $A.g^{-1}$, the performance reached 900 $F.g^{-1}$ in H_2SO_4 electrolyte. Even at high charge rates of 5 $A.g^{-1}$, the capacitance still remained at 770 $F.g^{-1}$. The material was cycled to show the extent of irreversible decay and found to be only 5% after 3000 cycles [104]. This result shows how important the short regular pore order, conductive support, and improved diffusion are to pseudocapacitive electrode materials, especially conducting polymers.

Yu et al. [105] created a high performance composite by using electrodeposition of PPy on free-standing graphene films (Figure 4.29a). Optimum composite performance was seen after 120 sec of deposition, leading to 240 $F.g^{-1}$ at $10mV.sec^{-1}$ in 1M KCl (Figure 4.29c). The substrate was a 20 μm thick graphene film that contained no binder or conductive additives. As a result,

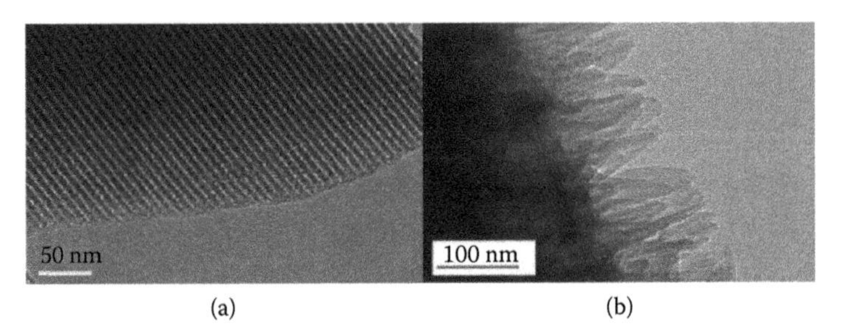

FIGURE 4.28
(a) TEM image of the crystal plane of composite PANI–mesoporous carbon electrode [100]. (b) SEM image of same material showing PANI nanowires deposited on carbon substrate. (*Source:* Wang, Y. G., H. Q. Li, and Y. Y. Xia. 2006. *Advanced Materials*, 18, 2619–2623. With permission.)

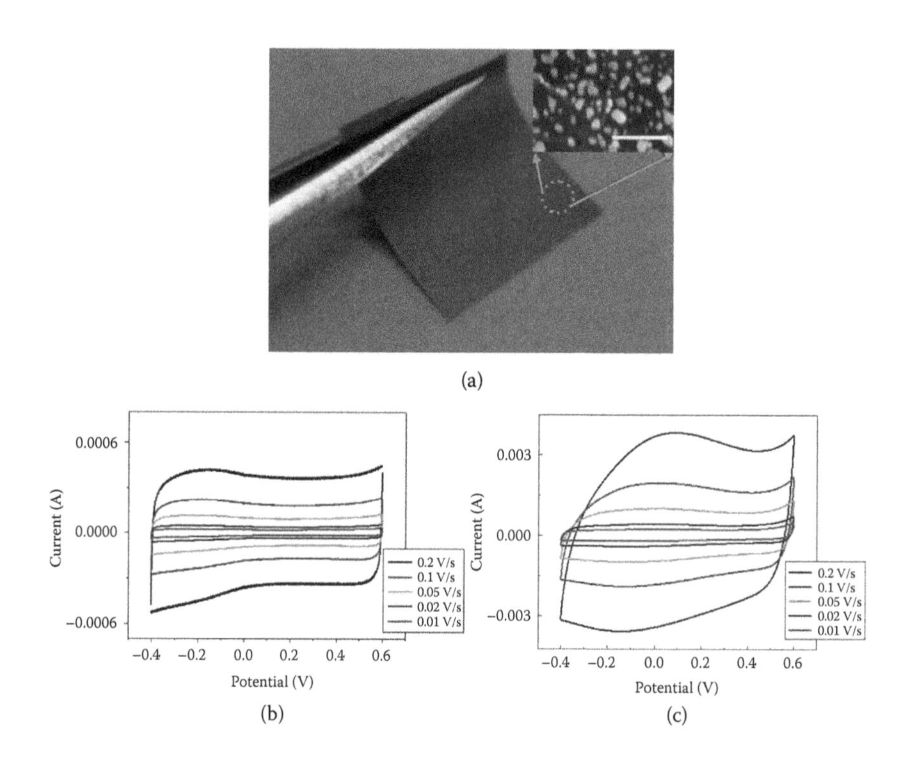

FIGURE 4.29
(See color insert.) (a) Graphene composite film with polypyrrole deposited for 120 sec. Inset shows SEM image at the observation area. White bar = 100 nm. (b) Cyclic voltammogram curves for pure graphene film. (c) Graphene with polypyrrole deposited for 120 sec in KCl solution between −0.4 and 0.6 V versus SCE at scan rates of 0.01, 0.02, 0.05, 0.1, and 0.2 V/sec. (*Source*: Davies, A. et al. 2011. *Journal of Physical Chemistry C*, 115, 17612–17620. With permission.)

the non-optimized graphene baseline exhibited moderate performance of 63 $F.g^{-1}$ (Figure 4.29b) [105]. After a 120 sec PPy deposition, the composite effectively quadrupled the overall performance compared to neat graphene film. Further, the flexible, free-standing nature of the composite suggests the composite electrode film could be used directly in supercapacitor cell construction or for other flexible storage devices.

4.2.10 Asymmetric Structures

Similar to the value proposition behind composite electrodes that use pseudocapacitive materials on high surface area carbon supports, hybrid structures offer improvements to the energy limitations of EDLC systems. Asymmetric systems utilize one EDLC electrode with the other made of either a pseudocapacitive material or a lithium electrode. Systems employing these techniques trade lower cycle life and power for an increase in energy density.

When designing a hybrid cell with optimal energy density, it is important to maintain durability and power, as long cycle life and high power represent two major advantages over battery devices. Another major advantage of the asymmetric design is the ability to match electrodes that exhibit overpotential reactions that block either oxygen or hydrogen gas evolution due to water decomposition. This manages to overcome the 0 V NHE hydrogen evolution and 1.2 V NHE oxygen evolution (water decomposition) that normally limit the potential window to 1 V (0.1 to 1.1 V NHE).

Belanger et al. [83] used a MnO_2-coated cathode with an EDLC carbon anode, leading to a system with a cell voltage of 2 V in neutral aqueous electrolyte. They achieved an energy density of 17.3 $Wh.kg^{-1}$ and a maximum power density of 19 $kW.kg^{-1}$, which is higher than the result from the symmetric MnO_2 device or the symmetric AC carbon cell tested. Further, the device was stable for over 5000 cycles.

The carbon anode showed an overpotential reaction, adsorbing hydrogen and preventing gas evolution until –0.65 V versus NHE. The proton absorption could block dihydrogen evolution until it became more thermodynamically feasible [83,86]. The reversible MnO_2 oxidation reactions at the cathode show oxygen overpotential to 1.4 V versus NHE, which allows the device's potential window to be extended.

Khomenko et al. [85] illustrated that correctly mating electrodes could extend potential range as high as 2 V for aqueous systems [28]. Results of their investigation can be seen in Table 4.4 [85]. We can see that conductive carbon materials can also extend the onset voltage for hydrogen evolution through their doping interactions. The best result (1.8 V, 13.5 $Wh.kg^{-1}$, low

TABLE 4.4

Electrochemical Characteristics of Symmetric and Asymmetric ECs Based on Different Active Materials

Electrode Material		EC Characteristic			
Positive	Negative	U (V)	E ($Wh.kg^{-1}$)	ESR ($\Omega.cm^2$)	P_{max} ($kW.kg^{-1}$)
PANI	PANI	0.5	3.13	0.36	10.9
PPy	PPy	0.6	2.38	0.32	19.7
PEDOT	PEDOT	0.6	1.13	0.27	23.8
Carbon Maxsorb[a]	Carbon Maxsorb	0.7	3.74	0.44	22.4
PANI	Carbon Maxsorb	1	11.46	0.39	45.6
PPy	Carbon Maxsorb	1	7.64	0.37	48.3
PEDOT	Carbon Maxsorb	1	3.82	0.33	53.1
MnO_2	MnO_2	0.6	1.88	1.56	3.8
MnO_2	PANI	1.2	5.86	0.57	42.1
MnO_2	PPy	1.4	7.37	0.52	62.8
MnO_2	PEDOT	1.8	13.5	0.48	120.1

Source: Khomenko, V. et al. 2005. *Applied Physics A*, 82, 567–573. With permission.

[a] Carbon Maxsorb is a high-surface-area activated carbon.

ESR, 120 kW.kg^{-1}) came from a combination of an amorphous MnO_2 cathode and PEDOT anode that contained 15 to 20 wt% CNT to maximize conductivity. The combination of high pseudocapacitance and high power carbon offers intriguing design possibilities. The trade-off is more complex systems and design criteria that can prevent the long cycle lives expected of ECs if not properly managed [28].

To increase energy density, another group of hybrid devices utilizing lithium storage with EDLC storage mechanisms [17] was explored. The major challenge in this approach is how to overcome the poor diffusion coefficient and poor electronic conductivity of lithium compared to EDLC and other fast redox materials. A lithium salt (Li^+BF_4 or Li^+PF_6) was used along with a pre-doped lithium (carbon electrode) anode and an AC cathode [106]. The pre-doped lithium source acts as the anode on discharge. The chemisorbed lithium charges oxidize and are released into the electrolyte. To compensate, Li^+ will temporarily adsorb onto the carbon cathode surface. A commercialized version of this design is made by Fuji Heavy Industries and its performance characteristics can be seen in Table 5.7.

Wang et al. [17] investigated an alternative design in which a composite titanium dioxide on CNT cathode was used to adsorb Li^+ ions released from a lithium-doped mesoporous templated carbon anode during discharge. Titania (TiO_2) was treated onto the porous CNT substrate in a thin layer and exhibited a high chemical activity toward lithium. The use of CNTs and mesoporous carbon could improve conduction, surface area, and ion conduction via ordered spacing. The resulting ES had an energy density of 25 Wh.kg^{-1} and a power density of 3 kW.kg^{-1} power and was stable for over 1000 cycles [17].

Naoi et al. [106] used a composite anode made of nanocrystalline lithium titanate (Li_4TiO_{14}, 5 to 20 nm) chemisorbed to carbon fibers (200 m^2.g^{-1}) and an AC cathode to achieve 55 Wh.kg^{-1} and 10.3 kW.kg^{-1}, respectively. Lithium titanate was chosen because it exhibits high Coulombic efficiency (95%) of its theoretical (175 mAh.g^{-1}) storage capacity even at higher power. In this case, a storage capacity of 158 mAh.g^{-1} was achieved. The composite also showed a low strain during charge discharge cycles that enhanced the cycle life.

Using composite electrodes, the challenges such as the power limitations created by the low Li^+ diffusion coefficient ($<10^{-6}$ cm^2.s^{-1}) and poor electronic conductivity ($<10^{-10}$ mS.cm^{-1}) of lithium titanate in bulk form could be addressed [106]. In a composite electrode, the nanocrystalline particles reduce diffusion problems, while the highly conductive nanofibers promote conductivity [106]. Along with higher energy density compared to other lithium doping techniques, the lithium titanate device maintained 90% energy density stability after 9000 cycles [106].

4.3 Electrolyte Structures and Materials

4.3.1 Electrolyte Overview

Electrolytes play an important role in overall ES performance. They exert critical effects on the development of the double-layer and accessibility of pores to electrolyte ions. Normally, electrolyte–electrode interactions and the ionic conductivity of the electrolyte play a significant role in internal resistance. Poor electrolyte stability at different cell operating temperatures and poor chemical stability at high rates can further increase resistances within an ES and reduce cycle life.

Electrolytes that exhibit high chemical and electrochemical stabilities allow larger potential windows without ruining performance characteristics. To ensure safe operation of ESs, electrolyte materials should have low volatility, low flammability, and low corrosion potential. Table 4.5, Table 4.6, and Table 4.7 show a range of different electrolytes, as well as several important operational properties [52]. Each solvent exhibits varying levels of ionic conductivity, voltage stability, size, and reaction concerns that must be considered when choosing an electrolyte. Solid polymer electrolytes are becoming

TABLE 4.5

Available Ion Sources for Organic and Inorganic Electrolytes

	Ion Size (nm)	
	Cation	Anion
Organic Electrolytes		
$(C_2H_5)_4N{\cdot}BF_4$ (TEA^+BF4^-)	0.686	0.458
$(C_2H_5)_3(CH_3)N{\cdot}BF_4$ ($TEMA^+BF_4^-$)	0.654	0.458
$(C_4H_9)_4N{\cdot}BF_4$ ($TBA^+BF_4^-$)	0.83	0.458
$(C_6H_{13})_4N{\cdot}BF_4$ ($THA^+BF_4^-$)	0.96	0.458
$(C_2H_5)_4N{\cdot}CF_3SO_3$	0.686	0.54
$(C_2H_5)_4N{\cdot}(CF_3SO_2)_2N$ (TEA^+TFSI^-)	0.68	0.65
Inorganic Electrolytes		
H_2SO_4		0.533
KOH	0.26[a]	
Na_2SO_4	0.36[a]	0.533
NaCl	0.36[a]	
$Li{\cdot}PF_6$	0.152[b]	0.508
$Li{\cdot}ClO_4$	0.152[b]	0.474

[a] Stokes diameter of hydrated ions
[b] The diameter of PC, depending on the solvent used
Source: Inagaki, M., H. Konno, and O. Tanaike. 2010. *Journal of Power Sources*, 195, 7880–7903. With permission.

TABLE 4.6

Basic Properties of Available Organic and Aqueous Solvents for ESs

Solvent	Melting Point (°C)	Viscosity (Pa·s^{-1})	Dielectric Constant (ε)
Acetonitrile	−43.8	0.369	36.64
γ-Butyrolactone	−43.3	1.72	39
Dimethyl ketone	−94.8	0.306	21.01
Propylene carbonate	−48.8	2.513	66.14
Water			

Source: Inagaki, M., H. Konno, and O. Tanaike. 2010. *Journal of Power Sources*, 195, 7880–7903. With permission.

TABLE 4.7

Electrolyte Resistances and Voltages of Various Electrolyte Solutions at Room Temperature

Electrolyte Solution	Density (g/cm^3)	Conductivity (mS/cm)	ΔU
Aqueous, KOH	1.29	540	1
Aqueous, KCl	1.09	210	1
Aqueous, sulfuric acid	1.2	750	1
Aqueous, sodium sulfate	1.13	91.1	1
Aqueous, potassium sulfate	1.08	88.6	1
Propylene carbonate, Et_4NBF_4	1.2	14.5	2.5 to 3
Acetonitrile, Et_4NBF_4	0.78	59.9	2.5 to 3
IL, Et_2MeIm^+ BF_4.	1.3 to 1.5	8 (25°C)	4
IL, Et_2MeIm^+ BF_4.		14 (100°C)	3.25

Note: See References 8, 73, and 109.

increasingly popular because of reduced leakage concerns and larger potential among other possible benefits.

4.3.1.1 Electrolyte Decomposition

Voltage in ECs is limited by the breakdown of materials within cells at higher voltages. As a result, the potential must be kept within a specific range. Experimentally, the evolution of side reactions at low or high voltages can be seen as sharp drifting current tails at either end of the voltage spectrum. By controlling the potential window, the redox tails due to decomposition (see Figure 4.30) can be avoided at either end of the potential spectrum utilized.

The figure illustrates the redox tails that occur through water decomposition when a large window is used for a three-electrode cell [107]. The decomposition potential is dependent upon the electrolyte and its

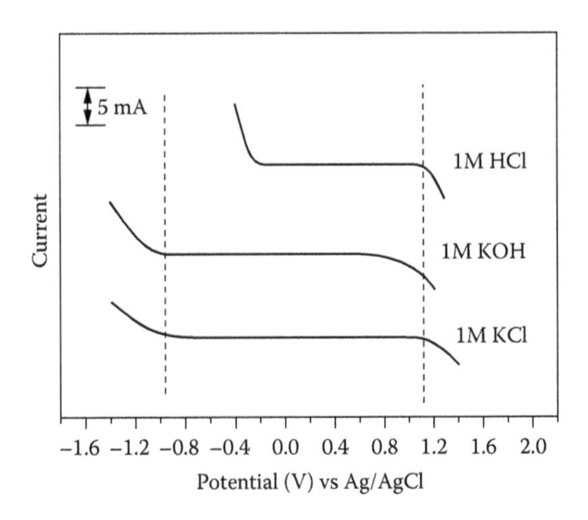

FIGURE 4.30
Gas evolution from various aqueous electrolytes during testing of stable platinum electrode. (*Source*: Hong, M. S., S. H. Lee, and S. W. Kim. 2002. *Electrochemical and Solid State Letters*, 5, A227. With permission.)

interaction effects with the solvent and the electrode material. In some cases, it is possible to utilize stabilizers to prevent decomposition reactions and increase potential. This concept is discussed in more detail for specific materials throughout this chapter. It is important to consider the effects of decomposition when testing and designing a cell to optimize performance and cycle life.

4.3.2 Aqueous Electrolytes

Aqueous electrolytes are used frequently due to low cost and availability. Ion sources include potassium hydroxide, potassium chloride, and sulfuric acid. Aqueous electrolytes are most commonly applied in the development stages of new ES materials. This is because of several key factors that include high ionic conductivity, mobility, and low hazard level. Further, aqueous electrolytes can be used in open environments and do not require water-free environments as organic electrolytes do.

The range of base, salt, and acid electrolytes makes it easier to tailor designs for electrode materials that require specific ion interaction mechanisms for optimal performance and avoid collector corrosion through undesirable redox reactions. For example, KCl is a safe, ionically conductive, neutral salt that has easy handling characteristics. Testing with KCl electrolyte and glassy carbon plates as the current collector works well and can be conducted safely. However, the chloride ions attack a large range of metals. This rules out low cost metal foils like stainless steel, nickel, and aluminum for collecting current.

The disadvantages of aqueous electrolytes involve corrosion and low stability-window (ΔV) issues that affect cell performance and stability. Acidic or basic pH conditions in a system can cause corrosion of collectors and packaging materials. Corrosive reactions detract from system performance and reduce cycle lives. Conversely, aqueous electrolytes exhibit water decomposition, resulting in hydrogen evolution at low cell potential (around 0 V) and oxygen evolution at high potential range (around 1.2 V) because of the poor voltage stability of water.

Rupturing cells threaten physical safety and reduce cycle life. Precautions must be taken with aqueous electrolyte systems to restrict the voltage window to avoid rupture. As a result, the potential window for most aqueous systems is limited to about 1 V. The low voltage stability of aqueous electrolytes greatly restricts the energy and power density possible in an ES. Conversely, the higher ionic conductivity and mobility of aqueous electrolytes seen in Table 4.7 translates into the best possible capacitance for an ES and lower internal cell resistance. The low internal resistance allows quick response time.

4.3.3 Organic Electrolytes

Organic electrolytes currently dominate the commercial ES market because of their potential window of operation in the range of 2.2 to 2.7 V. Table 5.7 provides a list of commercialized ES devices and their performance characteristics. Each device listed uses an organic electrolyte because of the enhanced potential window over aqueous electrolytes and moderate ion conduction provided. Most devices utilize acetonitrile, while others employ propylene carbonate solvents.

If organic electrolytes are used during peak operation periods, a dynamically controlled system could temporarily charge a cell to as high as 3.5 V [2]. The higher voltage window translates to the larger energy and power densities demanded by the consumer and industrial markets. The benefits are compounded when larger ES modules are used. Fewer components are needed to meet the module size requirements. Fewer cell balancing and connection components are required and less parasitic resistance arises from interfacing individual cells.

Acetonitrile is the current solvent standard, and is used to support the salt tetraethylammonium tetrafluoroborate (Et4NBF4, melting point >300°C) [108]. However, its continued use brings toxicity and safety concerns. A safer alternative is propylene carbonate, but it suffers from strong resistivity issues compared to acetonitrile.

Table 4.7 illustrates the resistivity and potential window properties of various electrolyte solvents [8,73,109]. The resistance of organic electrolytes is much higher than that of aqueous systems and negatively affects power and capacitive performances. The reduction in power performance is, however, balanced by the quadratic effect of the increased potential window [2].

TABLE 4.8

Evaluation of Capacitance and Resistance with Variations of Carbon A and Carbon B

Capacitor	Negative Electrode	Positive Electrode	Volumetric Capacity (F/cm³)	Internal Resistance (mΩ)
1	A	B	26.6	24
2	A	A	20.8	23
3	B	B	27.5	257
4	B	A	18.8	243

Source: Okamura, M. 1999. Electric double-layer capacitors and storage systems. U.S. Patent 6064562. With permission.

Note: Average pore diameters for carbon A and carbon B were 1.6 and 1.2 nm, respectively. Solvent used was tetraethylammonium tetrafluoroborate in acetonitrile.

As our understanding of pore and electrolyte ion interactions grows, it is clear that electrode materials should be developed with the intended electrolytes if possible. As an example, Table 4.8 correlates resistance to pore size and illustrates the importance of correctly matching ion and pore sizes [110]. Good design choices of electrolyte and pore size help optimize capacitance while minimizing the higher resistances seen in organic electrolyte systems. Even in optimized systems, the resistance of organic electrolytes still contributes to a much higher self-discharge current in ES devices. Self discharge stems from charge leakage across the double-layer interface. Water within an electrolyte can increase resistance and promote leakage. As a result, purification of electrolytes is necessary to prevent leakage and corrosion. The leakage across the double-layer interface of an EC ensures that long term energy storage is an inherent limitation of capacitive devices [111].

4.3.4 Ionic Liquids

Ionic liquids (ILs) begin to eliminate organic solvent safety issues and improve key parameters for use in ESs. ILs exist as viscous molten salts (gels) at ambient temperatures, allowing heavy concentrations in solvents or removal of solvents altogether. Low vapor pressure (rupture risks), low flammability, and low toxicity keep health risks low. High chemical stability of ILs allows operation at voltage windows as high as 5 V.

With the exception of the most studied ionic liquid, the $EtMeIm^+BF_4$ imidazolium salt, the major drawback of ionic liquids is their low conductivity at room temperature in aqueous and acetonitrile-based systems [112]. Table 4.7 illustrates that even $EtMeIm^+BF_4$ suffers from increased resistivity (lower conductivity) compared to aqueous or organic electrolytes. The correlation between increased stability ionic liquids and lowered conductivity can be seen in Table 4.9 [113].

IL electrolytes have high thermal stability that creates an opportunity for operation in high temperature environments. At high temperatures, the low conductivity that limits IL performance is overcome by increased ion mobility (kinetic energy), resulting in higher conductivity, greater device power, and better response time [113]. However, high heat reduces the potential window for ion stability and this negatively impacts the power and energy density. Another way to overcome the low conductivity of ILs is to balance the high potential windows of ionic liquids with the increased conductivity and power of organic electrolytes such as propylene carbonate and acetonitrile to optimize conductivity [112]. Utilizing such combinations can prevent safety issues, reduce toxicity, and result in a device with high energy density that maintains sufficient power performance [112].

4.3.5 Solid State Polymer Electrolytes

Gel and solid polymer electrolytes aim to combine the function of the electrolyte and separator into a single component to reduce the number of parts in an ES and increase the potential window through the higher stability offered by a polymer matrix. A gel electrolyte incorporates a liquid electrolyte into a microporous polymer matrix that holds in the liquid electrolyte through capillary forces, creating a solid polymer film. The chosen separator must be insoluble in the desired electrolyte and provide adequate ionic conductivity. Non-polar rigid polymers such as PTFE, PVA, PVdF, and cellulose acetate offer good ion conductivity when used as gel electrolytes [114]. Based on the data in Table 4.9, the ionic conductivity of $EtMeIm^+BF_4$ is 14 $mS.cm^{-1}$. Ionic conductivity of the same imidazolium salt used as a gel electrolyte in a PVdF matrix retains 5 $mS.cm^{-1}$[115].

Modern electrolytes need increased stability and ion mobility to operate at high potential windows. Gel electrolytes allow incorporation of aqueous, organic, and ionic liquids, depending on the requirements of the ES. Separators are used in conjunction with the electrolyte to help provide structured channels and prevent short circuits between the electrodes. The presence of solid electrolyte layers results in a reduced need for robust encapsulation technologies.

In order to combine the two structures, electrolyte is trapped within the polymer matrix during polymerization. The result is a solid, thin, flexible electrolyte. Gel electrolytes have clear manufacturing and assembly advantages because of their simplified forms and dual functionalities. However, the performance is a large factor in the ability of such ideas to proliferate.

Gel polymers offer slightly lower conductance than liquid electrolytes, but they provide structural improvement that improves the efficiency of ion transport mechanisms and cycle life [116,117]. Polyvinyl acetate (PVA) has been shown to offer good results in trapping aqueous electrolytes [116,118,119].

TABLE 4.9

List of Ionic Liquids and Their Parameters

	Electrochemical Stability					
Ionic Liquid	**Cathodic Limit (V)**	**Anodic Limit (V)**	**ΔU (V)**	**Working Electrode**	**Reference**	**Conductivity σ at 25°C (mS cm^{-1})**
Imidazolium						
$[EtMeIm]^+[BF_4]^-$	–2.1	2.2	4.3	Pt	Ag/Ag^+, DMSO	14.0
$[EtMeIm]^+[CF_3SO_3]^-$	–1.8	2.3	4.1	Pt	I^-/I_3^-	8.6
$[EtMeIm]^+[N(CF_3SO_2)_2]^-$	–2.0	2.1	4.1	Pt, GC	Ag	8.8
$[EtMeIm]^+[(CN)_2N]^-$	–1.6	1.4	3.0	Pt	Ag	–
$[BuMeIm]^+[BF_4]^-$	–1.6	3.0	4.6	Pt	Pt	3.5
$[BuMeIm]^+[PF_6]^-$	–1.9	2.5	4.4	Pt	Ag/Ag^+, DMSO	1.8
$[BuMeIm]^+[N(CF_3SO_2)_2]^-$	–2.0	2.6	4.6	Pt	Ag/Ag^+, DMSO	3.9
$[PrMeMeIm]^+[N(CF_3SO_2)_2]^-$	–1.9	2.3	4.2	GC	Ag	3.0
$[PrMeMeIm]^+[C(CF_3SO_2)_3]^-$		5.4	5.4	GC	Li/Li^+	–
Pyrrolidinium						
$[n\text{PrMePyrrol}]^+[N(CF_3SO_2)_2]^-$	–2.5	2.8	5.3	Pt	Ag	1.4
$[n\text{BuMePyrrol}]^+[N(CF_3SO_2)_2]^-$	–3.0	2.5	5.5	GC	Ag/Ag^+	2.2
$[n\text{BuMePyrrol}]^+[N(CF_3SO_2)_2]$	–3.0	3.0	6.0	Graphite	Ag/Ag^+	–
Tetraalkylammonium						
$[n\text{Me}_3\text{BuN}]^+[N(CF_3SO_2)_2]^-$	–2.0	2.0	4.0	Carbon		1.4
$[n\text{PrMe}_3\text{N}]^+[N(CF_3SO_2)_2]^-$	–3.2	2.5	5.7	GC	Fc/Fc^+	3.3
$[n\text{OctEt}_3\text{N}]^+[N(CF_3SO_2)_2]^-$			5.0	GC		0.33
$[n\text{OctBu}_3\text{N}]^+[N(CF_3SO_2)_2]^-$			5.0	GC		0.13

Pyridinium						
$[BuPyr]^+[BF_4]^-$	–1.0	2.4	3.4	Pt	Ag/AgCl	1.9
Piperidinium						
$[MePrPip]^+[N(CF_3SO_2)_2]^-$	–3.3	2.3	5.6	GC	Fc/Fc$^+$	1.5
Sulfonium						
$[Et_3S]^+[N(CF_3SO_2)_2]^-$			4.7	GC		7.1
$[nBu_3S]^+[N(CF_3SO_2)_2]^-$			4.8	GC		1.4

Source: Lewandowski, A. and M. Galinski. 2007. *Journal of Power Sources*, 173, 822–828. With permission.

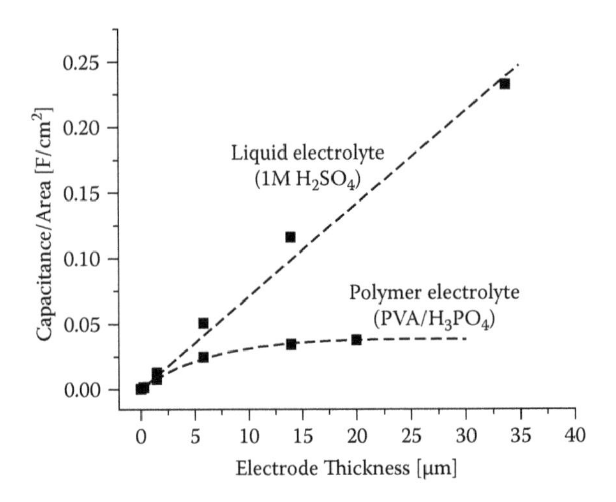

FIGURE 4.31
Thickness dependence of capacitance per area for CNT films comparing liquid (1 *M* H_2SO_4) and gel (PVA/H3PO4) electrolytes. (*Source:* Kaempgen, M. et al. 2009. *Journal of the American Chemical Society*, 9, 1872–1876. With permission.)

PVdF is also capable of providing structure and highly conductive channels for ion transport.

Polymer gel electrolytes suffer from a key restriction in electrode thickness that can be observed in Figure 4.31. Penetration of ions deep into highly porous electrodes is limited and saturation occurs in performance for thicker electrodes. The PVA-based gel electrolyte shown in Figure 4.31 illustrates a saturation around 10 μm and matched performance to aqueous systems at 2 to 3 μm. This suggests the market for gel electrolytes currently lies in flexible lower capacitance storage applications [120].

Solid polymer electrolytes made of polyethylene oxide (PEO) and polypropylene oxide (PPO) are considered because of their strong thermal conduction and electrochemical properties over a wide operating temperature range [114,115]. However, the low room temperature ionic conductivities exhibited by PEO and PPO solid state polymer electrolytes prevents successful application in ESs. When PEO was incorporated into a gel electrolyte to boost conductivity, the result indicated that PEO and PPO are actually found inferior compared to PVA and PVdF for gel electrolytes because the oxygen atoms in the polymer backbone limit ion mobility [115].

Another alternative solid state electrolyte under study is the use of solid-state proton conductors such as heteropoly acid (HPA) electrolytes. The two most common HPAs are $H_4SiW_{12}O_{40}$ (SiWA) and $H_3PW_{12}O_{40}$ (PWA) [121]. The HPA materials have high proton conductivities at room temperature (solid form of pure SiWA = 27 $mS.cm^{-1}$). The traditional problem with solid state proton conductors is their poor film making properties that make forming a separator difficult.

Lian et al. [121–123] investigated a composite solid state polymer that utilized PVA for its good film formability. The solid PVA–PWA and PVA–SiWA electrolytes have good film making properties and exhibit strong stability at high relative humidity. Nafion® is another proton conducting polymer that has good film forming properties and also exhibits high conductivity at room temperature, but its conductivity decreases significantly with decreasing humidity and cell moisture [122]. Stability means that HPA materials can be processed in ambient environments; they simplify packaging procedures and create leak-proof, corrosion-free cell designs. The PVA–SiWA solid state electrolyte exhibited 11 $mS.cm^{-1}$ and provided capacitance (50 $mF.cm^{-2}$) comparable to an aqueous H_2SO_4 electrolyte (70 $mF.cm^{-2}$) when tested with a symmetrical ruthenium dioxide cell (60 μm thick electrodes) [121]. Further optimization showed that an even mix of PWA and SiWA created a synergetic effect when combined with PVA and increased conductivity to 13 $mS.cm^{-1}$ [122].

4.4 Separator Structures

Separators play a role in preventing contact and electron transfer between anode and cathode. A separator must be mechanically strong to provide device durability and prevent migration of high carbon particles over time. It is important that the material possesses strong ion conductance and electronic insulating capability. High ionic conductivity is promoted by high porosity and low tortuosity [2]. Resistances to ion flow and interfacial contact resistance with an electrode can also be improved when separators exhibit sufficient wettability [114]. A separator film should be thin, while maintaining mechanical stability. Separators must be chemically resistant to corrosion from electrolytes and by-products of electrode degradation. It is also very important that separators prevent migration of active materials in order to eliminate short circuiting.

Batteries and ESs that operate near ambient temperatures often use materials such as cellulose paper, polymer, and glass wool. However, commercial separators vary based on electrolyte choice and temperature of operation. ESs represent a developing market that utilizes many common electrolytes used in battery systems. For this reason, separator choices closely mimic choices for batteries. Organics utilize microporous polymers and cellulose paper separators, whereas aqueous devices traditionally utilize glass, mica, and ceramic separators [124]. However, paper-based separators suffer from poor mechanical strength and durability in high temperature operation environments.

Polyolefin-based microporous separators (Figure 4.32) continue to displace natural materials such as glass and cellulose fibers because of high porosity, low cost, flexibility, corrosion resistance, and improved mechanical strength

FIGURE 4.32
Appearances and forms of polyolefin separators. (*Source:* Arora, P. and Z. J. Zhang. 2004. *Chemical Reviews*, 104, 4419–4462. With permission.)

[114]. Polyolefins continue to see increased utilization in aqueous systems as well. The problem with polyolefins for aqueous electrolytes is that they are hydrophobic and cannot be effectively wetted by an electrolyte. However, treatment of polyolefin films by graft polymerization enables modification of surface properties to increase surface hydrophilicity [114].

Separator structures fall into four main categories: microporous films, nonwovens, gel polymers, and solid polymers. Microporous films contain small pores (5 to 10 nm in diameter) and are often used for low temperature applications. They are made from nonwoven fibers such as cotton, polyester, glass, polyolefins (PP and PE), PTFE, and PVC. Microporous separators are commonly used with organic electrolytes and in acidic systems. Nonwovens are manufactured as mats of fibers and bind through frictional forces. They exhibit consistent weight, thickness, and degradation resistance but they show inadequate pore order and are difficult to make thinner than 25 μm. Nonwovens are generally made from cellulose, PTFE, PVC, PVdF, or a combination of polyolefins and receive preference in alkaline systems [114].

4.5 Current Collectors

Current collectors are used in ES devices to gather and feed electrical charges stored within the active capacitive material. In most cases, the conduction of charge throughout the active material of an electrode is insufficient and provides a large amount of resistance that can ruin performance characteristics of an otherwise acceptable ES material. Efficient contacts and additives are needed to effectively transport charge current and provide a system with sufficient power.

Typically metal and metal alloys with high electrical conductivities are used to move energy to a common ground for the capacitor stack. Metals of choice are aluminum, iron, copper, and steel thin films. In the case of a soft electrode, the material can be pressed into a metal wire mesh. The sheet resistance of the active material prevents effective conduction over long distances such as along the length of an electrode film. As a result, current conduction is applied through the thickness of the active material and then passes to a larger external end-plate collector through a low resistance solder joint. The metal collector film backs the active electrode material to minimize electron transport distance to the highly conductive collector.

Over prolonged cycling time, the connection between the electrode and the metal collector can be degraded due to shrinkage and loss of electrode material to the electrolyte. These losses translate to lowered capacitance and cycle life of the device and offer further reason to use other techniques to improve connections and reduce resistances. Another important concern on a system-wide level is the efficient transport of heat from the material. Using thermally conductive metals helps transport heat to the end plate, allowing better integration with heat sinks for passive cooling. Aluminum foils are common choices due to thermal transmission, low cost, high workability, and good conductivity.

In many cases, electrode materials or pastes are grown or deposited (e.g., spray coat, drop coat, spin coat, application of electrode paste) directly onto collector material to provide a good molecular contact created by drying and annealing of the electrode [125]. To further minimize resistance, a highly conductive coating that exhibits improved interfacial contact between components can be applied. Without low resistance connections, the electrode performance of a good material can otherwise be lost.

One common technique for generating a good contact is the use of silver paste that adapts to the surface roughness of the active material. Gold and lead thin films have been shown to significantly enhance interface integration of graphene (electrically and mechanically), resulting in higher device performance. Strong current collection allows efficient movement of charge and allows current density as high as 30 $A.g^{-1}$ (maintaining 70% of low current capacitance) and performance of 40 $Wh.kg^{-1}$ and 40 $kW.kg^{-1}$ [126].

Another technique used by Portet et al. to minimize contact resistance employs roughening the collector to bring ESR down from 50 to 5 Ω.cm2. Etching or surface roughening is an effective way to improve the contact area and increase the number of conductive sites. To further improve contact resistance, a conductive carbon gel was dried onto the collector surface. The carbon-coated collector exhibited a minimal interfacial boundary with the active carbon material and managed to reduce contact resistance to as low as 1 $\Omega.cm^2$ [127].

For free-standing electrodes, integration with collectors becomes even more important to avoid resistances because there is no intimate surface contact. This highlights the importance of high pressure lamination to reduce contact resistance between collector and active material when an intimate contact generated through drying or annealing is not possible or is insufficient to prevent power limitation.

Even with efficient current collection through the electrode–collector interface, an active material can exhibit large electronic resistances. Internal resistance within an electrode paste is caused by low conductivity and insufficient percolation to provide enough complete conduction pathways across the electrode layer thickness. These internal resistances are characteristic to a particular active material and can vary based on the quality of binding by the electrode production technique. Poor binding can also result in increased resistance and device failure during operation.

To reduce internal resistance, electrode pastes contain polymeric binding agents (PTFE, Nafion, PVdF, or PVB) and conductive carbon additives (acetylene black, Super P) [127–129]. Binders increase durability and prevent degradation of the collector–electrode interface over time. However, one downside of using these agents is an increase in internal electrode resistance that reduces device power [31]. Optimal balance of internal resistance and sufficient paste stability leads to loading of about 3 to 5% by weight.

Internal resistance within the electrode paste matrix is reduced by the enhanced percolation provided by 10 to 20% of conductive carbon filler. Collector metals and conductive pastes contribute to a non-negligible weight within a device. Dead weight is an important factor that cannot be overlooked when designing electrode materials. Any non-capacitive weight in a cell reduces device performance. As a result, optimization of cell design for minimal paste additives should be used and collector metals should be thin and lightweight.

With this in mind, a hybrid collector electrode material with CNTs was demonstrated. The highly conductive nanotubes provided sufficient current collection for pseudocapacitive [50] and EDLC devices [47]. In both cases, electrodes made of CNTs deposited on lightweight paper substrates replaced the heavy metal collectors and carbon paste additives and enhanced the specific capacitance of the overall system, compared with results from CNTs deposited on metal substrates. A CNT paper electrode of 30 $Wh.kg^{-1}$ and 200 $kW.kg^{-1}$ was demonstrated [47]. The downside is that CNTs offer low energy density compared to AC materials and cost more than metal collector substrates.

4.6 Sealants

Proper sealing is a key component of cell assembly. Gas evolution from electrolyte degradation, corrosion, and surface oxidation of electrode and packaging can be a problem over time [1]. Depending on the ion type and solvent, issues can vary. A proper seal will prevent water and gas from entering a cell. Another key reason for proper sealing is to prevent shunt resistances between neighboring electrodes and cells. A shunt current significantly increases self discharge and reduces device efficiency. A poor seal will break over time and cause degradation and short circuiting [2].

Rolled cells can contain separate sealant materials. The internal seal is a curable polymer placed onto the top of the metal casing before it is closed and mechanically crimped shut. Curable polymer sealants are also used to prevent short circuits across multiple layers in bipolar stacks (tops and bottoms of roll edges). The curable material composed of a thermosetting polymer, epoxy, polyurethane, polyester, polyacrylate creates a moisture-resistant and mechanically strong seal [130]. A second type is an insulating shrink wrap that seals the outer casing electrode, offers an external moisture barrier, and prevents unplanned discharges that create safety issues. Shrink wrap seals are made from loosely fitting preformed bags composed of polyolefins (PP, PE, reinforced PE) that surround the metal casing. Heat is then applied to cause the plastic to shrink and tightly seal around the casing insulating the cell.

Pouch cells have no metal casings. A multilayer polymer is placed around the cell and laminated on a production line to bind and seal the device between the bag layers. Before electrolyte infusion and bagging, the devices undergo a nitrogen purge of the environment. A vacuum is applied, then the electrolyte is fed through the bagging enclosure system under vacuum to create a tight seal. The vacuum applies even pressure along the contents of the film and drives the correct amount of electrolyte resin into the cells. The excess is removed and collected for reuse [131].

The process lends itself to lamination on a large scale. Many cells can be processed at one time within a single vacuum bag by careful arrangement and spacing to minimize wasted bagging material. After lamination, the laminated sheet can be cut into individual devices.

In smaller scale research applications, a polymer-coated foil bag (polybag) can be vacuumed and the edges heat-sealed by hand [47]. An ideal sealant bag is thermally conductive, electronically insulating, moisture resistant, and has a one-way gas permeability membrane to vent gases generated from the system.

Safety dictates that a seal must be strongly electronically insulating and be puncture resistant. While optimizing these properties, it is also desirable to minimize bag weight. The polybag used by Ioxus ESs is a patented laminate design [132]. LDPE sealant material is used as a first layer that re-forms at 120°C, insulates, and binds the cell surface to the packaging. A copper non-reactive foil layer provides moisture resistance and thermal conductivity away from the cell. The copper is sandwiched between another LDPE layer that provides further electrical insulation and binds an outer puncture-resistant layer of PET or Mylar® (biaxially aligned PET) that is mechanically strong. After aligning the films, they are laminated in production and sent to module assembly.

4.7 Summary

In this chapter, several important aspects of component materials and requirements and cell and stack construction of electrochemical supercapacitors (ESs) were discussed to enable readers to: (1) understand the functional differences of capacitors, batteries, and ECs and how they impose different material and performance restrictions; (2) recognize the importance of pore sizes and their interactions with electrodes, separators, and electrolyte materials, with a focus on existing and new carbon electrode materials; (3) observe the role of pseudocapacitance in boosting energy density while overcoming durability issues; (4) illustrate the importance of making strong design choices that meet the energy, safety, and power needs of a system and looking to the future and economies of scale; (5) gain an overview of electrode composites choices to overcome material disadvantages, particularly the composite devices that integrate components to simplify, and in some cases improve, device mechanics.

References

1. Simon, P. and Y. Gogotsi. 2008. Materials for electrochemical capacitors. *Nature: Materials*, 7, 845–854.
2. Kötz, R. 2000. Principles and applications of electrochemical capacitors. *Electrochimica Acta*, 45, 2483–2498.
3. Linden, D. and T. Reddy, *Handbook of Batteries*, 4th ed. New York: McGraw Hill.
4. Burke, A. and M. Miller. 2011. The power capability of ultracapacitors and lithium batteries for electric and hybrid vehicle applications. *Journal of Power Sources*, 196, 514–522.
5. Winter, M. and R. J. Brodd. 2004. What are batteries, fuel cells, and supercapacitors? *Chemical Reviews*, 104, 4245–4269.
6. Worlds of David Darling Encyclopedia (online). Lead–acid battery. http://www.daviddarling.info/encyclopedia/L/AE_lead–acid_battery.html [accessed April 4, 2012].
7. Georgia State University. 2012. Lead–acid battery: hyperphysics (online). http://hyperphysics.phy–astr.gsu.edu/hbase/electric/leadacid.html [accessed April 9, 2012].
8. Davies, A. and A. Yu. 2011. Material advancements in supercapacitors: From activated carbon to carbon nanotube and graphene. *Canadian Journal of Chemical Engineering*, 89, 1342–1357.
9. Qu, D. and H. Shi. 1998. Studies of activated carbons used in double-layer capacitors. *Journal of Power Sources*, 74, 99–107.
10. Kim, Y. et al. 2004. Correlation between the pore and solvated ion size on capacitance uptake of PVDC-based carbons. *Carbon*, 42, 1491–1500.

11. Chmiola, J. et al. 2008. Desolvation of ions in subnanometer pores and its effect on capacitance and double-layer theory. *Angewandte Chemie*, 47, 3392–3395.
12. Chmiola, J. et al. 2006. Anomalous increase in carbon capacitance at pore sizes less than 1 nanometer. *Science*, 313, 1760–1763.
13. Largeot, C. et al. 2008. Relation between the ion size and pore size for an electric double-layer capacitor. *Journal of the American Chemical Society*, 130, 2730–2731.
14. Huang, J. B. Sumpter, and V. Meunier. 2008. Universal model for nanoporous carbon supercapacitors applicable to diverse pore regimes, carbons, and electrolytes. *European Journal of Chemistry*, 14, 6614–6626.
15. Huang, J. B. Sumpter, and V. Meunier. 2008. Theoretical model for nanoporous carbon supercapacitors. *Angewandte Chemie*, 47, 520–524.
16. Feng, G. et al. 2010. Atomistic insight on the charging energetics in subnanometer pore supercapacitors. *Society*, 114, 18012–18016.
17. Liu, C. et al. 2010. Advanced materials for energy storage. *Advanced Materials*, 22, E28–E62.
18. Geng, D. et al. 2011. Nitrogen doping effects on the structure of graphene. *Applied Surface Science*, 257, 9193–9198.
19. Zhao, L. et al. 2010. Nitrogen-containing hydrothermal carbons with superior performance in supercapacitors. *Advanced Materials*, 22, 5202–5206.
20. Lin, Z. et al. 2011. Surface engineering of graphene for high performance supercapacitors. *Synthesis*, 236–241.
21. Jeong, H. M. et al.2011. Nitrogen-doped graphene for high performance ultracapacitors and the importance of nitrogen-doped sites at basal planes. *Nanoletters*, 11, 2472–2477.
22. Algharaibeh, Z. and P. G. Pickup, 2011. An asymmetric supercapacitor with anthraquinone and dihydroxybenzene modified carbon fabric electrodes. *Electrochemistry Communications*, 13, 147–149.
23. Pognon, G. et al. 2011. Performance and stability of electrochemical capacitor based on anthraquinone modified activated carbon. *Journal of Power Sources*, 196, 4117–4122.
24. Pognon, G. T. Brousse, and D. Bélanger. 2011. Effect of molecular grafting on the pore size distribution and the double-layer capacitance of activated carbon for electrochemical double-layer capacitors. *Carbon*, 49, 1340–1348.
25. Conway, B. E. 1999. *Electrochemical Supercapacitors*, New York: Plenum.
26. Gryglewicz, G. et al. 2005. Effect of pore size distribution of coal-based activated carbons on double-layer capacitance. *Electrochimica Acta*, 50, 1197–1206.
27. Smith, P. and T. Jiang. 2009. High Energy Density Ultracapacitors. NAVSEA-Caderock Division. http://www1.eere.energy.gov/vehiclesandfuels/pdfs/merit_review_2009/energy_storage/esp_22_smith.pdf
28. Frackowiak, E. 2007. Carbon materials for supercapacitor application. *Physical Chemistry–Chemical Physics*, 9, 1774–1785.
29. Wu, F. C. et al. 2004. Physical and electrochemical characterization of activated carbons prepared from fir woods for supercapacitors. *Journal of Power Sources*, 138, 351–359.
30. Wang, X. and D. Wang. 2003. Performance of electric double-layer capacitors using active carbons prepared from petroleum coke by KOH and vapor re-etching. *Journal of Materials Science and Technology*, 19.

31. Obreja, V. 2008. On the performance of supercapacitors with electrodes based on carbon nanotubes and carbon activated material: A review. *Physica E*, 40, 2596–2605.
32. Otowa, T. R. Tanibata, and M. Itoh. 1993. Production and adsorption characteristics of Maxsorb, a high surface area active carbon. *Gas Separation and Purification*, 7, 241–245.
33. Wen, Z. et al. 2009. An activated carbon with high capacitance from carbonization of a resorcinol–formaldehyde resin. *Electrochemistry Communications*, 11, 715–718.
34. Ania, C. O. et al. 2007. The large electrochemical capacitance of microporous doped carbon obtained by using a zeolite template. *Advanced Functional Materials*, *17*, 1828–1836.
35. Fuertes, A. et al. 2005. Templated mesoporous carbons for supercapacitor application. *Electrochimica Acta*, 50, 2799–2805.
36. Zhang, L. L. and X. S. Zhao. 2009. Carbon-based materials as supercapacitor electrodes. *Chemical Society Reviews*, 38, 2520–2531.
37. Wang, H. et al. 2009. High performance of nanoporous carbon in cryogenic hydrogen storage and electrochemical capacitance. *Carbon*, 47, 2259–2268.
38. Portet, C. et al. 2009. Electrical double-layer capacitance of zeolite-templated carbon in organic electrolyte. *Journal of the Electrochemical Society*, 156, A1–A6.
39. Nishihara, H. et al. 2009. Investigation of the ion storage/transfer behavior in an electrical double-layer capacitor by using ordered microporous carbons as model materials. *Chemistry*, 15, 5355–5363.
40. Vixguterl, C. et al. 2004. Supercapacitor electrodes from new ordered porous carbon materials obtained by a templating procedure. *Materials Science and Engineering B*, 108, 148–155.
41. Kim, N. D. et al. 2008. Electrochemical capacitor performance of N-doped mesoporous carbons prepared by ammoxidation. *Journal of Power Sources*, 180, 671–675.
42. Lufrano, F. et al. 2010. Mesoporous carbon materials as electrodes for electrochemical supercapacitors. *International Journal of Electrochemical Science*, 5, 903–916.
43. Beck, J. S. et al. 1992. New family of mesoporous molecular sieves prepared with liquid crystal templates. *Journal of the American Chemical Society*, 114, 10834–10843.
44. Vix-Guterl, C. et al. 2005. Electrochemical energy storage in ordered porous carbon materials. *Carbon*, 43, 1293–1302.
45. Lei, Z. et al. 2011. Mesoporous carbon nanospheres with an excellent electrocapacitive performance. *Journal of Materials Chemistry*, 21, 2274.
46. Wang, D. W. et al. 2008. A 3-D aperiodic hierarchical porous graphitic carbon material for high rate electrochemical capacitive energy storage. *Angewandte Chemie*, 47, 373–376.
47. Hu, L. et al.2009. Highly conductive paper for energy storage devices. *Proceedings of National Academy of Sciences of the United States of America*, 106, 21490–21494.
48. Yoon, B. 2004. Electrical properties of electrical double-layer capacitors with integrated carbon nanotube electrodes. *Chemical Physics Letters*, 388, 170–174.
49. Honda, Y. et al. 2007. Aligned MWCNT sheet electrodes prepared by transfer methodology providing high power capacitor performance. *Electrochemical and Solid State Letters*, 10, A106–A110.

50. Zhou, R. et al. 2010. High performance supercapacitors using a nanoporous current collector made from super-aligned carbon nanotubes. *Nanotechnology*, 21, 345701.
51. Collins, P. 2000. Nanotubes for electronics. *Scientific American*, 67–69.
52. Inagaki, M., H. Konno, and O. Tanaike. 2010. Carbon materials for electrochemical capacitors. *Journal of Power Sources*, 195, 7880–7903.
53. Hasan, T. and V. Scardaci. 2007. Stabilization and "debundling" of single wall carbon nanotube dispersions in N-methyl-2-pyrrolidone (NMP) by polyvinylpyrrolidone (PVP). *Journal of Physical Chemistry C*, 111, 12594–12602.
54. An, K. H. et al. 2001. Electrochemical properties of high power supercapacitors using single-walled carbon nanotube electrodes. *Advanced Functional Materials*, 11, 387–392.
55. Niu, C. et al. 1997. High power electrochemical capacitors based on carbon nanotube electrodes. *Applied Physics Letters*, 70, 1480.
56. Li, J. et al. 2007. Correlations between percolation threshold, dispersion state, and aspect ratio of carbon nanotubes. *Advanced Functional Materials*, 17, 3207–3215.
57. Liu, C. 2005. The electrochemical capacitance characteristics of activated carbon electrode material with a multiwalled carbon nanotube additive. *New Carbon Materials*, 20, 205–210.
58. Hata, K. et al. 2004. Water-assisted highly efficient synthesis of impurity-free single walled carbon nanotubes. *Science*, 306, 1362–1364.
59. Hiraoka, T. et al. 2010. Compact and light supercapacitor electrodes from surface-only solid by opened carbon nanotubes with 2200 m^2/g surface area. *Advanced Functional Materials*, 20, 422–428.
60. Pech, D. et al. 2010. Ultrahigh power micrometer-sized supercapacitors based on onion-like carbon. *Nature: Nanotechnology*, 5, 651–654.
61. Rao, C. N. R. et al. 2009. Graphene, the new two-dimensional nanomaterial. *Angewandte Chemie*, 48, 7752–7777.
62. Stankovich, S. et al. 2007. Synthesis of graphene-based nanosheets via chemical reduction of exfoliated graphite oxide. *Carbon*, 45, 1558–1565.
63. Park, S. and R. S. Ruoff. 2009. Chemical methods for the production of graphenes. *Nature: Nanotechnology*, 4, 217–224.
64. Wang, Y. et al. 2011. Preventing graphene sheets from restacking for high capacitance performance. *Journal of Physical Chemistry C*, 115, 23192–23197.
65. Wang, Y. et al. 2009. Supercapacitor devices based on graphene materials. *Journal of Physical Chemistry C*, 113, 13103–13107.
66. Yu, A. et al. 2010. Ultrathin, transparent, and flexible graphene films for supercapacitor application. *Applied Physics Letters*, 96, 253105.
67. Zhu, Y. et al. 2011. Carbon-based supercapacitors produced by activation of graphene. *Science*, 1537.
68. Yang, X. et al. 2011. Bioinspired effective prevention of restacking in multilayered graphene films: toward the next generation of high performance supercapacitors. *Advanced Materials*, 23, 2833–2838.
69. Barranco, V. et al. 2010. Amorphous carbon nanofibers and their activated carbon nanofibers as supercapacitor electrodes. *Carbon*, 114, 10302–10307.
70. Kim, C. and K. S. Yang. 2003. Electrochemical properties of carbon nanofiber web as an electrode for supercapacitor prepared by electrospinning. *Applied Physics Letters*, 83, 1216.

71. Kim, C. et al. 2004. Supercapacitor performances of activated carbon fiber webs prepared by electrospinning of PMDA-ODA poly(amic acid) solutions. *Electrochimica Acta*, 50, 883–887.
72. Yan, X. et al. 2011. Fabrication of carbon nanofiber–polyaniline composite flexible paper for supercapacitor. *Nanoscale*, 3, 212–216.
73. Burke, A. 2007. R&D considerations for performance and application of electrochemical capacitors. *Electrochimica Acta*, 53, 1083–1091.
74. Conway, B. E. V. Birss, and J. Wojtowicz. 1997. The role and utilization of pseudocapacitance for energy storage by supercapacitors. *Journal of Power Sources*, 66, 1–14.
75. Conway, B. E. 1991. Transition from supercapacitor to battery behavior in electrochemical energy storage. *Journal of the Electrochemistry Society*, 138, 1539–1548.
76. Zhang, L. L. et al. 2009. Manganese oxide–carbon composite as supercapacitor electrode material. *Microporous and Mesoporous Materials*, 123, 260–267.
77. Zheng, J. P. P. J. Cygan, and T. R. Jow. 1995. Hydrous ruthenium oxide as an electrode material for electrochemical capacitors. *Journal of the Electrochemical Society*, 142, 2699–2703.
78. Jayalakshmi, M. et al. 2007. Hydrothermal synthesis of SnO_2–V_2O^5 mixed oxide and electrochemical screening of carbon nanotubes (CNTs), V_2O_5, V_2O_5 CNTs, and SnO_2-V_2O_5-CNT electrodes for supercapacitor applications. *Journal of Power Sources*, 166, 578–583.
79. Hu, C. C., W. C. Chen, and K. H. Chang. 2004. How to achieve maximum utilization of hydrous ruthenium oxide for supercapacitors. *Journal of the Electrochemical Society*, 151, A281–A290.
80. Raistrick, I. D. 1992. Electrochemical capacitors. In *Electrochemistry of Semiconductors and Electronics: Processes and Devices*, New York: Noyes, 297–365.
81. Zheng, J. P. 1995. A new charge storage mechanism for electrochemical capacitors. *Journal of the Electrochemical Society*, 142, L6–L8.
82. Hu, C. 2004. Effects of substrates on the capacitive performance of $RuO_x{\cdot}nH_2O$ and activated carbon–RuO_x electrodes for supercapacitors. *Electrochimica Acta*, 49, 3469–3477.
83. Cottineau, T. et al. 2005. Nanostructured transition metal oxides for aqueous hybrid electrochemical supercapacitors. *Applied Physics A*, 82, 599–606.
84. Rajendra Prasad, K. and N. Miura. 2004. Electrochemically synthesized MnO_2-based mixed oxides for high performance redox supercapacitors. *Electrochemistry Communications*, 6, 1004–1008.
85. Khomenko, V. et al. 2005. High voltage asymmetric supercapacitors operating in aqueous electrolyte. *Applied Physics A*, 82, 567–573.
86. Tomko, T. et al. 2011. Synthesis of boron–nitrogen substituted carbons for aqueous asymmetric capacitors. *Electrochimica Acta*, 56, 5369–5375.
87. Toupin, M. T. Brousse, and D. Belanger. 2004. Charge storage mechanism of MnO_2 electrode used in aqueous electrochemical capacitor. *Chemical Materials*, 16, 3184–3190.
88. Toupin, M. T. Brousse, and D. Belanger. 2004. Charge storage mechanism of MnO_2 electrode used in aqueous electrochemical capacitor. *Journal of the American Chemical Society*, 16, 3184–3190.
89. Wang, Y. and I. Zhitomirsky/ 2011. Cathodic electrodeposition of Ag-doped manganese dioxide films for electrodes of electrochemical supercapacitors. *Materials Letters*, 65, 1759–1761.

90. Li, J. and I. Zhitomirsky. 2009. Electrophoretic deposition of manganese dioxide–carbon nanotube composites. *Journal of Materials Processing Technology*, 209, 3452–3459.
91. Fan, Z. et al. 2006. Preparation and characterization of manganese oxide/CNT composites as supercapacitive materials. *Diamond and Related Materials*, 15, 1478–1483.
92. Yang, Y. and C. Huang. 2009. Effect of synthetical conditions, morphology, and crystallographic structure of MnO_2 on its electrochemical behavior. *Journal of Solid State Electrochemistry*, 14, 1293–1301.
93. Zhang, J. et al. 2011. Synthesis, characterization and capacitive performance of hydrous manganese dioxide nanostructures. *Nanotechnology*, 22, 125703.
94. Liu, T. C. 1998. Behavior of molybdenum nitrides as materials for electrochemical capacitors. *Journal of the Electrochemical Society*, 145, 1882.
95. Choi, D. G. E. Blomgren, and P. N. Kumta. 2006. Fast and reversible surface redox reaction in nanocrystalline vanadium nitride supercapacitors. *Advanced Materials*, 18, 1178–1182.
96. Choi, D. and P. N. Kumta. 2006. Nanocrystalline TiN derived by a two-step halide approach for electrochemical capacitors. *Journal of the Electrochemical Society*, 153, A2298–A2303.
97. Zheng, J. P. 2004. Resistance distribution in electrochemical capacitors with a bipolar structure. *Journal of Power Sources*, 137, 158–162.
98. Rudge, A. et al. 1994. Conducting polymers as active materials in electrochemical capacitors. *Journal of Power Sources*, 47, 89–107.
99. Zhang, K. et al. 2010. Graphene–polyaniline nanofiber composites as supercapacitor electrodes. *Chemistry of Materials*, 22, 1392–1401.
100. Frackowiak, E. et al. 2006. Supercapacitors based on conducting polymer–nanotube composites. *Journal of Power Sources*, 153, 413–418.
101. Yan, J. et al. 2010. Preparation of graphene nanosheet/carbon nanotube/polyaniline composite as electrode material for supercapacitors. *Journal of Power Sources*, 195, 3041–3045.
102. Pan, L. et al. 2010. Conducting polymer nanostructures, template synthesis and applications in energy storage. *International Journal of Molecular Sciences*, 11, 2636–2657.
103. Liu, R. and S. B. Lee. 2008. MnO_2–poly(3,4-ethylenedioxythiophene) coaxial nanowires by one-step coelectrodeposition for electrochemical energy storage. *Journal of the American Chemical Society*, 130, 2942–2943.
104. Wang, Y. G., H. Q. Li, and Y. Y. Xia. 2006. Ordered whisker-like polyaniline grown on the surface of mesoporous carbon and its electrochemical capacitance performance. *Advanced Materials*, 18, 2619–2623.
105. Davies, A. et al. 2011. Graphene-based flexible supercapacitors: pulse electropolymerization of polypyrrole on free-standing graphene films. *Journal of Physical Chemistry C*, 115, 17612–17620.
106. Naoi, K. et al. 2010. High rate nanocrystalline $Li_4Ti_5O_{12}$ attached on carbon nanofibers for hybrid supercapacitors. *Journal of Power Sources*, 195, 6250–6254.
107. Hong, M. S., S. H. Lee, and S. W. Kim. 2002. Use of KCl aqueous electrolyte for 2 V manganese oxide–activated carbon hybrid capacitor. *Electrochemical and Solid State Letters*, 5, A227.
108. 2007. Basic research for electrical energy storage. *Physical Review Letters*.

109. Lide, D., Ed. 2009. *CRC Handbook of Chemistry and Physics*, 89th ed., Boca Raton: CRC Press, 800–900.
110. Okamura, M. 1999. Electric double-layer capacitors and storage systems. U.S. Patent 6064562.
111. Ricketts, B. W. Self discharge of carbon-based supercapacitors with organic electrolytes. *Journal of Power Sources*, 89, 64–69.
112. Lewandowski, A. et al. 2010. Performance of carbon–carbon supercapacitors based on organic, aqueous and ionic liquid electrolytes. *Journal of Power Sources*, 195, 5814–5819.
113. Lewandowski, A. and M. Galinski. 2007. Practical and theoretical limits for electrochemical double-layer capacitors. *Journal of Power Sources*, 173, 822–828.
114. Arora, P. and Z. J. Zhang. 2004. Battery separators. *Chemical Reviews*, 104, 4419–4462.
115. Hashmi, S. A. Ionic liquid incorporated polymer electrolytes for supercapacitor application. *Indian Journal of Chemistry*, 49, 743–751.
116. Staiti, P. and F. Lufrano. 2010. Investigation of polymer electrolyte hybrid supercapacitor based on manganese oxide–carbon electrodes. *Electrochimica Acta*, 55, 7436–7442.
117. Kalpana, D., N. G. Renganathan, and S. Pitchumani. 2006. A new class of alkaline polymer gel electrolyte for carbon aerogel supercapacitors. *Journal of Power Sources*, 157, 621–623.
118. Meng, C. et al. 2010. Highly flexible and all solid state paper-like polymer supercapacitors. *Nanoletters*, 10, 4025–4031.
119. Yoo, J. J. et al. 2011. Ultrathin planar graphene supercapacitors. *Nanoletters*, 11, 1423–1427.
120. Kaempgen, M. et al. 2009. Printable thin film supercapacitors using single-walled carbon nanotubes. *Journal of the American Chemical Society*, 9, 1872–1876.
121. Lian, K. and C. M. Li. 2009. Solid polymer electrochemical capacitors using heteropoly acid electrolytes. *Electrochemistry Communications*, 11, 22–24.
122. Gao, H., Q. Tian, and K. Lian. 2010. Polyvinyl alcohol heteropoly acid polymer electrolytes and their applications in electrochemical capacitors. *Solid State Ionics*, 181, 874–876.
123. Lian, K. and C. Li. 2008. Heteropoly acid electrolytes for double-layer capacitors and pseudocapacitors. *Electrochemical and Solid State Letters*, 11, A158–A162.
124. Schneuwly, A. and R. Gallay. 2000. Properties and applications of supercapacitors: From the state of the art to future trends. 1–10.
125. Wu, H. C. et al. 2009. High performance carbon-based supercapacitors using Al current collector with conformal carbon coating. *Materials Chemistry and Physics*, 117, 294–300.
126. Ku, K. et al. 2010. Characterization of graphene-based supercapacitors fabricated on Al foils using Au or Pd thin films as interlayers. *Synthetic Metals*, 160, 2613–2617.
127. Portet, C. 2004. Modification of Al current collector surface by sol–gel deposit for carbon–carbon supercapacitor applications. *Electrochimica Acta*, 49, 905–912.
128. Demarconnay, L., E. Raymundo-Piñero, and F. Béguin. 2010. A symmetric carbon–carbon supercapacitor operating at 16 V by using a neutral aqueous solution. *Electrochemistry Communications*, 12, 1275–1278.

129. Hong, M. S., S. H. Lee, and S. W. Kim. 2002. Use of KCl aqueous electrolyte for 2 V manganese oxide–activated carbon hybrid capacitor. *Electrochemical and Solid State Letters*, 25, A227.
130. Shiue, L, R. Cheng, and C. Chun-Shen. 2004. Supercapacitor with High Energy Density. U.S. Patent 6,762,926.
131. *Epoxy Vacuum Bagging Techniques*, 7th ed. 2010, Bay City, MI: Gougeon Brothers, 53.
132. Eilerstsen, T. 2011. Multi-Electrode Series Connected Arrangement Supercapacitor. U.S. Patent 7,830,646.

5

Electrochemical Supercapacitor Design, Fabrication, and Operation

5.1 Introduction

The design, operation, and manufacturing of electrochemical ES devices are important in supercapacitor technologies. However, the industry advancements of electrode fabrication and materials design largely remain proprietary to individual developers and manufacturers. Ensuring the security of their intellectual property is necessary to maintain a competitive edge in the fast growing field of energy storage. As a result, a thorough guide containing specific details of current advancements in electrochemical supercapacitor materials, fabrication, and design remains largely inaccessible. Therefore, the discussion about these commercial technologies is limited by information that is freely accessible to the public sector.

The performance of electrochemical supercapacitors is strongly dependent on the design and integration of active electrodes, electrolytes, and current collectors. Many of their production features are similar to the processes for manufacturing batteries. Both ES devices and batteries require electrolyte accommodations, separators, proper sealings, and casings. Therefore, supercapacitor manufacturers apply battery engineering and production machinery when designing and producing supercapacitors to save money and improve performance.

Supercapacitor technology has advanced in recent decades and devices are classified as economically viable and unviable options. In general, academic endeavors allow the pursuit of both options, but commercial research and development (R&D) firms are not interested in unviable research. As a result, much of their focus has centered on improving supercapacitor performance and developing inexpensive materials for mass production. Commercial R&D is driven by the need to be innovative, decrease manufacturing costs that limit competition, and improve sales. In an effort to achieve these goals, the electrode and electrolyte materials required for ESs have been identified as major challenges, as discussed in Chapter 4.

This chapter incorporates academic literature, conference proceedings, and seminar publications issued over the past 10 years to provide up-to-date information about electrochemical supercapacitor design, fabrication, and operation. A review of publicly accessible patents is also presented, along with information made available by manufacturers.

5.2 Design Considerations

In general, the first consideration in designing any device, including an electrochemical supercapacitor (ES), is the intended application. An ES is used for energy storage and delivery across a wide range of applications. The important design considerations include: (1) cell voltage, (2) frequency response, (3) lifetime and cycle charging, (4) polarity, (5) heat and temperature effects, and (6) humidity. For cell voltage considerations, an ES must be suitable for low voltage, high voltage, or both types of applications. Since the application systems require immediate high power (supplied within less than a millisecond to a few seconds) or delaying the desired power for longer periods ranging from a few seconds to minutes, the operational voltage and its frequency response must be considered.

Other factors such as cycle life, cell polarity, and heat management must combine to achieve synergy and optimization to ensure adequate performance and efficiency. All these factors can also be governed by additional intrinsic and extrinsic aspects of cell materials and their manufacturing processes. This section will discuss each design factor along with material and manufacturing requirements and implementation methods. It is important to note that these considerations are general guidelines for carbon-based electric double-layer ESs.

5.2.1 Cell Voltage

As discussed in Chapter 3, the operating voltage of a single cell ES is largely limited by the electrolyte used. According to the equation of ES energy density ($E = \frac{1}{2}CV^2$), the energy density is proportional to the square of the cell voltage, determined mainly by the electrolyte voltage window. Therefore, electrolyte selection has a direct impact on energy storage capabilities and the number of single-stacked cells to achieve a rated operating voltage. In commercial manufacturing, some organic electrolytes are preferred due to their wider voltage windows compared to aqueous electrolytes. For example, tetraethylammonium tetrafluoroborate (Et_4NBF_4) as the electrolyte salt dissolved in an acetonitrile or propylene carbonate solvent is the most popular choice to attain high operating potentials near 3V. A typical ES stack for applications requiring high load operating potentials, such as uninterrupted power supplies or hybrid electric vehicles, contains numerous ~2.7 V single cells in series.

In ES stack design and manufacturing, maximizing the safe operating voltage of each cell is a priority, along with reducing costs by minimizing material requirements, using cost-effective materials, and simplifying manufacturing processes. In general, a wide ES operating voltage improves the efficiency of storing energy and supplying power. For example, electronic systems generally possess a threshold for a minimum operating voltage that directly relates to the usable power stored in an electrochemical capacitor cell. At the current technological state, utilization of 75% of the stored energy can be achieved when an ES is discharged from an initial charge potential $V_{initial}$ to one half of $V_{initial}$.

In ES design and manufacturing, other factors related to the operation should be considered: unforeseen voltage spikes, electrolyte decomposition, effects of applying excessive voltage, and cell performance decay. Electrolyte decomposition is caused by continued operation at potentials greater than the acknowledged limits of the dielectric or the electrolyte's thermodynamic limitations. Excessive voltage can lead to gas production, mechanical swelling, and cell bursting. Finally decay in cell performance through a reduction in capacitance and increase in equivalent series resistance should also be considered in relation to sustainable energy and power supplies to application devices.

5.2.2 Frequency Response

The frequency response of an electrochemical capacitor is dependent upon its measured time constant τ ($\tau = R_{ESR}C$), which represents the time required to achieve 63.2% of the full charge from a noncharged state or discharge 32.8% of its charge from a fully charged state. Generally, the frequency response ranges from milliseconds to seconds, depending on the cell configuration, electrode materials, and electrolyte that all contribute to the equivalent series resistance and capacitance of a cell. Minimizing the resistances and maximizing the capacitance can increase the efficiency and the power rating, allowing an ES to play a more critical role in pulse power applications that require rapid responses with minimal losses of power.

5.2.3 Lifetime and Cycle Charging

Theoretically, the lifetime of an ES is unlimited because no final event indicates that it is dead. However, the continuous charging and discharging of an ES at a constant current can actually result in exponential decay of the capacitance, leading to an increase in internal resistance. The end of an ES life cycle is defined as maximum acceptable loss in relative capacitance. The life cycle of an ES (Figure 5.1) details the initial dramatic loss generally found in all cycle testing, a steady linear decline in relative capacitance over time,

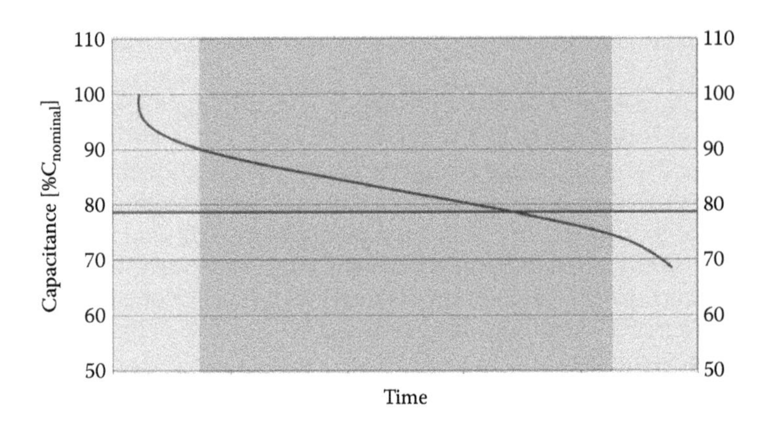

FIGURE 5.1
Life cycle profile of electrochemical capacitor operating within moderate parameters where time is 10^6 cycles. The horizontal line indicates acceptable loss in relative capacitance.

and then a rapid decline in capacitance. The horizontal line indicates the acceptable loss in relative capacitance.

In ES lifetime testing, it is often observed that resting the device between cycles offers a short-term reprieve and recovery of capacitance but the device will eventually return to its pre-rested performance. Operating temperatures, voltages, and maintained charge voltages over a period of time contribute to an increase in internal resistance and a decline in capacitance. Figure 5.2 shows the temperature effects on capacitance; a mild increase of only 5°C can cause a significant lifetime degradation. An optimal electrochemical supercapacitor normally experiences a loss of less than 10% initial capacitance after extended cycling (>1,000,000 cycles). The cycling life, even for commercial devices, is generally greater than 500,000 cycles with a minimal loss in performance. As most commercial products are packed in

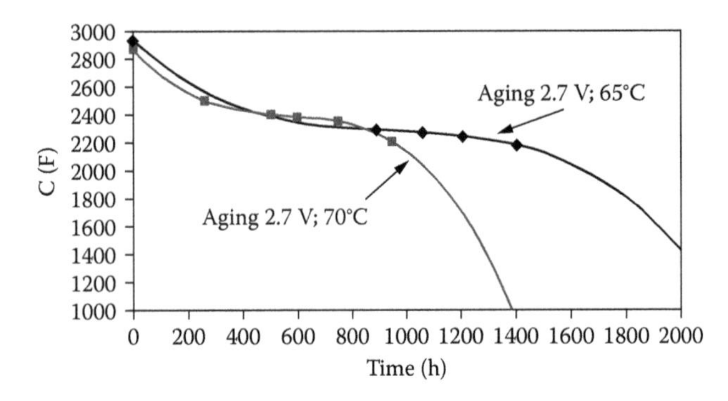

FIGURE 5.2
Supercapacitor maintained at 2.7 V to evaluate aging affects at constant temperatures of 65 and 70°C. (*Source:* Gualous, H. et al. 2010. *Microelectronics Reliability*, 50, 1783–1788. With permission.)

vacuums, a product should be used within 6 months of opening the vacuum seal to prevent the soldered leads from oxidizing.

The Coulombic efficiency of ESs, defined as the ratio of the discharge current to the charging current, can be expected to approach 100% even at high currents. In practical use, the charge and discharge current values are controlled to be the same, so the Columbic efficiency equals 100%. This efficiency applies to the current, and does not directly correspond to high energy efficiency.

5.2.4 Polarity

Electric double-layer ESs are commonly designed to be symmetrical, with both the anode and cathode using the same materials such as high surface carbon particles. Therefore, they do not have a theoretical polarity. However some manufacturers use stainless steel casings that indicate the intended polarity for use. Labeling is used to avoid the reverse corrosion potential of the stainless steel that leads to subsequent loss of life. Manufacturers recommend that users maintain the polarity, but no catastrophic failure will occur if an ES is reverse charged. However, if the EC has been conditioned for charge in a certain direction and the direction is changed, the lifetime of the device will be reduced.

The lifetime of an EC will be shortened considerably when the cell connected to the positively marked terminal is connected to the negative terminal of the charging source. Reversing the voltage may not show any ill effects on performance, but the cell will show degraded lifetime and performance characteristics after it is reconnected correctly. The negative effects on lifetime and performance can be avoided if cells are properly connected based on their polarities.

For asymmetric ESs, as discussed in Chapter 2, polarity is very important because the anode material is different from the cathode material. Particularly, if an electrochemically active material is used as the electrode material, the corresponding electrochemical reaction or process on that electrode may be different from the activities of the other electrode, resulting in different electrode potentials and different polarities. Therefore, the polarity must be marked on each electrode in an asymmetric ES to indicate how each material should perform within a specific operating potential.

5.2.5 Heat and Temperature Effects

The ability of ESs to operate at temperature ranges from –40 to 65°C is favorable for increasing their applications across different environments. This range, in particular temperatures below ice point, favor organic rather than aqueous electrolytes because of water's freezing point. However, ES performance can be affected by heat generated from its proximate operating environment, high power operations, or current resistances due to continual cycling at high frequencies.

A trade-off exists between increasing and decreasing temperatures. An increase in operating temperature will decrease the resistance contribution from the electrolyte as a result of improved electrolyte mobility and increased dissociation; decreasing the operating temperature exerts the opposite effect. However, high temperature operation causes declines in capacitance and performance and may upset the positive effect of R_{ESR} reduction.

The temperature effects are even more severe in an ES stack than a single cell because the surface area to volume ratio of an ES is smaller. Therefore, heat management design and optimization are important and necessary in maintaining an ES's desired performance.

5.2.6 Humidity

ES cells can operate effectively at high levels of humidity. However, storage humidity during production can have ill effects on ESs. ES cells are usually shipped in vacuum-packed containers to prevent exposure of cells to moisture. The terminal pins of the cells are susceptible to oxidation and eventually corrosion due to high levels of humidity over prolonged periods. If an original vacuum pack is opened and not all the cells are used, the package must be resealed to shield the remaining cells from humidity. Failure to consider this important storage requirement will lead to oxidation of the pins, reduced solderability of the pins, and possible damage to the terminal pins. Oxidized terminations can be removed cautiously with an abrasive emery cloth.

5.3 Single Cell Manufacturing

Fabrication techniques for electrode materials are diverse in the range of materials available and the synthesis conditions required. These issues must be determined before production starts. Popular ES electrode materials include activated carbon; carbon foams, fibers, and nanotubes; and metal oxides. For detailed information about the synthesis of these materials and their corresponding technologies, please see Chapter 4.

5.3.1 Electrode Materials

In the current technological state, most commercially available products are electric double-layer ESs. Their electrodes are manufactured using an identical material, such as a high surface carbon powder, to serve as both anode and cathode materials. Using carbons as electrode materials is advantageous because they are widely available, inexpensive, and diverse in both structure and composition. Over the years, research and development focused heavily on investigating carbons. In general, raw carbons are not favorable for the development of significant capacitance. However, procedures involving pyrolysis (carbonization) in a non-oxidizing environment introduce high surface area and enhance porosity, increasing capacitance. As a result, in the manufacturing of carbon-based ES cells, the choice of precursor material and carbonization and activation techniques are critical to achieve a suitable electroactive material for well performing devices.

5.3.2 Electrode Fabrication

Fabricating ES electrodes in commercial applications is proprietary information and normally not disclosed to the public domain. However, the laboratory and commercial techniques used for the fabrication of electrodes are similar. The process involves mixing several ingredients, including the electrode material such as a carbon powder (carbon black or CNT), conducting particles, and a fluorine-containing polymer (often PTFE) binding agent with a solvent to obtain a paste or slurry. This is followed by a period of ball milling or ultrasonication to ensure a good dispersion of the active material within a homogeneous composition. The paste or slurry is then rolled, heat pressed, and dried to form an electrode layer film. Alternatively, a spray deposition of the slurry or spreading of the paste onto a substrate acting as the current collector can also be used to prepare the electrode layer, followed by drying, annealing, and possibly press procedures to achieve the required film.

5.3.3 Electrolyte Preparation

Both aqueous electrolytes, including ionic liquids [e.g., 1-ethyl-3-methylimidazolium tetrafluoroborate ($EMIMBF_4$)] and non-aqueous electrolytes (e.g., Li or Na tetrafluoroborate, tetraethylammonium perchlorate dissolved in acetonitrile, or propylene carbonate) can be used for saturation of the electrode layer and the separator. This process must be performed in a glove box if a moisture-sensitive electrolyte, such as a non-aqueous electrolyte, is used because any water inside a non-aqueous electrolyte will cause gas to form during operating potentials intended to reach 2.5 to 3.5 V. Furthermore, water contamination can disrupt the cell sealant during cycling and cause higher internal resistance, lower power ratings, higher leakage current, and loss of cycle life. To effectively maintain performance over the intended operating life, commercially available ESs are generally equipped with pressure safety valves along the sides or ends of the cells to release any gas produced by contamination or overheating.

5.3.4 Current Collector Preparation

Metal foils of aluminum, copper and nickel with thicknesses between 20 and 80 μm are largely used as current collectors or electrode substrates due to high conductance properties and low costs relative to gold or platinum. During preparation, a foil should be etched by acid or other chemicals to introduce higher surface area and irregularities along the surface to improve the contact between the current collector and electrode layer. Three ways to assemble the electrode layer and current collector are: (1) bind the previously prepared electrode layer film onto the current collector, (2) deposit the electrode layer onto the current collector, and (3) spray the electrode layer onto the current collector. If the electrode layer is not insufficiently adhered to the current collector, a higher internal resistance will arise. After that, the assembly obtained can be cut into a desired shape such as a pellet. If the pellet is sufficiently dry, it will take several hours for the electrolyte to sufficiently saturate the pores of the electrodes and separator.

5.3.5 Single Cell Structure and Assembly

The design, assembly, and packaging of a supercapacitor depends strongly on the desired rating and application, and these parameters vary in commercial products and research devices. Figure 5.3 shows a basic single supercapacitor cell. Figure 5.3a shows a single custom-made supercapacitor cell for fundamental studies—a complete cell composed of a pair of electrodes that are electrically insulated from each other by a separator made of thin, porous, non-conducting material. This assembly is then sandwiched between current collector electrodes under constant pressure. A Swagelok cell system (Figure 5.3b) is frequently employed in research settings to evaluate electroactive materials in a complete cell configuration.

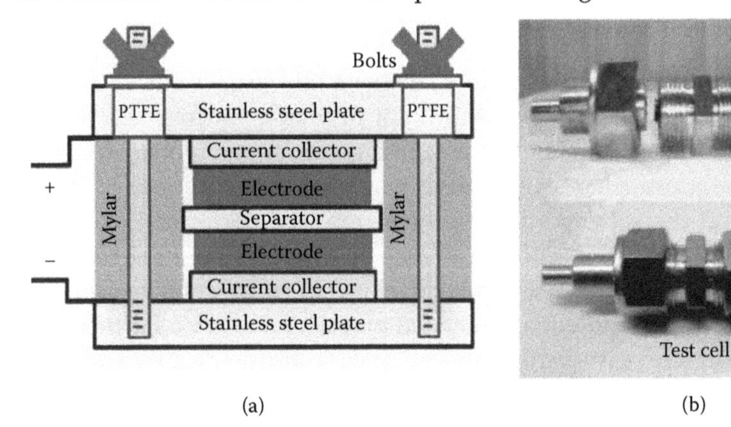

FIGURE 5.3
Basic cell diagrams. (a) Simple stainless steel framework housing. [2] (b) Swakelog system. (*Source*: Reddy, A. L. M. and S. Ramaprabhu. 2007. *Journal of Physical Chemistry* C, 111, 7727–7734. With permission.)

Commercially produced supercapacitors often take stacked cylindrical or coin forms. Each manufacturer has its own propriety packaging systems but the fundamentals are consistent for most commercial applications. Cylindrical types, typically called "jelly rolls" because of their shape, require formation of the electrode layer film by rolling or spraying a carbon material on both sides of a separator. An outer separator is applied to ensure the layers are electrically isolated from each other. Coin-type constructions are usually manufactured by companies that use construction schematics (Figure 5.4). In coin assemblies, the cells are uniformly pressed to ensure good electrical contact and proper sealing to prevent electrolyte leakage. Coin cells have also been fabricated in research laboratories based on their simple construction. Figures 5.4b and d demonstrate a typical cylinder cell manufactured by Maxwell.

Both forms of single supercapacitor cells can use aqueous or non-aqueous electrolytes. As noted above, if a non-aqueous electrolyte is used, the assembly must occur in a moisture-free environment such as a dry box. In addition,

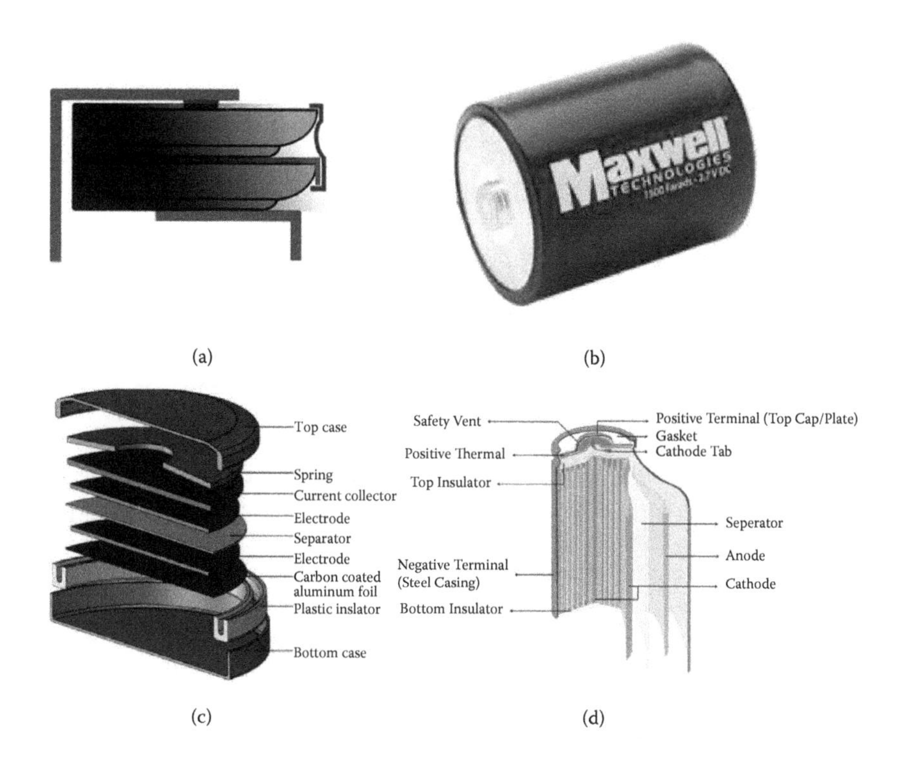

FIGURE 5.4
(See color insert.) (a) Image of a manufactured two-coin cell stack covered by shrink wrap sleeve. (b) Manufactured rolled cell in metal can design. (c) Coin cell. (d) Rolled cell designs for ESs and batteries. See References 4 through 7.

good contact between the current collector and active material layer must be maintained to minimize internal resistance. They must be pressed together uniformly to avoid pressing out a critical volume of electrolyte.

5.3.5.1 Coin Cells

Coin cells provide good media for small capacitance devices. The casing of a coin cell is made from conductive metal end caps and the fitting is designed to be crimped under high pressure. The top plate sits in a hydrophobic insulating rubber or Teflon gasket seal that prevents water moisture from entering the device and disrupting performance. However, the result is a thin device with a low mass of active material. To create larger capacitors, it is important to better utilize cell volume and produce larger cells. Rolled cells are used in larger capacitance devices where the electrode material can be synthesized in long sheets and spirally wound into rolls. A high level of compression is needed for good contact with the aluminum support. However it is important not to remove electrolyte to a level that will limit ion availability.

If too little electrolyte is present within a soaked separator, ion saturation will occur and performance will suffer. The presence of an aluminum collector allows for strong conduction. The external contact tabs are soldered onto the collector foil during the rolling process. Rolled cells provide uniform rolling and compression on a large scale. However, they exhibit poor space optimization and the thick nature of the cell makes designing of efficient heat sink and cooling designs difficult.

5.3.5.2 Cylindrical Cells

Alternatively, a roll can be wound to fit a prismatic case as commonly found in lithium ion batteries. The roll is placed in a fitted conductive metal casing that has a Teflon or rubber gasket seal separating the outer can and top button contact. The cylindrical metal rod around which the film is wound becomes the internal contact. Contacts are ensured by soldering and then electrolyte is injected into the cell followed by a curable polymer sealant. The top of the casing is fitted with a vent, gasket layer, and top contact plate. To seal the device, the top of the can is mechanically crimped and the outer metal casing acts as the other contact and provides mechanical stability and rupture resistance.

To prevent discharge and ensure safety, an external sealant or insulator must be applied to the electrode as a solid casing or as a shrink wrap to finish the manufacture. In stiff metal cases for which gas evolution is a concern, a one-way safety vent must be incorporated into the device. The safety vent prevents moisture from entering but will rupture and release gas if the capacitor enters a regime in which pressure or heat build-up is a concern. Another assembly concern for organic electrolytes in commercial devices is contamination via penetration of water moisture into the electrolyte. Cells made with

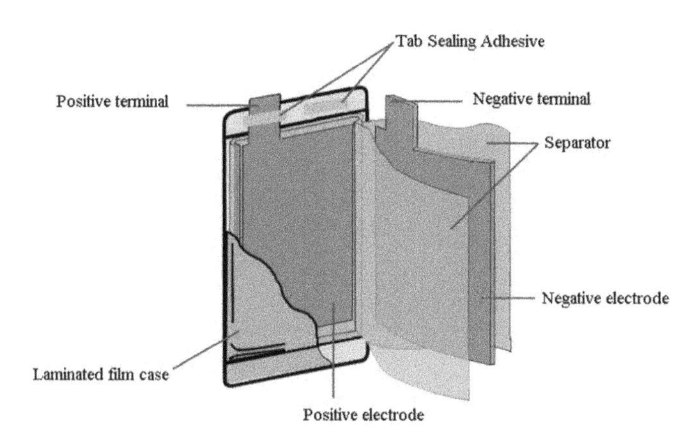

FIGURE 5.5
(See color insert.) Pouch cell design for solid electrolyte lithium polymer battery. Design is easily adaptable to ECs made with organic electrolyte. (*Source: Electropedia: Cell Construction* (online). http://www.mpoweruk.com/cell_construction.htm [accessed April 5, 2012]. With permission.)

organic electrolytes must be assembled under vacuum to prevent electrolyte degradation.

5.3.5.3 Pouch Cells

Pouch cells (Figure 5.5) present an alternative for packaging and hermetically sealing solid electrolyte batteries. The process is also applicable for ES systems utilizing organic electrolytes. In a pouch cell, electrodes can easily be stacked cell designs or spirally wound prismatic rolls. Soldered contact tabs are placed in contact with the appropriate film layers to complete the design. A prismatic roll is placed between layers of polymeric sealant materials that are laminated together to create a fitted polybag around the cell. The cells are flexible and polymer based, making for easy incorporation of melt shutdown and vents that allow the cells to expand before rupture. This reduces risks more effectively than poorly vented metal casings.

The simplicity of the design removes the need for many of the components of rolled cells. Only the metal tabs, the electrode roll, and the laminate sealant are needed. The pouch cell design inherently offers low weight because it needs fewer components and the metal casing improves overall pack performance density. Cells can also be made very thin and offer optimal space usage, resulting in improved volumetric performance versus cylindrically rolled cells when scaled up to module size. The thin rectangular profile ensures that cell modules can be solid stacks with very short interconnects. This lowers pack ESR and further improves power performance over other small-sized rolled designs (Figure 5.6). The flexible prismatically wrapped cells cannot reach the same large size or large total capacitance of cylindrically rolled cells.

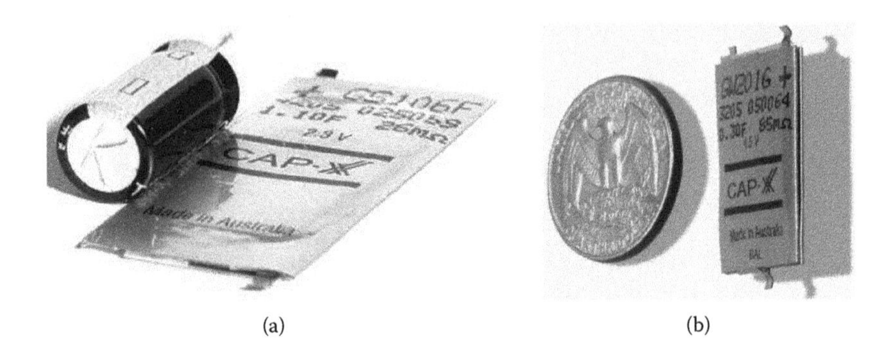

FIGURE 5.6
Pouch cell form factor for CAP-XX© EC made from activated carbon and organic electrolyte. (a) Single cell, 2.5V. (b) Two stacked cells with short interconnects that bring voltage to 5 V. (*Source: Cap-XX Photo Gallery* (online). http://www.cap-xx.com/news/photogallery.htm [accessed April 5, 2012]. With permission.)

5.3.6 Considerations for Contact Area and Positioning

The contact space of a solder contact tab or the outer casing is very small compared to the long lengths of the films in rolled cylindrical, pouch, and prismatic designs. As discussed in Section 5.5, the active carbon material in an EDLC electrode is not a strong enough conductor to prevent large electron transport resistances in batteries. It is logical that even the low sheet resistance of highly conductive metals such as steel and aluminum foils can considerably reduce performance over long lengths in the meter range, such as the lengths present in a wound cell.

The situation can be exacerbated because sheet resistance increases when metal foils are chosen to be as thin as possible to reduce device weight. When a single contact tab or outer layer casing is considered for a wound film, the distance for conduction is many times more than the width of the film. Contact methodology for wound cells is therefore critical if the intent is to reduce system losses of a large cell.

As shown in Figure 5.7, the variation in reaction rate arising from inadequate current collection can cause more of the battery electrode to be used near the tabs [9]. Temperature variation within the device also indicates hot spots that negatively impact cycle life, efficiency, and performance characteristics, if the device has insufficient contact tabs. The hot spots and electrode usage problems will only be exacerbated in ESs that run at far higher power densities than batteries. Utilizing continuous tab design (Figure 5.8) by offsetting the electrode materials to create a single end cap tab can improve the performance of a device [4,9]. One downside of the continuous tab design is the loss in active material volume to accommodate the electrode overlap. Also, continuous tab design requires care when applying the end plate to prevent short circuiting the electrodes.

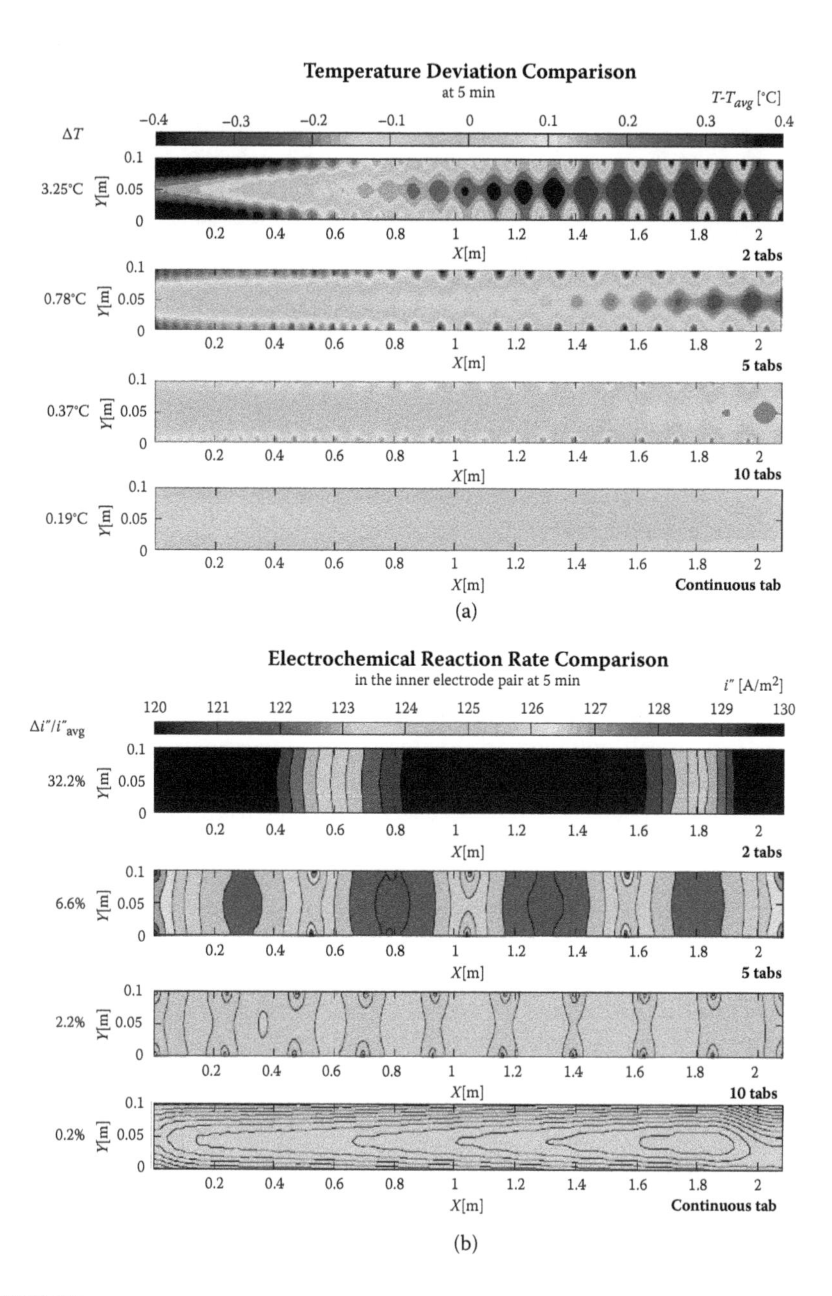

FIGURE 5.7
(See color insert.) (a) Temperature and (b) electrode utilization profiles over area of a battery electrode for increasing number of solder tabs within roll of wound cell. (*Source:* Lee, K. J., G. H. Kim, and K. Smith. 2010. In *Proceedings of 218th ECS Meeting*, NREL/PR-5400-49795. With permission.)

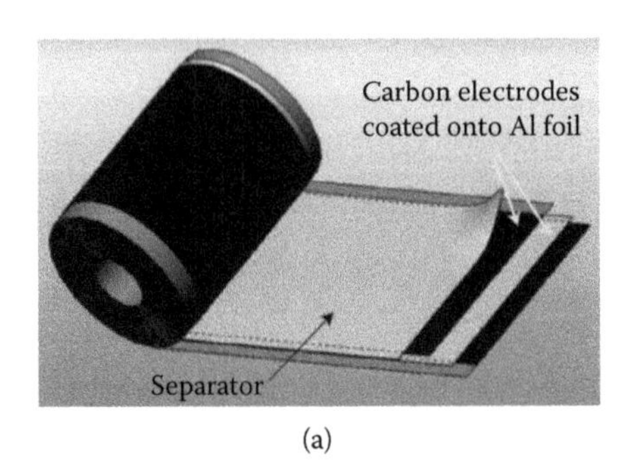

(a)

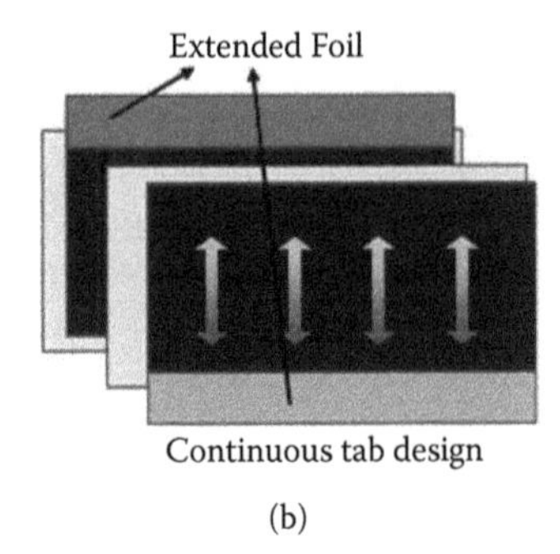

(b)

FIGURE 5.8
(a) Capacitive rolled cell. (b) Continuous tab design scheme applied in some batteries and electrochemical capacitors. (*Sources:* Simon, P. and Y. Gogotsi. 2008. *Nature: Materials, 7,* 845–854; Lee, K. J., G. H. Kim, and K. Smith. 2010. In *Proceedings of 218th ECS Meeting,* NREL/PR-5400-49795. With permission.)

5.4 Supercapacitor Stack Manufacturing and Construction

5.4.1 Cell Stacking to Form Modules

A number of identical supercapacitors can be stacked in series to form a supercapacitor stack whose voltage is the sum of all individual supercapacitors' single cell voltages. In this way, a high voltage can be reached for desired applications. If more charge or capacitance is wanted, a number of identical supercapacitors can be stacked in parallel to form a parallel supercapacitor stack. The combination of series and parallel supercapacitor stacks can achieve an overall capacitance at an intended operating voltage.

Most ES stacks are manufactured for the transport industry and focus on high load and current applications to meet operating voltages, temperatures, and power requirements of the intended system. The design and manufacturing of both ES stacks and ES banks should consider the materials used in the arrangement of capacitor electrodes, sealants, and casings, and incorporate a voltage balancing system to avoid cell decay and performance loss. In general, an ES stack design should meet both the energy and power requirements of an application. In general, the two types are: (1) the series ES stack in which all individual cells are connected in series, and (2) the parallel stack in which all individual cells are connected in parallel. The details of stacking are described in Chapter 2.

In designing and manufacturing an ES stack, the application target must be well defined. According to the requirements of targeted applications, the materials, volumes, stack sizes, shapes, and modes of packaging can be selected, designed, and manufactured. In any case, minimizing stack weight and volume is always a great concern to improve energy and power densities while reducing costs.

During ES stack construction and packaging, series stacking is more practical than parallel stacking. In series stacking, individually packaged cells are serially integrated into a multiunit stack through external metal bars or soldering the cells onto a PCB board. Alternatively, the uncased jelly rolls could be compartmentalized to reduce dead space and packaging weight and increase the intimacy of contacts within the module [10]. However, when module designs are scaled, the combination of interparticle electrode resistances and parasitic intercell contacts can cause increasingly large pack resistances. Large scale resistances can cause great effects on power performance [11]. Pack voltage is often tailored by creating external lateral contacts that also affect resistances. During use, interconnected cells begin to experience non-uniform charge distribution [12]. These charge distributions can be mitigated by using active and passive balancing components discussed later in this chapter.

5.4.2 Utilizing Bipolar Design

Use of bipolar electrodes to form an ES stack is shown in Figure 5.9. The bipolar arrangement can effectively minimize the volume of the stack and circumvent the use of additional materials and external connections. In addition, the intimate surface level connection can help overcome macroscopic resistances generated from solder joints, long interconnects, and tabs that contact only part of the collector foil. The reduction in packing material (grid weight) for a module also improves cell performance.

Bipolar electrodes are commonly used in fuel cell, battery, and ES applications. The construction of bipolar electrodes requires the use of a highly conductive paper or plate substrate (e.g. titanium, carbon fiber paper, etc.). During electrode preparation, one side of the conductive bipolar electrode sheet is attached to an electroactive layer by pressing a previously prepared electrode film or depositing an active layer. The other side of this sheet is attached by another electrode layer. In the stack, one side of the bipolar electrode acts as the anode of a cell and the other side acts as the cathode of the adjacent cell (Figure 5.9b). It is very important that no ion conductivity through the bipolar plate is possible in order to prevent short circuits. A gasket, sealant, or non-conducting porous material is placed between each bipolar plate to act as a separator to avoid short circuiting of the cells. An external pressure is then applied to the outer casing of the cell to improve material contact and reduce resistances.

Figure 5.10 shows a series sequence of bipolar plates. Each plate has a double-layer capacitance on both surfaces. However, each side maintains a polarity opposite that of the electrolyte solution. The spaces between each bipolar plate are filled with an electrolyte in liquid or hydrated solid polymer form. From the current collecting plate E_1 to the first side of bipolar plate A_1, a voltage drop occurs to represent the voltage of the first cell in the three-cell stack.

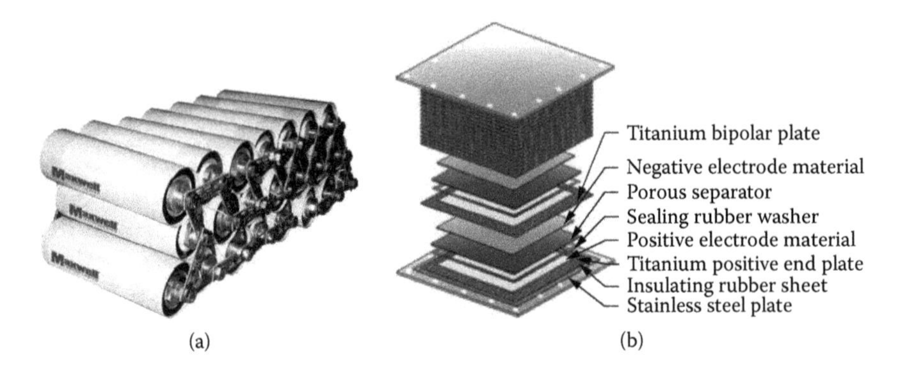

FIGURE 5.9
(a) Configuration for larger scale series stacking of individual supercapacitor cells. [13] (b) Encased multi-cell bipolar connected stack. Note contrast in volume of each method. (*Source:* Zhou, X., C. Peng, and G. Z. Chen. 2012. *AIChE Journal*, 58. With permission.)

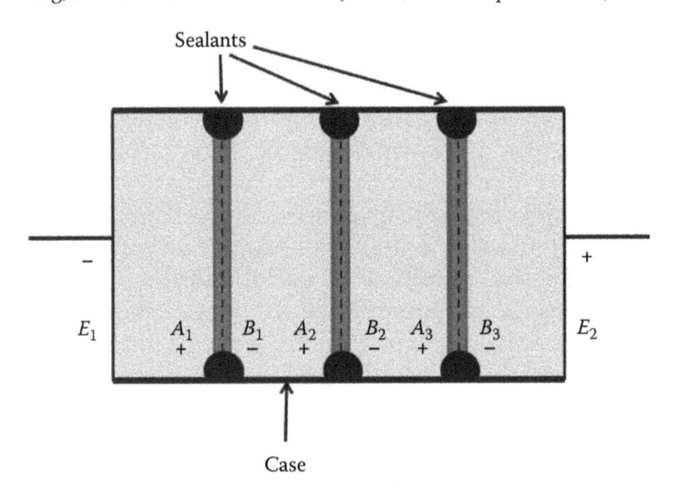

FIGURE 5.10
Bipolar electrode cell design adapted from U.S. Patent 4022952 [accessed February 12, 2012]. A1, A2, and A3 are positive electrodes of supercapacitor; B1, B2, and B3 are negative electrodes of supercapacitor.

Following in succession, A_1 is in direct contact to electrode B_1 as they are both represented by the single bipolar electrode. As a result, for n bipolar electrodes, the stack will contain $n + 1$ cells in series. The current collectors on both ends of the stack labeled E_1 and E_2 are critical in transmitting the power density of a stack and therefore, require excellent intrinsic conductivity and good contact with the active material.

Metal end plates are normally used as the current collectors because they possess sound mechanical properties that maintain the stack's structural stability and provide a facile means of connecting external leads. However, the current collector surface must be treated to enhance the contact of the two materials and reduce series resistances. Normally, acid or anode etching is used to treat the collector surface.

As noted earlier, bipolar electrode stacking arrangements have a volumetric efficiency. Unfortunately, a drawback for bipolar electrode arrangements is that the heat dispersion is more difficult than occurs with an external connection due to the internal I^2R heat generation within the stack. If this heat is not effectively managed, over-heating will lead to cell degradation. Further, cooling surface area is low within the bipolar stack, making it difficult to incorporate passive cooling without high thermal conduction properties [15]. This concern is more significant in large-scale supercapacitor packs. Therefore an additional cooling system is needed, but will cause a parasitic loss of stack power. In addition, for a bipolar electrode-based ES stack, proper sealing of the stack is crucial to circumvent any form of electrolyte leakage that ultimately creates undesirable shunt currents that cause self discharge behavior.

5.5 Voltage Cell Balancing

In an ES stack containing a number of single cells, it is impossible to have the same voltages in all cells when operating under a load. Some cells have high voltages and others have low voltages. This creates an unbalanced stack and results in low efficiency. In general, low cell voltage is caused by high internal resistance and high voltage is caused by low capacitance. For an ES stack, the damage from high cell voltage is more severe than that caused by low cell voltage because it could lead to gas evolution and stack rupture

If all cells in a stack are in series, they will experience an equal current. The cell voltage is relatively dependent on capacitance and charging current, so incorporating a single cell with a smaller capacitance will lead to a higher voltage across the cell. A cell with a larger than average capacitance will experience a smaller voltage. This will become even worse during stack charging, as the voltage distributed across a stack is initially dependent on the capacitance. The cells having lower capacitances will experience voltages higher than the limited ratings of individual cells within the stack, leading to aging and performance degradation. Unequal capacitances of cells can arise from several factors aside from manufacturing variances including cell aging and temperature gradients.

It is important in stack design to balance voltages to prevent unequal voltage distributions. This requires consistency in the manufacturing of single cells to make every cell have the same capacitance and internal resistance or make the capacitance and internal resistance as close as possible. An engineering concern is that the material mass and mode of construction should be equivalent in each manufactured cell. To further balance the voltage distribution to avoid damaging effects, two types of voltage equalization circuits developed for this purpose are passive balancing and active balancing.

5.5.1 Passive Balancing

Passive balancing is classified into two subtypes: (1) resistance balancing, and (2) Zener diode balancing.

5.5.1.1 Resistance Balancing

Resistance balancing is a simple method of maintaining control of equalized charges across the stack. This can be done by regulating the voltage of each cell using a bypass resistor connected in parallel to each single cell, as demonstrated in Figure 5.11. The parallel resistor draws a current proportional to the voltage and induces a cell discharge when the cell voltage is higher than a balanced voltage. The disadvantage to such an approach is that the energy from surplus charging is dissipated and wasted as thermal energy through the resistors. In addition, the voltages across each cell are not precisely regulated and the charging of the entire stack is less efficient.

In this circuit, each single cell is charged separately. After the first cell C_1 is fully charged, the charging current of the subsequent C_2 capacitor must pass through the first parallel resistor R_1. This process continues throughout the system, resulting in a lengthy charge process with a significantly low ratio of energy stored in comparison to the energy required for charging.

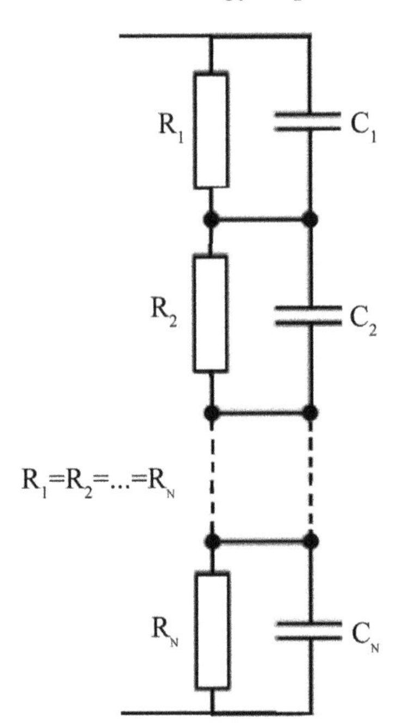

FIGURE 5.11
Circuit of resistive balancing method for electrochemical supercapacitor stack. (*Source:* Linzen, D. et al. 2005. *Electronics*, 41, 1135–1141. With permission.)

For example, a simulation by Barrade et al. [16] tested an ES stack containing four ESs of 800 F and a fifth 1000 F capacitor connected in series and charged to 12.5 V. In the stack, five 0.1 Ω parallel resistors were used to bypass each cell for voltage balancing. It was found that an extended time of 400 seconds was necessary to charge an equalized voltage across the entire stack. Furthermore, the total energy needed to charge the stack was about 120 kJ, but the stored energy was only about 15 kJ, so the charging efficiency was only 12.5%. Therefore, more efficient balancing methods are necessary.

5.5.1.2 Zener Diode Balancing

This balancing method uses Zener diodes instead of resistors as the bypassing elements to maintain a threshold voltage across each capacitor equal to the threshold potential of the diode(s). The current passes through a connected capacitor and can be effectively limited upon reaching a defined potential. This method results in an energy efficiency as high as 92%. For example, if 16.3 kJ are used to charge an ES stack, 15 kJ are stored [1]. Zener diodes can also dissipate the power beyond the local threshold voltage.

5.5.2 Active Balancing

As discussed above, passive balance schemes utilize shunts (resistors and diodes) to dissipate energy to regulate cell voltages. The resulting circuits are mainly suited for applications requiring low power or low current charging and discharging rates. High charging and discharging rates within a short time require a minimal loss in efficiency (non-dissipative equalization) to optimize performance. Some active balancing methods have been developed to achieve this.

Figure 5.12 shows an active balancing circuit scheme. In the figure, buck–boost transistors or diodes can provide an equalizing current I_{eq} for charging current I. The direction and magnitude of the equalizing currents depend on the local voltage across the corresponding single cell. For example, if a local voltage exceeds the limiting voltage, a reverse current can be applied until the voltage is balanced. This process actively proceeds along the entire stack until equivalent voltages are established. Using this active balancing circuit, the energy efficiency may be as high as 97%.

5.6 Cell Aging and Voltage Decay

An ES has an unlimited shelf life when stored in a discharged state, but aging does occur and typically decreases the capacitance and increases the resistance. The life of an ES specified by industry standards is a 20% decrease

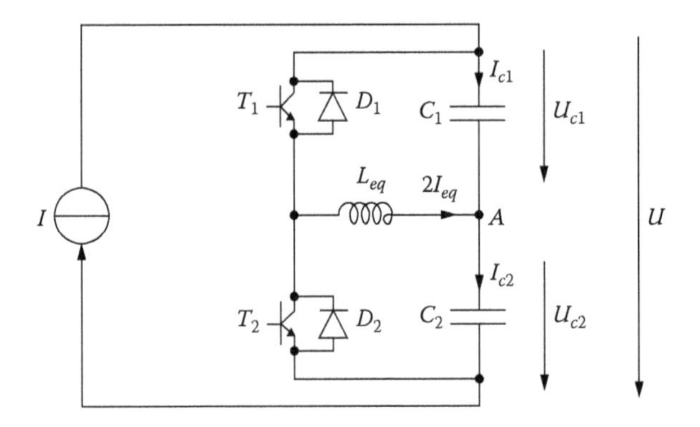

FIGURE 5.12
Active balancing circuit for ES stack using reversible buck–boost converters optimized to regulate voltage across each cell. (*Source:* Sharma, P. and T. S. Bhatti. 2010. *Energy Conversion and Management*, 51, 2901–2912. With permission.)

in capacitance and/or 200% increase in resistance. ES lives do not suddenly end. Rather their performance continually degrades over their lifetimes. End of life for an ES means only that its performance no longer meets the application requirements.

A large number of carbon materials have been developed in recent years in an effort to improve the performance of ES electrode materials (see Chapter 4). Among carbon materials, activated carbon-based materials remain the predominant choices for electrode production due to low cost and good cycle lives. However, some performance deterioration can be observed after prolonged use with organic electrolytes, particularly at high temperatures and at overcharging or discharging voltages. These factors are recognized as the main contributors to the detrimental behavior caused by cell aging.

Cell or stack aging is indicated by a loss in overall performance due to diminishing capacitance, slower charging and discharging rates, and increased series resistance. In addition, these signs can also be associated with macroscopic phenomena, for example, localized detachment of electrode materials from the metallic collector through electrode swelling, gas evolution, and loss of elements involved in faradic reactions.

The typical degradation behavior of an ES resembles exponential decay. Voltage decay in power sources or energy storage devices results from aging effects. The most dramatic effect of life degradation is on the internal resistance of the device because it is a direct indication of aging. Voltage drops of a single cell or stack can also be induced by self discharging, but they are not treated as voltage decay caused by cell aging. Charging voltages and temperatures above a cell rating limit have been identified as predominant aging factors that are generally observed during operation. An increase in either of these factors can exponentially accelerate the rates of the electrochemical reactions responsible for cell aging. For example, an increase of 10 K above the rated temperature or an increase of 100 mV above the rated cell voltage

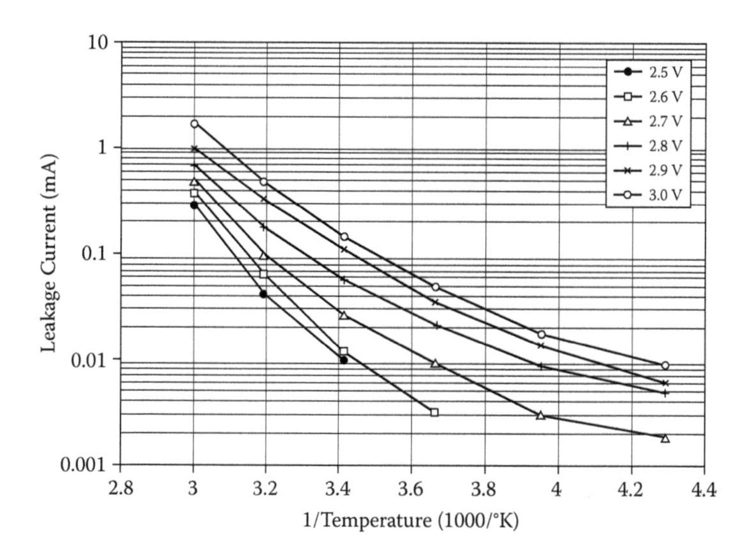

FIGURE 5.13
Arrhenius plots of measured leakage currents of EDLC evaluated at a range of voltages. Initially maintained at 60°C for 100 hours at each voltage, the temperature then stepped down from −20 to −40°C. Measurements taken after 10 hr constant conditions. (*Source:* Kötz, R., M. Hahn, and R. Gallay. 2006. *Journal of Power Sources*, 154, 550–555. With permission.)

will double a cell's aging rate as observed by Kotz et al. [19] in their experiments with a BCAP0305 supercapacitor from Maxwell Technologies.

Figure 5.13 shows Arrhenius plots of the logarithmic leakage current at temperatures ranging from 60 to −40°C in a voltage range from 2.5 to 3.0 V to determine the activation energies of the degradation process. It can be seen that a 0.1 V increase over the maximum operating voltage or a 10 K increase in temperature can increase the aging factors within 1.5 to 2 and 1.7 to 2.5, respectively. Bohlen et al. [20] provide a more comprehensive review of the effects of temperature and voltage on ES produced from leading manufacturers. They prepared a detailed analysis of ES aging behavior in terms of impedances (Table 5.1).

Changes in the chemical properties of an ES material are largely responsible for the aging and voltage decay, in particular, by increasing the temperature

TABLE 5.1

Commercially Available Electrochemical ESs Used for Cell Aging Analysis

Type	Manufacturer	Capacity (F)	Rated Voltage (V)	Geometry
A	Epcos	600	2.5	Cylindrical
B	Nesscap	600	2.7	Prismatic
C	Maxwell	350	2.5	Cylindrical

Source: Bohlen, O., J. Kowal, and D. U. Sauer. 2007. *Journal of Power Sources*, 172, 468–475. With permission.

or voltage. For example, some adverse chemical activities such as electrode and electrolyte decomposition can be enhanced by temperature or voltage increases and can increase cell resistance.

Additionally, the metal- and oxygen-containing surface-functional groups on the active carbon are also partially responsible for performance degradation from participating in electrolyte decomposition reactions that ultimately reduce the microporosity of the electrode. For example, in the presence of oxygenated and/or residual water, the decomposition of BF_4^- anion to $BF_xO_y^-$ ions can occur. Solid products formed from the redox decomposition of the electrolyte can block the pores on carbon particles, causing a loss in capacitance. Decomposition products also block pores within the cell separator, causing an increase in electrolyte resistance. Water impurity in organic electrolytes is believed to be partially responsible for the macroscopic effects and capacitance fading observed during cell aging. In addition, interaction of the electrode functional groups with organic electrolyte $TEABF_4$ (tetraethylammonium tetrafluoroborate) in acetonitrile also causes capacitance fading and increased resistance.

In general, the positive electrode (anode) is more susceptible to loss in surface area. Aging studies found that fluoride could be bonded covalently to the anodic carbon support. The nitrogen species including pyridinic (-C = N-C-) and amine ($C\text{-}NH_2$) moieties can also bond covalently to the graphitic structure of the anode, facilitating the polymerization of the acetonitrile solvent and causing aging of the anode [21,22].

To study aging, a high operating voltage above the nominal rate of 2.5 to 2.7 V was employed using quaternary ammonium salts in propylene carbonate (PC) or acetonitrile. Gas degradation products at both electrodes from an elevated voltage (2.6 to 4 V) are CO, CO_2, ethene, propene, and H_2, resulting in pressure increase, then losses in electrolyte ion and electrode cohesion. Evolution of CO_2 at a nominal voltage (2.5 V) and rated temperature limit (70°C) comes from the reaction of acetonitrile and water.

The electrode kinetics and effects on the ES electrochemical performance have also been studied. At an extremely low kinetic rate, capacitance loss accompanied by an increase in series resistance could occur, leading to aging. This aging is accelerated at both high temperatures and voltages.

Understanding the mechanisms of ES cell aging is important in developing mitigation strategies to improve device lifetimes. Furthermore, the knowledge of cell aging can held find ways to increase the temperature or charge conditions that currently limit the use of ES systems.

5.7 Self Discharging

Self-discharging phenomena are common in all ESs but their origins are not well understood. Self discharging can reduce performance in terms of energy and power densities. For example, the observed loss in the open circuit voltage of an ES over time after charging is normally caused by self discharging. Even when operating with a load, fast self discharging can cause a significant power loss. Several factors affect the self-discharging process, including charging history, chemistry and electrochemistry of the system, purity of the electrode materials and electrolyte, operating temperature and voltage, and pore structure. The voltage decay caused by self discharging is determined by the self-discharging mechanism(s). In general, self discharging can be induced by five processes:

Faradic decomposition reactions of electrolyte solvent — If the charging voltage is beyond the thermodynamic limitations of the solvent, the solvent will be electrochemically oxidized or reduced. For example, if water is used as the solvent, it could be oxidized to O_2 at the positive electrode and reduced to H_2 at the negative electrode.

Parasitic redox reactions involving impurities of electrode materials and electrolytes — During the charging process, if the electrode material contains electrochemically active surface groups, oxidation will occur on the positive electrode and a reduction will occur on the negative electrode, forming a complete electrochemical reaction. If the charging voltage is beyond the reversible redox voltage, a faradic charge leakage current can result, especially when the concentrations of the corresponding redox couples are high. For metal impurities in carbon materials, a similar mechanism can also cause self discharging. If the electrolyte contains impurities that are oxidized on the positive electrode or reduced on the negative electrode over the potential range of ES cell voltage, a faradic charge leakage current will cause self discharging.

Reduction of dissolved oxygen in electrolyte — Under the driving force of cell voltage, oxygen dissolved from the air or produced by an overcharged cell voltage can be reduced to peroxide or water on the negative electrode. Simultaneously, other oxidizable compounds such as electrode impurities or H_2 produced by overcharge voltage can be oxidized on the positive electrode, forming a self-discharging current that consumes the charge stored in the ES. Although the dissolved oxygen can be removed before cell assembly to reduce the effect of self discharging, some degree of discharge may remain due to carbon's affinity to adsorb oxygen.

Charge redistribution within deep pores of electrode material — A dominant effect of self discharge can be the redistribution of charge within the deep pores of an electrode material. When an ES is fully charged, the deep pores cannot be fully accessed by the charge in a short time. After charging stops, charge carriers will begin to travel further down the depths of the pores to develop a more uniform distribution of charge and additional ions will continue to enter the pores. Depending on the branching and tortuosity of the pore structure and the geometry of the pores, ~50 hr are required to ensure uniform distribution of the charge along the pore depth.

This process causes a slow decline of cell voltage, like a self-discharging process.

Leaking current through short circuit pathway — Self discharging in a series stack can occur through a short circuit pathway induced by improper isolation of electrodes. In addition, if the cells in the stack do not have equal capacitances, overcharging can occur on some cells resulting in self discharging. This problem requires optimized manufacturing procedures to ensure quality control [23].

5.8 Patent Review

This section will review the patents filed in the past decade (Table 5.2) for electrochemical ES issues such as electrode materials, electrolytes, and procedures. A wide range of electrolytes and derivative electrode materials including electric double-layer capacitive carbons, pseudocapacitive metals (metal oxides), and electrically conducing polymers (ECPs) will be discussed.

5.8.1 Patents on Electrode Materials

Carbon materials are widely used across electrochemical systems such as batteries, fuel cells, and ESs and are of research interest in both academic and industrial settings. A tremendous number of patents claim their preparation and production procedures, including the variables of carbon precursors, carbonization, and possible activation methods at various incremental temperature rates, exposure times, and repeatable steps.

One noted patent presented by the U.S. Department of Energy (US 6,299,850 B1) describes an activation method for the carbonization of a polyimide or other polymeric material in which an aqueous HNO_3 oxidizing agent followed by multiple thermal treatments is used. The resulting activated carbon material possesses a high volumetric capacitance of 300 F/cm^3. Several commercial firms developed similar proprietary methods in the synthesis of activated carbon.

Since the early 1990s, activation processes have expanded to include fibrous materials to enhance accessible surface area, for example, carbon precursor polyacrylonitrile and poly(vinylidene fluoride) (PVDF) used to develop fibers for use in electrodes. Carbon aerogels were also demonstrated by the U.S. military to enhance capacitance through electropolymerization of various electroactive polymers [polyaniline, polyarylamine, polypyrrolepoly(o-methoxyaniline)]. The production of thin polymer films (< 10 nm), particularly with poly(-methoxyaniline), produces a negligibly changed porosity considered responsible for improved performance.

Advancements in nanomaterials have yielded recent patents that focus on single walled and multi-walled carbon nanotubes (CNTs) and graphene

TABLE 5.2

Survey of Patents for Use as Electrode Materials in Electrochemical Capacitors

	U.S. Patent No. (Date) Author/Assignee	Electrode Materials
1	WO 2011078901 (06/30/2011) Raghurama, Raju/ Honeywell International Inc.	Composite synthesized from MnO_2 and TiO_2 in preferable weight ratio of 95:5 is provided. Testing of material yields preferred wt% values of electrodes as 75, 20, and 5% for TiO2–MnO2 composite, polymer binder, and carbon black, respectively.
2	20110051316 (03/03/2011) Jun, Liu; Aksay, Iihan A.; Kou, Rong; Wang, Donghai/Battelle Memorial Institute, Princeton University, U.S.	Functionalized graphene sheets are coated with mesoporous metal oxides. A mesoporous silica layer was coated using this nanocomposite as electrode material. A 171 F/g double-layer capacitance was exhibited by the graphene sheets alone at scan rate of 2mV/s.
3	20100233496 (09/16/2010) Kim, Hak-Kwan; Ra, Seung-Hyun; Bae, Jun-Hee; Jung, Hyun-Chul/Samsung Electro-Mechanics Co., Ltd., South Korea	Metal oxide developed from Cu, Ag, Au, Ni, Cr, Sn, Cd, Pb, Rd, Pt, In, Ru, Mn, Zn, and Co achieved via immersion technique. Suggested transition metal oxides for electrode use are MnO_2, RuO_2, CoO, and NiO.
4	20100238607 (09/23/2010) Park, Jin-hwan; Park, Sung-ho; Shin, Tae-yeon/ Samsung Electronics Co., Ltd., South Korea	Metal nanostructures formed for electrode design on conductive substrate pointing perpendicular to substrate. Metals used include at least one from a group consisting of Au, Ag, Ni, Cu, Pt, Ru, Mn and Li. Metal oxides shown to integrate are chosen from a group consisting of RuO_2, MnO_2, IrO_2, NiO_x $(0 < x < 2)$, and CoO_x $(0 < x < 2)$. A porous polymer (Nafion, Aciplex, Flemion or Dow) may be coated on metal oxide layer to bind layer to nanometal structure.
5	(WO 2010120560) (10/21/2010) Risser, Stephen M.; Mcginniss, Vincent; Tan, Bing; Spahr, Kevin; Castenada-Lopez, Homero/Battelle Memorial Institute, U.S.	Composite electrode material of Mn and Fe metal oxides with ratio ranging from 3.5:1 to 4.5:1 for use in ESs. Metal oxides react to form gel and are subsequently dried in supercritical CO_2 to form powder. An electrode with a weight composition of 15 to 60% carbon and 40 to 80% metal oxide nanoparticles achieved minimum specific capacitance of 500 F/g at 1 mV/s in 1*M* KOH electrolyte.
6	WO 2010028162 (03/11/2010) Gruner, George/University of California, U.S.	Multilayer electrodes use an array of metals and metal oxides [Zn, Co, Ni, Li, TiO_2, RuO_2, Fe_3O_4, WO_3, V_2O_5,Ni(OOH)], carbon and carbon derivatives (graphite, carbon black, carbon nanotubes, activated carbons, and aerogels) and conducting polymers (polyaniline, polythiophene, polypyrrol, and PEDOT). Specific use of carbon networks and polyaniline coating was found to increase capacitance significantly; this configuration showed capacitance of 0.016 F/m^2.

(continued)

TABLE 5.2 (CONTINUED)

Survey of Patents for Use as Electrode Materials in Electrochemical Capacitors

	U.S. Patent No. (Date) Author/Assignee	Electrode Materials
7	20100203391 (08/10/2010) Sergey D. Lopatin; Robert Z. Bachrach; Dmitri A. Brevnov; Christopher Lazik; Miao Jin; Yuri S. Uritsky/Applied Materials, Inc.	Mesoporous carbon hybrid matrix with 2 to 50 nm pore sizes is constructed by interconnection of spherical carbon fullerene onions and a CNT hybrid matrix. The mesoporous carbon material forms on conducting substrate by a CVD-like process that allows carbon atoms to undergo continuous nanoscale self assembly used as an electrode.
8	20100021819 (01/28/2010) Aruna, Zhamu; Bor Z., Jang; Jinjun, Shi/U.S.	Graphene nanoplatelets less than 100 nm in length and 2 nm in thickness are dispersed in binder or matrix material such as PAN, furfuryl alcohol, phenolic formaldehyde, polyacrylonitrile, and cellulosic polymers. Carbonization of sample slurry results in increased specific surface area from 492 to 1560 m^2/g.
9	20090320253 (12/31/2009) Joseph; St Clair, Todd P.; Nadjadi, Andrew; Bourcier, Roy Schneider, Vitor Marino/Corning Inc., U.S.	Electrodes made of conductive carbon bonded onto graphite-based paper; carbonizable resin for binder. Polymer beads are used as thermally insulting material. Number, spacing, and sizes of beads depend on electrode and material composition.
10	20090316336 (12/24/2009) Fan, Qinbai/Technology Institute, U.S.	Electrode material composed of graphite slurry (in isopropanol and Teflon solution) table cast onto aluminum substrate. Bipolar membrane then constructed from spraying polyolefin amine-based anion exchange membrane (5% Nafion) solution onto cation exchange membrane.
11	WO 2009148977 (10/12/2009) Mohit, Singh/SEEO Inc.	Nanostructures (nanocrystals, nanoparticles, nanowires, nanorods) are made from semiconductive ZnS, ZnO, ZnSe, ZnTe, HgSe, MgSe, CaSe, SrSe, BaS, BaTe, GaN, GaP, InP, PbS, PbSe, AlS, and AlSb and used as film electrodes.
12	7623340 (11/24/2009) Lulu Song; Aruna Zhamu; Jiusheng Guo; Bor Z. Jang/Nanotek Instruments, Inc.	Fully separated graphene platelets dispersed within matrix binder (polyaniline, carbonaceous material, amorphous or glassy carbon, or combination) solution and treated by activation or carbonization to increase total surface area. Suspension is then frozen and solvent removed by vacuum sublimation leaving a porous nanographene platelet–polymer composite.
13	WO 2009106842 (09/03/2009) Tobias, James; Alan Daniel/ Nano-Tecture Ltd.	Capacitor device consisting of electrodes of mesoporous particles (nickel hydroxide; nickel oxide; nickel oxy-hydroxide; manganese dioxide; nickel–manganese oxides and lithiated forms; titanium dioxide and lithiated forms; tin, tin alloys and lithiated forms) with at least 75 wt% of the particles larger than 15μm is used. To increase structural strength, particles may be deposited on a substrate. Binders and other inactive materials that improve electrical conductivity may be added.

TABLE 5.2 (CONTINUED)

Survey of Patents for Use as Electrode Materials in Electrochemical Capacitors

	U.S. Patent No. (Date) Author/Assignee	Electrode Materials
14	20090136834 (05/28/2009) Coowar, Fazlil Ahmode; Blackmore, Paul David/ Qinetiq Limited, UK	Carbon monofluoride, conductivity additive, and polyvinylidene fluoride binder were used to create a paste and subsequently cast on aluminum foil to create cathode.
15	20080287607 (11/20/2008) Lien Tai Chen; Shu-Hui Cheng; Tun-Fun Way; Tzu Hsien Han/Industrial Technology Research Institute	Precursor polyacrylonitrile dispersed in dimethylforamide or dimethylacetamide undergoes nanospinning process. Spun fibers are stabilized at 270°C, and activated through heating at 1000°C. Evaluation of capacitance using aqueous electrolyte yields >300 F/g.
16	20100134954 (10/25/2008) Chris Wright; Jennifer Sweeney; Daniel Peat; Phillip Andrew Nelson/ Nano-Tecture Ltd.	Supercapacitor electrode is constructed of mesoporous metal (nickel and its oxides, hydroxides, and oxyhydroxides) or metal compound using carbon; volume ratios are preferably 15:1 to 50:1. Capacitance of Ni electrode measured at current density of 2092 mA/cm^2 to be 1635 F/cm^3. Capacitance of carbon electrode measured at same current density was 106 F/cm^3.
17	20080232028 (09/25/2008) Zhao Xin/College of William & Mary, U.S.	Carbon nanosheet of <2 nm thickness and specific surface area between 1000 and 2600 m^2/g attached to smooth carbon fibers by thermal chemical deposition or microwave plasma chemical deposition. Composite is then used as electrode in full cell supercapacitor.
18	20080218939 (09/11/2008) Marcus, Matthew S.; Gu, Yuandong/Honeywell International, U.S.	Nickel or gold nanowire electrode between 0.02 and 0.2 µm wide is created using porous alumina or silica membrane. Dissolving membrane leaves metal nanowires to be used as electrode.
19	20080010796 (01/17/2008) Pan, Ning; Du, Chunsheng/ University of California, U.S.	Functionalized CNTs are attached to metal substrate (nickel, copper, aluminum, gold, or platinum) by electrophoretic deposition after dispersion and charging in a high concentration colloidal suspension take place. The composite is baked onto metal substrate at 500°C in H_2.
20	20070148335(06/28/2007) Tanaka, Hideki; Katai, Kazuo/Tdk Corporation, Japan	Electrode is processed by spreading a paste containing activated carbon, polytetrafluoroethylene (binder), and carbon black (conducting agent) mixed in a solvent (n-methyl-2-pyrolidone) onto a current collector such as aluminum, then solvent is evaporated and material compressed into an electrode.

(continued)

TABLE 5.2 (CONTINUED)

Survey of Patents for Use as Electrode Materials in Electrochemical Capacitors

	U.S. Patent No. (Date) Author/Assignee	Electrode Materials
21	20070139865 (06/21/2007) Sakamoto, Ryuichi; Takeishi, Minoru; Koshimizu, Masahiro/ TDK Corporation, Japan	Porous active carbon particles derived from various precursors and polytetrafluoroethylene are used as electrode material and binder material, respectively. Fluoro rubbers such as vinylidene fluoride, hexafluoropropylene, and tetrafluoropropylene can be used as binder. Specific surface area of porous particles is preferably between 2000 and 2500 m^2/g to obtain high volume capacity.
22	20070134151 (06/14/2007) Jo, Seong-Mu; Kim, Dong-Young; Chin, Byung-Doo; Hong, Sung-Eun/Korea Institute of Science and Technology, South Korea	Halogenated polymers such as poly(vinylidene fluoride and poly(vinylidenefluoride-co-hexafluoropropylene) with preferred diameter of 1 to 1000 nm are mixed with graphitizing metal nanograin catalyst (Pt, Ru, Cu, Fe, Ni, Co, Pd, W, Ir, Rh, Sr, Ce, Pr, Nd, Sm, Re, B, Al), catalyst precursor (metal chlorides: $CuCl_2$, $COCl_2$, $OsCl_3$, $CrCl_2$,$FeCl_3$), metal nitrides: $Fe(NO_3)_3$ and $Pd(NO_3)_2$) or organic metal compounds such as iron acetylacetonate. Chosen mixture is electrospun at charge of 9 kV, and resulting fiber treated and heated at carbonizing temperature between 800 and 1800°C in N_2. The resulting excellent ultrafine carbon fiber can then be processed as electrode material.
23	20070095657 (05/03/2007) Dong-Young Kim; Seong-Mu Jo; Byung-Doo Chin; Young-Rack Ahn/ Korea Institute of Science and Technology, South Korea	Ultrafine metal fiber matrix of titanium dioxide deposited electrochemically acts as double-layer electrode with wide specific surface area. Next, a metal oxide such as ruthenium oxide, rubidium oxide, iridium oxide, nickel oxide, cobalt oxide, manganese oxide, or vanadium oxide can be deposited onto substrate by electrodeposition.
24	WO 2006135439 (12/21/2006) Ma, Jun; Chishti, Asif; Ngaw, Lein; Fischer, Allen; Braden, Robert/Hyperion Catalysis International Inc.	Nanotubes reacted with ozone at temperatures from 0 to 100°C are used develop aggregates. Dispersing product into polyethyleneimine cellulose forms mixture. Pyrolyzing product at temperatures above 650°C results in rigid pure carbon structure as electrode material. Additional solids can be added to mix before pyrolyzation to increase capacitance.
25	20060250750 (11/06/2006) Yoshinori Yonedu; Michihiro Sugou; Yoshinori Ogawa/ Shin-Etsu Chemical Co.	Polyimide silicone from polyamic acid precursor is mixed with conductive material, preferably carbon or transition metal oxide to create a minimal resistance active electrode. Mass ratio of conductive material to polyimide silicone is preferably 99:1 to 80:20.
26	20060098389 (05/11/2006) Liu, Tao; Kumar, Satish/ Georgia Tech Research Corporation	Single wall carbon nanotubes mixed with copolymer such as acrylonitrile or methylacrylate are made into a suspension and cast on glass plate for film development. Subsequent activation of films by CO_2 and Ar at 700°C for 8 min. tested at current density of 01 mA to yield 167 F/g.

TABLE 5.2 (CONTINUED)

Survey of Patents for Use as Electrode Materials in Electrochemical Capacitors

	U.S. Patent No. (Date) Author/Assignee	Electrode Materials
27	20050153130 (07/14/2005) Long, Jeffrey W.; Rolison, Debra R./Secretary of the Navy representing U.S.	Carbon aerogel or porous carbon electrode is further developed by pore coating process using an electroactive polymer such as polyaniline, polyarylamine, or polypyrrole to increase capacitance.
28	20050142898 (06/30/2005) Kim, Kwang Bum/Hyundai Motor Company, Kia Motors Corporation	Porous alumina template with pore diameters between 20 and 200 μm is submerged in solution of nickel salt of concentration 0.05 to 0.5*M* where electroplating of nickel metal layer occurs. Porous alumina template is then dissolved in NaOH to leave only nickel metal electrode.
29	20040214081 (10/28/2004) Tomoki Nobuta; Hiroyuki Kamisuki; Masaya Mitani; Shinako Kaneko; Tetsuya Yoshinar; Toshihiko Nishiyama; Naoki Takahashi/Japan	Powdery conducting polymer serves as both cathode and anode in supercapacitor. Examples of doped or undoped proton compounds include polyaniline, polythiophene, polypyrrole, polyacetylene, polyperinaphthalene, polyfuran, polyfiurane, polythienylene, polypyridinediyl, polyindole, indoletrimer compounds, polyaminoanthraquinone, polyimidazole, quinones such as benzoquinone, and anthraquinone where a quinone oxygen can be converted to a hydroxyl group.
30	WO 2004032162 (04/15/2004) Mikhailovich, Alexander; Anatolijevich, Sergey; Ivanovich, Dmitry; Viktorovna, Svetlana/ Gen3 Partners, Inc.	Proposed electrode uses polymer deposited onto conducting substrate via electrochemical polymerization. Polymer is a poly-[Me(R-Salen) type. Me represents complex compound of transition metals (Ni, Pd; Co, Cu, and Fe) with at least two different oxidation states. R represents electron donating substituent such as CH_3-O-, C_2H_5O–, HO–, –CH_3 radicals. Salen represents residue of bis(salicylaldehyde) ethylenediamine in Schiff's base. Preferences for conducting substrate include carbon fiber, carbon materials with metal coatings, and metal electrodes with high specific surface areas.
31	WO 2003088374 (10/23/2003) Philips, Jeffrey; Hewson, Donald/Powergenix Systems, Inc., Canada	Nickel-based positive electrode made of 3 to 95 wt% NiOH; balance consists of either nickel powder, cobalt powder, or carbon powder, possibly in combination. Carbon-based negative electrode made of 10 to 95 wt% carbon, with balance consisting of at least one metal oxide from group of bismuth oxide, iridium oxide, cobalt oxide, iron oxide, and iron hydroxide, and hydride from Groups IIIA, IIIB, IVA, IVB, VB, VIB, VIIB, and VIIIB.

(continued)

TABLE 5.2 (CONTINUED)

Survey of Patents for Use as Electrode Materials in Electrochemical Capacitors

	U.S. Patent No. (Date) Author/Assignee	Electrode Materials
32	20050089754 (08/28/2003) Lang, Yoel/Cellergy Ltd., Israel	Small amounts of inorganic fillers such as fumed silica, high surface area alumina, bentonites, glass spheres and ceramics are mixed with polyols such as propylene glycol to increase viscosity for printed electrodes. Proposed printed electrodes are carbon black, graphite, metallic or plated metallic particles.
33	20030086238 (05/08/2003) Bendale, Priya; Malay, Manuel R.; Dispennette, John M.; Nanjundiah, Chenniah; Spiess, Frederic/Bendale, Priya; Malay, Manuel R.; Dispennette, John M.; Nanjundiah, Chenniah; Spiess, Frederic	Proposed electrode is made from activated carbon combined with small amount of conducting carbon and binder to yield slurry that is applied to first layer of conducting carbon (carbon powder) and binder polyvinylpyrrolidone initially deposited on aluminum current collector.
34	6493209 (10/23/2003) Kamath, Hundi P.; Rasmussen, Paul S.; Manoukian, Daniel M./ Powerstor Corporation, U.S.	Compounds such as resorcinol formaldehyde, phenol resorcinol formaldehyde, catechol formaldehyde and phloroglucinol formaldehyde are mixed with catalyst and possibly a metal to create aquagel that is subsequently dried and pyrolyzed to produce carbon aerogel.
35	5993996 (11/02/2002) Firsich, David W/Inorganic Specialists, Inc.	Carbon nanofibers undergo hydrogenation at temperatures from 400 to 800°C followed by treatment with fuming sulfuric acid at temperature between 110 and 150°C to create high performance carbon electrodes. Untreated carbon nanofibers exhibited capacitance of 22 F/g; treated carbon fibrils obtained capacitance of 95 F/g.
36	20020163773 (09/25/2002) Niiori, Yusuke; Katsukawa, Hiroyuki/NGK Insulators, Ltd., Japan	Carbon material undergoes mechanical stress process to increase its partial oxidation, after which the oxidized carbon is mixed with a conducting agent and binder to form a sheet placed onto acid-etched aluminum substrate.
37	6299850 (10/09/2001) Doughty; Daniel H.; Eisenmann, Erhard T./ Department of Energy representing U.S.	Polymeric material with high molecular directionality is carbonized, treated in 1 to 5 *M* aqueous nitric acid and heated at 350°C in non-oxidizing atmosphere; process is repeated 7 to 10 times to yield carbon supercapacitor electrode with volumetric capacitance up to 300 F/cm^3.

TABLE 5.2 (CONTINUED)

Survey of Patents for Use as Electrode Materials in Electrochemical Capacitors

	U.S. Patent No. (Date) Author/Assignee	Electrode Materials
38	20010026850 (10/04/2001) Shah, Ashish; Muffoletto, Barry C./U.S.	Substrate selected from metals group (tantalum, titanium, nickel, molybdenum, niobium, cobalt, stainless steel, tungsten, platinum, palladium, gold, silver, copper, chromium, vanadium, aluminum, zirconium, hafnium, zinc, and iron) undergoes ruthenium oxide spray coating. Resulting electrode has a thin layer coating ranging between 10 nm and 1 mm thick and internal surface area between 10 and 1500 m^2/g. Patent predicts capacitance within range of 50 to 900 F/g.
39	WO 00/19461 (04/06/2000) Niu, Chun-Ming/Hyperion Catalysis International, Inc., U.S.	Electrochemically active materials such as activated carbons, carbon aerogels, and carbon foams (all derived from polymers) oxides, hydrous oxides, carbides, and nitrides are used to form composites with carbon nanofibers. Additional active materials such as oxides, hydrous oxides, and carbides can be combined to form a composite. Process requires dispersion in water with carbon nanofiber and subsequent filtration and washing. Capacitance of 249 F/g was measured from a RuO_2 xH_2O metal oxide and carbon nanofiber composite.
40	6181545 (08/26/1999) Amatucci, Glenn G.; Du Pasquier, Aurelien; Tarascon, Jean-Marie/ Telcordia Technologies, Inc., U.S.	Activated carbon powder is dispersed in a plasticized copolymer matrix solution which is then dried to form membrane. Examples of copolymers are poly(vinylidene fluoride-co-hexafluoropropylene) and poly(vinylidene fluoride-co-chlorotrifluoroethylene. The composite is thermally laminated to aluminum current collector as electrode.
41	20010001194 (02/23/1999) Jow, T. Richard; Zheng, Jian-Ping/Department of the Army representing U.S.	Proton inserted ruthenium oxide ($HRuO_2 \cdot xH_2O$) electrodes are created by reducing RuO_2. Energy is stored through this reversible reaction. At 1 V operating range in aqueous H_2SO_4, observed capacitance per unit mass and observed capacitance per unit area for this electrode are 380 F/g, and 200-300m F/cm^2, respectively.
42	6310765 (12/23/1998) Tanahashi, Masakazu; Igaki, Emiko/Matsushita Electric Industrial Co., Ltd., Japan	Carbon particles were roll pressed onto nickel foil, proceeded by the electro-polymerization of polypyrrole to form a 20 µm layer. The electrolytic capacitor had a capacitance of 71 µF on the average.
43	20030030963 (11/20/1997) Tennent, Howard; Moy, David; Niu, Chun-ming/ Hyperion Catalysis International, Inc., U.S.	Cylindrical carbon nanofibers are coated with thin layer of pyrolyzed carbonaceous polymer. Coating layer is comprised of one or more polymers selected from group consisting of phenolic formaldehyde, polyacrylonitrile, styrene DVB, cellulosic polymers, and H-resin.

(continued)

TABLE 5.2 (CONTINUED)

Survey of Patents for Use as Electrode Materials in Electrochemical Capacitors

	U.S. Patent No. (Date) Author/Assignee	Electrode Materials
44	5585999 (12/17/1996) De Long, Hugh C.; Carlin, Richard T./Department of the Air Force representing U.S.	Thin film of palladium as positive electrode undergoes reversible faradic reaction to palladium chloride. A metallic aluminum cathode is used. A capacitance of 550 μF/cm^2 was obtained from this configuration.
45	5527640 (06/18/1996) Rudge, Andrew J.; Ferraris, John P.; Gottesfeld, Shimshon/University of California, U.S.	Supercapacitor electrodes consist of *p*-doped conducting polymer for positive electrode and *n*-doped conducting polymer for negative electrode. Conducting polymers are selected from a group consisting of polythiophene (thiophene is in 3 position) polymers having aryl and alkyl groups attached to thiophene in 3 and 4 positions and polymers synthesized from bridged dimers having polythiophene backbone.
46	5402306 (05/28/1995) Mayer, Steven T.; Kaschmitter, James L.; Pekala, Richard W./ University of California, U.S.	Carbon aerogels are used as electrodes with densities between 0.3 and 0.9 g/cm^3 and surface areas from 400 to 1000 m^2/g. Xerogel–aerogel hybrid is made by laminating xerogel and aerogel layers together.

materials. Suspensions of CNTs for film casting have been used in an effort to effectively utilize their surface area (<500 m^2 g^{-1}) as a counterpart to their high conductivity and mechanical strength. Similarly, a recent U.S. patent by Applied Materials Inc. (20100203391) claims to integrate carbon-like onions (carbon fullerene onions) and CNTs by a CVD process to develop a self assembled mesoporous structure for electrode use.

Despite the benefits of CNTs, graphene is more practical because of synthesis costs, purification methods, and surface area. Similar efforts to utilize the high surface area of CNTs (theoretical surface area > 2500 m^2 g^{-1}) through solution mixing and subsequent carbonization are claimed by Aruna, Bor Z., and Jinjun to achieve 1560 m^2 g^{-1} (U.S. 20100021819). Integration with pseudocapacitive materials is also thought to complement the EDLC of a two-dimensional carbon material for high gravimetric or volumetric capacitance.

A focus on patent applications for metal oxides including oxidized ruthenium, nickel, cobalt, manganese, and vanadium yielded nanowire, fiber, and nanoparticle geometries with high electroactive surface areas. This was achieved through a variety of methods including the use of alumina, zeolite, mesoporous SBA-15 templating, and controlled growth by electrochemical deposition. Alloy oxides that prevalently incorporate manganese oxide are cited in two patents claiming beneficial integration with titanium oxide (WO

2011078901) and iron oxide (WO 2010120560). The latter is claimed to achieve a gravimetric capacitance of 500 F g^{-1} at 1 mV s^{-1} in 1 *M* KOH.

5.8.2 Patents on Electrolytes

The preferential usage of acetonitrile (AN) in commercial applications is currently challenged by its intrinsic hazards of flammability and toxicity. Propylene carbonate (PC) is less dangerous but is hindered by a threefold decrease in electrolyte conductivity. Novel ionic liquids show promise as non-toxic, highly stable next generation electrolytes if their electrolyte conductivity can be enhanced. A combination of organic quaternary salts with a preferred solvent [ethylene carbonate (EC)] in U.S. Patent 20080137265 claims to be an electrolyte that can be used over a wide temperature range (–40°C) and a preferred potential window from 0 to 4.0 V. Ionic molten salts in pure form have also undergone patenting (U.S. Patent 20040106041), and mixtures of these in optimum combinations with non-aqueous solvents are recognized to improve upon their deficient electrolyte conductivity.

Solid state films that have been developed to utilize solid polymer electrolytes without requiring safety sealing and additional packaging. Patents claim that organosilicon compounds (U.S. Patent 20070076349) and polyoxyalkylene-modified silanes (U.S. Patent 20070048621) are suitable with the additions of varying electrolyte salts (and separators if needed) for use as solid film electrolytes. Table 5.3 lists recent patents on electrolytes.

5.8.3 Patents on ES Designs

Commercialized supercapacitor designs commonly employ bipolar electrodes configured into a stack to increase the operating voltage and efficiently minimize cell volume. These cell designs commonly include jelly roll fabrication. U.S. Patent 6762926 [10] emphasizes this configuration. Jelly roll designs highlight supercapacitor portability, and multiple patents stress their use in module construction.

Module housing compartments imparting robust designs are critical to overall bank designs for high voltage applications. A patent by Maxwell Technologies claims to have engineered the design of a module housing that is unique in its ability to use tongue-and-groove connectors to extend the housing if required to accommodate more cells; cells sizes may vary (U.S. Patent 20070053140). The use of supercapacitor modules in photovoltaic cell windmill technologies and their significance in auto industry applications reveals their growing importance. Table 5.4 lists some of these designs.

A further review of patents, specifically hybrid or full electric vehicle applications indicates that some corporations hold more than 20 patents pertaining to this technology (Figure 5.14). Matsushita Electric holds more than 100 patents. To complement the tabulation of maximum patents assigned,

TABLE 5.3

Patent Survey of Electrolyte Developments for Use in Electrochemical Capacitors

	U.S. Patent No. (Date) Author/Assignee	Electrolyte Materials
1	20110076572 (03/31/2011) Amine, Khalil; Chen, Zonghai; Zhang, Zhengcheng/University of Chicago/Argonne, LLC, U.S.	Ionic electrolyte salt and non-aqueous electrolyte solvent described. Solvent includes mixture of siloxane, silane, or both; a sulfone and a fluorinated ether, fluorinated ester, or both; and ionic liquid or carbonate.
2	20080137265 (06/12/2008) Venkateswaran, Sagar N./U.S.	Organic compounds of quaternary fluoroborateomnium in mixture of ethylene carbonate, ethylmethyl carbonate, diethylmethyl carbonate, and γ-butyrolactone are used in combination to create high voltage, non-toxic electrolytes. Mixtures can be used down to –40°C; are flame retardant when (di)-ethylmethyl carbonate is omitted.
3	20100046142 (09/13/2007) Aitchison, Phillip Brett; Nguyen, Hung Chi/ Cap-XX Ltd., Australia	Created electrolyte comprised of solvent and ionic species. Possible ionic species include lithium tetrafluoroborate, tetrabutylammonium perchlorate, and tetraethylammonium tetrafluoroborate. Salt most preferred is ethyltriethylammonium tetrafluoroborate; preferred solvent is non-aqueous. Examples include ethylene carbonate, propylene carbonate, N-methypyrrolidione, 1,2-dimethoxyethane, methyl formate, sulfuryl chloride, and tributyl phosphate. The most preferable solvent is a nitrile, specifically propionitrile. A 1 *M* solution of MTEATFB in propionitrile yielded conductivity of 48 mS/cm at 95°C and 28 mS/cm at 23°C.
4	20070076349 (04/05/2007) Dementiev, Viacheslav V.; West, Robert C.; Hamers, Robert J.; Tse, Kiu-Yuen/ Wisconsin Alumni Research Foundation, U.S.	Electrolyte is a solution of lithium salt in an ion-conducting organosilicon compound such as oligoethyleneoxide-substituted organosilane. Preferred materials are organosilanes and organosiloxanes that can have linear, branched, hyper-branched, or cross-linked structures. At room temperature, preferred organosilicon electrolytes can have conductivity of 1 mS/cm.
5	20070048621 (03/01/2007) Kashida, Meguru; Nakanishi, Tetsuo; Miyawaki, Satoru; Aramata, Mikio/Shin-Etsu Chemical Co., Ltd., Japan	Polyoxyalkylene-modified silane (SiHX) is combined with non-aqueous solvent and electrolyte salt to form non-aqueous solution. Preferred SiHX is 0.001 to 0.1% by volume of solution. Light metals such as Li, Na, K, Mg, Ca, and Al are used as electrolyte salts in concentration of 0.5 to 2 *M*. Suitable solvents include aprotic dielectric constant compounds such as ethylene carbonate, γ-butyrolactone, and aprotic low viscosity solvents such as dimethylcarbonate, sulfolane, and acetic acid esters.

TABLE 5.3 (CONTINUED)

Patent Survey of Electrolyte Developments for Use in Electrochemical Capacitors

	U.S. Patent No. (Date) Author/Assignee	Electrolyte Materials
6	20080045615 (12/08/2005) Best, Adam S.; Viale, Sebastien M. J.; Picken, Stephen J./Dutch Polymer Institute	Polymer electrolyte is composed of polymer having ion exchangeable functional groups. The polymer consists of ionic liquid functional groups claimed to increase ion conduction at elevated temperatures and can be derivative of pyridinium, imidazolium, and alkyl-ammonium. Ion exchange functional groups include polymer-bound anionic groups such as sulfonates, carboxylates, and phosphonates.
7	20040214078 (10/27/2004) Mitani, Masaya; Nobuta, Tomoki; Kamisuki, Hiroyuki; Yoshinari, Tetsuya/NEC Tokin Corporation, Japan	Electrolyte is created with proton-conducting compounds such as polyaniline, quinone, and benzoquinone for both the anode and cathode. The electrolytic solution consists of a polymeric compound having an atom with an unpaired electron in the principal chain and an acid. Examples of polymeric compounds are methylene oxide, ethylene oxide, and propylene oxide that may be substituted by a hydroxide group.
8	7070706 (07/04/2006) Chu, Po-jen; Chiang, Chin-yeh/National Central University, Taiwan	Polymer electrolyte is composed of polymer substrate and metal salt. Possible polymer substrates include polyethylene oxide, polyvinylidene fluoride, poly(methyl methacrylate), polyvinylidene chloride, and polyacrylonitrile. Examples of cations of metal salt are Li, Na, K, and Mg. Possible anions are BF_4, SCN, SO_3CF_3, AsF_6, PF_6, and $N(CF_3SO_3)$. An inorganic modifier such as nanoparticle TiO_2 can be added to increase mechanical properties of supercapacitor. Mass is divided so that polymer substrate is 30 to 90 wt%, metal salt is to 2 to 30 wt%, and nanotube modifier is 3 to 30 wt%.
9	20040106041 (06/03/2004) Reynolds, John R.; Zong, Kyukwan; Stenger-Smith, John D.; Anderson, Nicole; Webber, Cynthia K.; Chafin, Andrew P./ Secretary of the Navy representing U.S.	Molten salt with 1-ethyl-3-methyl-1-H-imidazolium cation is used in electrolyte solution to increase charge and discharge speeds, operating up to 300°C and 4.3 V.

(continued)

TABLE 5.3 (CONTINUED)

Patent Survey of Electrolyte Developments for Use in Electrochemical Capacitors

	U.S. Patent No. (Date) Author/Assignee	Electrolyte Materials
10	20050211136 (09/25/2003) Drummond, Calum John; Nguyen, Hung Chi; Wade, Timothy Lawrence/ Energy Storage Systems Pty Ltd., Australia	Non-aqueous solvent system for electrolyte use includes low and high boiling point components. The low boiling component is a nitrile, preferably acetonitrile. The high boiling component is a lactone such as y-butyrolactone or an organic carbonate such as ethylene carbonate. Non-aqueous solvent includes acetonitrile, γ-butyrolactone, and ethylene carbonate in a mole ratio of 3:1.72:1. The described supercapacitor reported an ESR of no more than 784 $m\Omega \cdot cm^2$ at 85°C and no more than 946 $m\Omega \cdot cm^2$ at 23°C.
11	6669860 (04/04/2001) Maruyama, Satoshi; Suzuki, Hisashi; Sakurai, Kozo; Kujira, Masakatsu; Yamamoto, Takamasa/ TDK Corporation/Toyo Roshi Kaisha, Ltd., Japan	Solid electrolyte using polyvinylidene fluoride polymer dissolved in solvent such as dimethylacetamide, dimethylformamide, N-methyl-2-pyrrolidone, and dimethyl sulfoxide Is then cast into film. Preferred porosity of films is 70 to 80% and preferred pore diameter is 0.02 to 2 μm. Film is then immersed in solution of ethylene carbonate, propylene carbonate, or 2-methyltetrahydrofuran.
12	6219222 (03/29/2000) Shah, Ashish; Scheuer, Christina; Miller, Lauren; Muffoletto, Barry C./ Wilson Greatbatch Ltd., U.S.	Water and ethylene glycol are mixed to create solvent used to dissolve ionic salt with ammonium cation. Acid or acids can be used as additive to achieve desired supercapacitor conditions. Electrolyte is composed 0 to 85 wt% deionized water, 0 to 95 wt% ethylene glycol, 0 to 80 wt% acetic acid, 0 to 6 wt% phosphoric acid, and 0 to 50 wt% ammonium acetate.
13	5973913 (02/18/1999) McEwen, Alan B.; Evans, David A.; Blakley, Thomas J.; Goldman, Jay L./ Covalent Associates, Inc., U.S.	Non-aqueous electrolytes containing salts of alkyl substituted cyclic delocalized aromatic cations. Certain polyatomic anions, preferably PF_6^- are dissolved in high conducting organic liquids such as alkyl carbonate solvents or liquid SO_2 at concentrations >1 *M*. Salt of 1-ethyl-3-methylimidazolium and hexafluorophosphate (PF_6^-) is reported to have high conductivity of less than 13 mS/cm, electrochemical stability >2.5 V, and high thermal stability >100°C. High capacitance of >100 F/g is obtained when used with activated carbon electrodes.
14	5986878 (09/25/1997) Li, Changming; Reuss, Robert H./Motorola, Inc., U.S.	Electrolyte is formed from a poly (isopoly or heteropoly) acid. Specific examples are phosphotungstic acid (PWA) and phosphomolybdic acid (PMA) for which conductivities of 17 and 18 mS/cm, respectively, were reported.

TABLE 5.4

Design Patents for Supercapacitor Application

	U.S. Patent No. (Date) Author/Assignee	Design
1	WO 2010028162 (03/11/2010) Gruner, George/University of California, U.S.	ESs of both double-layer and battery-like materials such as porous carbon and RuO_2, respectively, form multilayer electrode structures. The two electrodes either combine the two materials or use them separately. Device configuration consists of basic electrode, separator with electrolyte, and electrode design.
2	20090320253 (12/31/2009) Bourcier; Roy Joseph/ Corning Incorporated, U.S.	Planar electrodes are arranged in series and conductive carbon material is applied to an electrode when it is between at least two alternating electrodes to provide efficient electrical contact between them. Stacks of discrete sheet electrodes have electrically insulating material disposed between electrodes. Electrodes can be of any shape or size and have first face, opposing second face, and defined thickness from first face to second.
3	20090316336 (12/24/2009) Fan, Qinbai/Gas Technology Institute	Bipolar membrane separator with first side having plurality of anions to face positive electrode and second side having plurality of cathodes facing negative electrode. Membrane builds its own electrical field to become a sub-capacitor with respect to external electric field to prevent recombination through maintained charge separation.
4	20100302708 (10/30/2008) Shiue, Lih-Ren; Goto, Masami/Linxross, Inc., Japan	Two end electrodes connect to power source, with at least one electrode having no connection at all. If the number of intervening electrodes exceeds one, separators must be placed so intervening electrodes are not in contact with each other. Intervening electrodes are wound concentrically on cylindrical element. A cylindrical 10 V sample measured at a 1 A discharge rate showed capacitance of 1.56 F, IR drop of 0.38 mm, and measured ESR of 192 mΩ at 1 KHz.
5	20080013255 (01/17/2008) Schneuwly, Adrian/ Maxwell Technologies Inc., U.S.	Two-sided electrode film element with single top side punched to create several cavities. Electrode foil element is then affixed onto bottom side. Second electrode film undergoes same process and is affixed to other side of foil element.
6	20070139865 (06/21/2007) Sakamoto, Ryuichi; Takeishi, Minoru; Koshimizu, Masahiro/ TDK Corporation, Japan	Structure for electrochemical device that enables easy passage of electrolyte solution into interior of member of stack or bank. Expansion of polarized electrode layers is effectively suppressed via a pair of fixed collecting plates. Each plate has a depressed portion to exert pressure on other stacked members, while the protruding portion forms a gap to permit electrolyte passage.

(continued)

TABLE 5.4 (CONTINUED)

Design Patents for Supercapacitor Application

	U.S. Patent No. (Date) Author/Assignee	Design
7	20070053140 (03/08/2007) Soliz, Ray/Maxwell Technologies, Inc., U.S.	Customizable housing for ES of one, two, and three dimensions that can accommodate various cell numbers is created by laser welding jelly roll ES in aluminum housing on one side and having a collector plate on other side. Customizable enclosures using tongue-and-groove connectors create slidable interference joint. Cylindrical cells of jelly roll design are connected by bus bars within housing.
8	20090201629 (02/01/2007) Lang, Joel/Cellergy Ltd., Israel	Energy storage device composed of one or more cells. Cell is defined by a pair of electrodes around a separator. Each cell is bounded by two current collectors. Each electrode is printed on the face of one of the two current collectors and the peripheral region where the second electrode is printed on is sealed.
9	6762926 (07/13/2004) Shiue, Lih-ren; Cheng, Chun-shen; Chang, Jsung-his; Li, Li-ping; Lo, Wan-ting; Huang, Kun-fu/ Luxon Energy Devices Corporation, Taiwan	Bipolar supercapacitor created by winding three electrode sheets and three separators in a jelly roll to create unitary roll. The six sheets are heat sealed to create one body. A conventional box-like container with compartments the size of the supercapacitor roll is used to connect unitary rolls in series to increase working voltage.
10	6493209 (12/10/2002) Hundi P. Kamath; Paul S. Rasmussen; Daniel M. Manoukian/Powerstor Corporation	Design includes a cell container, pair of electrodes within the container, separator placed between the electrodes and an electrolyte within the cell container. Numerous openings expose the electrode for direct electrical contact. Cell containers are easily stackable.
11	20020182349 (01/17/2002) Pynenburg, Rory Albert James/Storage Systems Pty. Ltd., Australia	Laminate package stores supercapacitor within cavity. Polyethylene inner barrier defines cavity to contain device. Sealant layer of Nucrel™ resin is dispersed between inner barrier layer and terminals; outer barrier plastic layer is bonded to metal layer and inner barrier layer.

Table 5.5 lists specific Department of Energy-funded patents with the highest number of citations for this technology[24].

5.9 Major Commercial ES Products

Development of and demands for supercapacitors arose from the heightened awareness of alternative energies and a conservative outlook on energy efficiency. Both factors helped reduce costs of these devices to ~99% per farad.

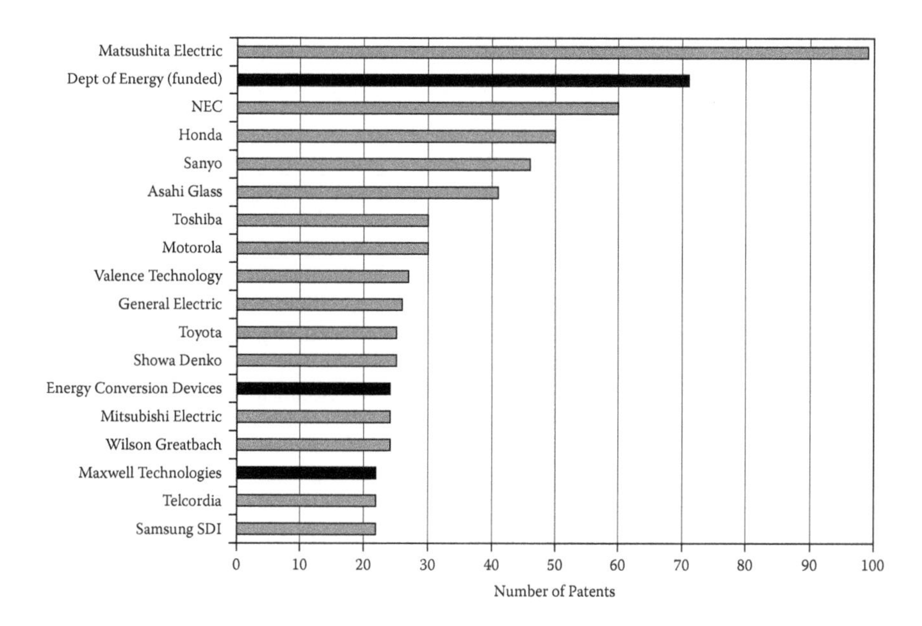

FIGURE 5.14
Organizations holding largest numbers of battery and ultracapacitor patents pertaining to HEV/PHEV/EV. (*Source:* Ruegg, R. and P. Thomas. 2008. Ultracapacitors for Hybrid, Plug-In Hybrid, and Electric Vehicles. U.S. Department of Energy.)

The remarkable decline in price is complemented by an increase in the number of producing companies and their forecasted product lines intended to meet the increasing demands for ES devices in the future [26].

A number of manufacturers around the world develop ES products including Maxwell Technologies and Epcos in North America; Cap-XX in Australia; NEC Tokin and Panasonic in Japan; and Nesscap in South Korea. See Table 5.6 for a complete list.

Several of these companies strive to manufacture cells over a range of capacitance values, voltages, and dimensions to address the integration of ESs in portable electronics (cameras, laptops, mobile PDAs, and cell phones) and meet demands for high loads of uninterrupted power and load leveling. Table 5.7 lists the performance parameters of these manufacturers. The largest market for ES is expected to be the automotive industry. Regenerative brake energy systems for micro, mini, mid, and full size hybrids and the required stabilization of power train systems against voltage sags currently drive the demand for ES innovation. The major markets of three companies are described below to demonstrate conditions in the current market.

Maxwell Technologies — Maxwell has an extensive history of developing and manufacturing power products. It produces a diverse range of small (prismatic and cylindrical) and large cells (cylindrical) to match competitors' products. Cell capacitances range from 5 to 3000 F, with larger bank

TABLE 5.5
Department of Energy-Funded Patents Receiving Largest Numbers of Citations

Citing Patent	Issue Year	DOE Patents Cited	Assignee	Title
6,962,613	2005	19	Cymbet Corp.	Low-temperature fabrication of thin-film energy-storage devices
6,643,119	2003	10	Maxwell Technologies	Electrochemical double-layer capacitor having carbon powder electrodes
6,168,884	2001	10	Lockheed Martin	Battery with in situ activation plated lithium anode
6,955,694	2005	10	Maxwell Technologies	Electrochemical double-layer capacitor having carbon powder electrodes
6,402,795	2002	8	Polyplus Battery	Plating metal negative electrodes under protective coatings
6,723.140	2004	8	Polyplus Battery	Plating metal negative electrodes under protective coatings
6,706.449	2004	8	Sion Power	Lithium anodes for electrochemical cells
6,572.993	2003	8	Visteon Corp.	Fuel cell systems with controlled anode exhaust
7,012.124	2006	6	Arizona State University	Solid polymeric electrolytes for lithium batteries
5,358.802	1994	6	University of California	Doping of carbon foams for use in energy storage devices
6,630.262	2003	6	Greenstar Corp.	Metal–gas cell battery with soft pocket
6,428.933	2002	6	3M	Lithium ion batteries with improved resistances to sustained self heating
7,170.260	2007	5	Maxwell Technologies	Rapid charger for ultracapacitors
5,336.274	1994	5	University of California	Method for forming cell separator for use in bipolar stack energy storage devices

packaging options. Maxwell's use of an engineered technique employing penetration welding bonds the cells to reduce series resistance and increase heat extraction efficiency. By focusing on large power requirements at high load potentials, its multiple modules are manufactured for industry use within the 16 to 125 V range. Maxwell emphasizes product applications for power leveling and back-up energy storage for trains, trams, and cranes. The successful integration of its modules in hybrid buses and trains is expected to expand its focus on the HEV niche market.

Cap-XX — Founded in 1997, Cap-XX focuses on thin, flat, single or dual (prismatic) cells that produce short term high power required for pulse

TABLE 5.6

Global Supercapacitor Manufacturers, Device Names, and Capacitance and Voltage Ratings

Company	Country	Device Name	Capacitance Range (F)	Voltage Range (V)	Web Address
AVX	U.S.	Bestcap	0.022 to 0.55	3.5 to 12	http://www.avx.com
Cap-XX	Australia	Supercapacitor	0.09 to 2.5	2.25 to 4.5	http://www.cap-xx.com
Cooper	U.S.	Powerstor	0.47 to 50	2.3 to 5	http://www.powerstor.com
E LNA	U.S.	Dynacap	0.033 to100	2.5 to 6.3	http://www.elna-america.com/index.htm
ESMA	Russia	Capacitor modules	100 to 8000	12 to 52	http://www.esma-cap.com/?lang = English
Epcos	U.S.	Ultracapacitor	5 to 5000	2.3, 2.5	http://www.epcos.com
Evans	U.S.	Capattery	0.01 to 1.5	5.5, 11	http://www.evanscap.com
Kold Ban	U.S.	KAPower	1000	12	http://www.koldban.com
Maxwell	U.S.	Boostcap	1.5 to 2600	2.5	http://www.maxwell.com
NEC	Japan	Supercapacitor	0.01 to 6.5	3.5 to 12	http://www.nec-tokin.net
Nesscap	Korea	EDLC	10 to 3500	3	http://www.nesscap.com
Panasonic	Japan	Gold capacitor	0.1 to 2000	2.3 to 5.5	http://www.pct.panasonic.co.jp
Tavrima	Canada	Supercapacitor	0.13 to 160	14 to 300	http://www.tavrima.com

loads. The company claims that its small devices possess the lowest RESR values for their size ($<0.1\ \Omega$) and time constant τ levels as fast as 20 msec. Its intended market targets high volume mobile and consumer electronic devices (cameras, laptops, and mobile phones).

Nesscap — Initially established as NESS in 1988, this Korean supercapacitor manufacturer changed its name to Nesscap in 2002. It offers a similar diversity in small and large cells in competition with Maxwell Technologies. Its cell capacitances range from a few farads to 5,000 F within a similar voltage range. Its large capacitance cells have prismatic designs for efficient module stacking. This led to its entry into the automotive market. Nesscap also

TABLE 5.7

Commercial Electrochemical Devices

Device	V Rating	C (F)	R (mΩ)	RC (sec)	Wh/kg	W/kg (95%)[a]	W/kg Matched Impedance	Weight (kg)	Volume (L)
Maxwell[b]	2.7	2,800	0.48	1.4	5.97	900	8,000	0.475	0.32
Maxwell	2.7	650	0.8	0.52	3.29	1281	11,390	0.20	0.211
Maxwell	2.7	350	3.2	1.1	5.91	1068	9,492	0.06	0.05
Ness	2.7	1,800	0.55	1.0	4.80	975	8,674	0.38	0.277
Ness	2.7	3,640	0.3	1.1	5.67	928	8,010	0.65	0.514
Ness	2.7	5,085	0.24	1.22	5.78	958	8,532	0.89	0.712
Ashai Glass (propylene carbonate)	2.7	1,375	2.5	3.4	6.63	390	3,471	0.21 (estimated)	0.151
Panasonic (propylene carbonate)	2.5	1,200	1.0	1.2	3.06	514	4,596	0.34	0.245
Panasonic	2.5	1,791	0.3	0.54	5.02	1890	16,800	0.31	0.245
Panasonic	2.5	2,500	0.43	1.1	5.49	1035	9,200	0.395	0.328
EPCOS	2.7	3,400	0.45	1.5	5.74	760	6,750	0.60	0.48
LS Cable	2.8	3,200	0.25	0.8	5.53	1400	12,400	0.63	0.47
Power Systems (activated carbon, propylene carbonate)	2.7	1,350	1.5	2.0	6.51	650	5,875	0.21	0.151
Power Systems (advanced carbon, propylene carbonate)	3.3	1,800	3.0	5.4	12.96	825	4,320	0.21	0.15
ESMA-Hybrid (C/NiO/aqueous electrolytes)	1.3	10,000	0.275	2.75	2.13	156	1,400	1.1	0.547
Fuji Heavy Industries (C–metal oxide hybrid)	3.8	1,800	1.5	2.6	15.56	1025	10,375	0.232	0.143

Source: Burke A. 2010. Ultracapacitor technologies and application in hybrid and electric vehicles, *International Journal of Energy Research*, 34,133–151. With permission.

[a] Power is based on $P = (9/16)*(1 - EF)*V^2/R$. EF = efficiency of discharge.

[b] All devices use acetonitrile electrolyte other than those noted (Burke, 2007).

offers high energy density cells developed from pseudocapacitive materials that have greater energy storage capabilities but shorter cycle lives. Their cells are popular in solar-powered applications.

5.10 Summary

Problems pertaining to the current distribution of supercapacitor devices arise from cell arrangement and assembly. Dominant effects in scale-up are the ohmic drops caused by interparticle resistance within a carbon or oxide matrix and outer particle surface contacts. For large-scale industry applications, the detrimental effects from these dynamics can substantially affect power performance.

The use of bipolar electrodes in devices requiring multiple electrodes for high voltage can avoid macroscopic resistance caused by the lateral contacts of separate electrode matrices and non-uniform current distributions. Excellent contact of endplates and bipolar electrode surfaces are still required to maintain minimal ESR during charge and discharge, and can be accomplished through surface treatments of the current collector and application of an optimal compressive force across the cell.

References

1. Gualous, H. et al. 2010. Supercapacitor ageing at constant temperature and constant voltage and thermal shock. *Microelectronics Reliability*, 50, 1783–1788.
2. Stoller, M. D. et al. 2008. Graphene-based ultracapacitor. *Nanoletters*, 8, 3498–3502.
3. Reddy, A. L. M. and S. Ramaprabhu. 2007. Nanocrystalline metal oxide dispersed multiwalled carbon nanotubes as supercapacitor electrodes. *Journal of Physical Chemistry C*, 111, 7727–7734.
4. Simon, P. and Y. Gogotsi. 2008. Materials for electrochemical capacitors. *Nature: Materials*, 7, 845–854.
5. *Electropedia: Cell Construction* (online). http://www.mpoweruk.com/cell_construction.htm [accessed April 5, 2012].
6. *How Coin and Multilayer Electric Double Layer Capacitors Are Manufactured* (online). http://www.elna.co.jp/en/capacitor/double_layer/manufacture.html [accessed April 5, 2012].
7. *Ultracapacitor k2 Series* (online). http://www.maxwell.com/products/ultracapacitors/products/k2-series
8. *Cap-XX Photo Gallery* (online). http://www.cap-xx.com/news/photogallery.htm [accessed April 5, 2012].

9. Lee, K. J., G. H. Kim, and K. Smith. 2010. A 3-D thermal and electrochemical model for spirally wound large format lithium ion batteries. In *Proceedings of 218th ECS Meeting*, NREL/PR-5400-49795.
10. Shiue, L., R. Cheng, and C. Chun-Shen. 2004. *Supercapacitor with High Energy Density*. U.S. Patent 6,762,926.
11. Conway, B. E. 1999. *Electrochemical Supercapacitors*, New York: Plenum.
12. *Tectate Ultracap 162V 25F Module* (online). http://www.tecategroup.com/store/index.php?main_page = product_info&cPath = 26_30_73&products_id = 1225 [accessed April 5, 2012].
13. Maxwell Technologies (online). www.maxwell.com [accessed February 12, 2012].
14. Zhou, X., C. Peng, and G. Z. Chen. 2012. 20-V stack of aqueous supercapacitors with carbon (2), titanium bipolar plates, and CNT–polypyrrole composite (1). *AIChE Journal*, 58.
15. Fritts, D. 1977. Electrode Assembly for Bipolar Battery. U.S. Patent 4,022,952.
16. Barrade, P. Series connection of supercapacitors: comparative study of solutions for active equalization of voltages. *Main*. 1–6.
17. Linzen, D. et al. 2005. Analysis and evaluation of charge balancing circuits on performance, reliability, and lifetime of supercapacitor systems. *Electronics*, 41, 1135–1141.
18. Sharma, P. and T. S. Bhatti. 2010. Review of electrochemical double-layer capacitors. *Energy Conversion and Management*, 51, 2901–2912.
19. Kötz, R., M. Hahn, and R. Gallay. 2006. Temperature behavior and impedance fundamentals of supercapacitors. *Journal of Power Sources*, 154, 550–555.
20. Bohlen, O., J. Kowal, and D. U. Sauer. 2007. Aging behaviour of electrochemical double-layer capacitors. *Journal of Power Sources*, 172, 468–475.
21. Ruch, P. W. et al. 2010. Aging of electrochemical double-layer capacitors with acetonitrile-based electrolyte at elevated voltages. *Electrochimica Acta*, 55, 4412–4420.
22. Zhu, M. et al. 2008. Chemical and electrochemical aging of carbon materials used in supercapacitor electrodes. *Carbon*, 46, 1829–1840.
23. Conway, B. E. 1999. *Electrochemical Supercapacitors*, New York: Plenum.
24. U. S. Department of Energy. 2007. *Basic Research Needs for Electrical Energy Storage*.
25. Ruegg, R. and P. Thomas. 2008. Linkages of DOE's Energy Storage R&D to Batteries and Ultracapacitors for Hybrid, Plug-In Hybrid, and Electric Vehicles.
26. Ahern, C. 2009. *Market Overview: Supercapacitors* (online). http://www.foresightst.com/
27. Davies, A. and A. Yu. 2011. Material advancements in supercapacitors: from activated carbon to carbon nanotube and graphene. *Canadian Journal of Chemical Engineering*, 89, 1342–1357.

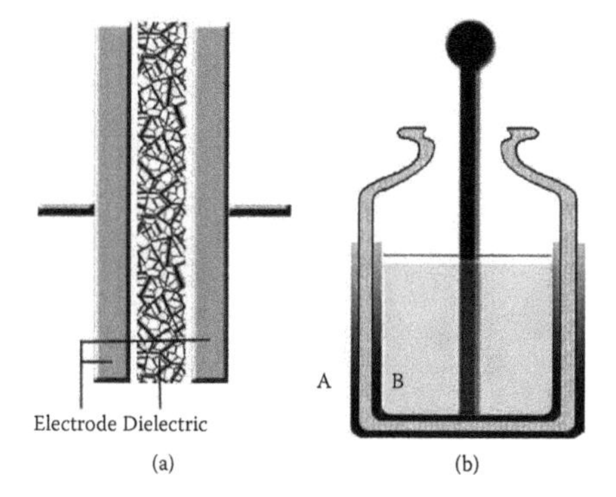

FIGURE 1.1
(a) Simplified schematic of capacitor design. (b) Cross-sectional schematic of Leyden jar (water-filled glass jar containing metal foil electrodes on its inner and outer surfaces, denoted A and B).

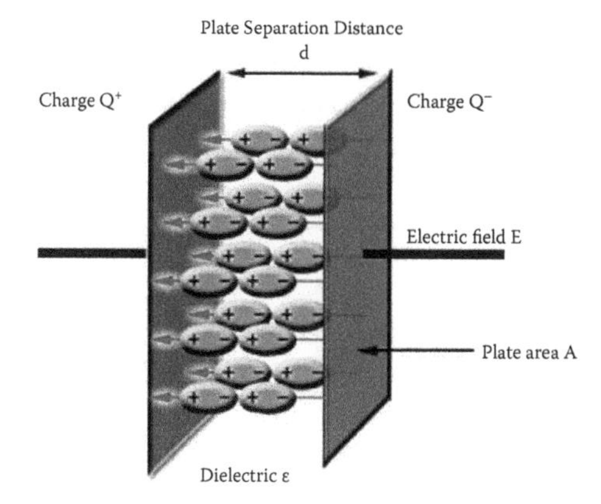

FIGURE 1.3
Charged capacitor device. (*Source:* Halliday, D., R. Resnick, and J. Walker. 2008. *Fundamentals of Physics*, 8th ed. New York: John Wiley & Sons, pp. 600–900. With permission.)

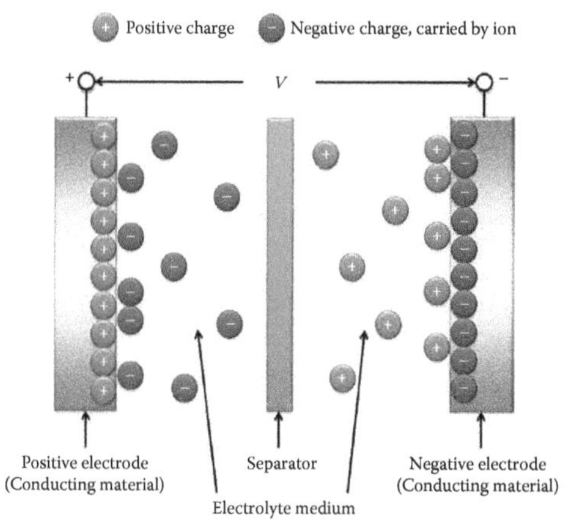

FIGURE 2.1
Electric double-layer supercapacitor.

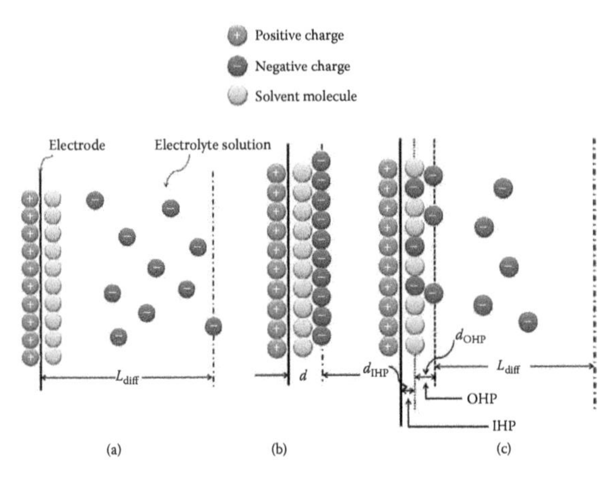

FIGURE 2.2
Electric double-layer models at interface of electrode and electrolyte solution. (a) Diffuse layer or Gouy–Chapman model. (b) Helmholtz layer or model; the d represents the double-layer thickness. (c) Stern–Grahame layer or model in which the IHP represents the inner Helmholtz plane and the OHP represents the outer Helmholtz plane.

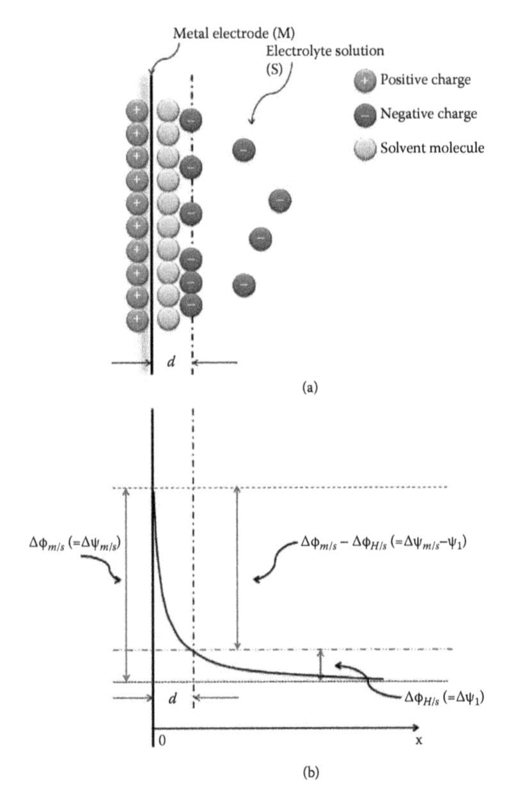

FIGURE 2.3
Drop of double-layer potential.

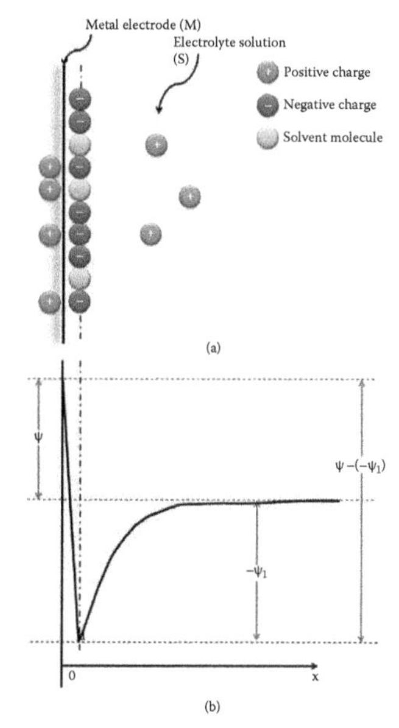

FIGURE 2.10
Double-layer with specific ion adsorption and its corresponding potential distribution.

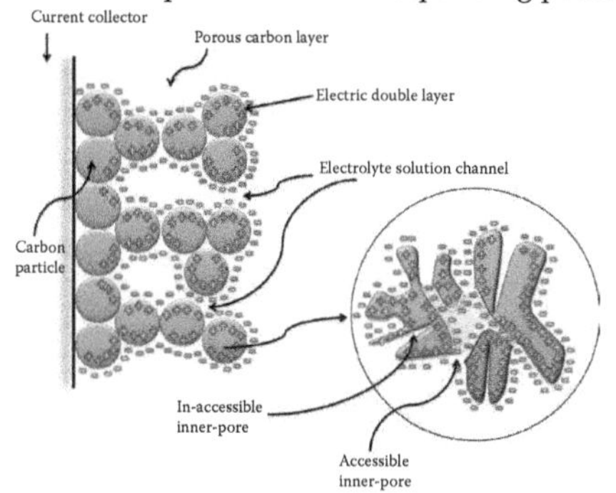

FIGURE 2.12
Electrode for electrochemical double-layer supercapacitor.

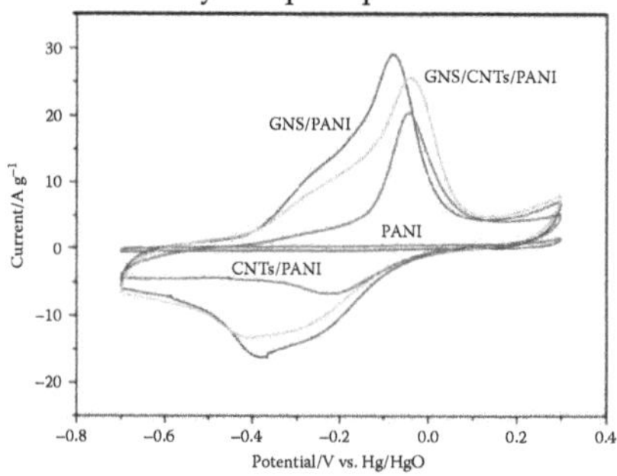

FIGURE 3.10
Cyclic voltammograms of different PANI composites and their effects on performance. (*Source:* Zhang, K. et al. 2010. *Chemistry of Materials*, 22, 1392–1401. With permission.)

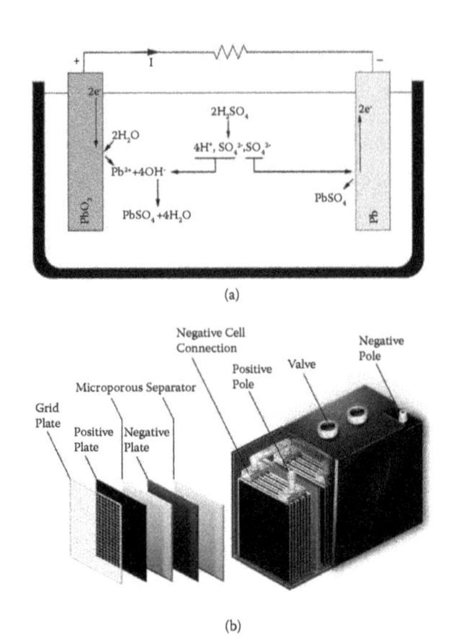

FIGURE 4.2
(a) Lead acid battery showing anode, cathode, and sulfuric acid electrolyte. (b) Cross section of lead acid battery pack. Separation of plates is created by a nonconductive separator; cells are stacked within battery module. (*Sources: Worlds of David Darling Encyclopedia* (online). Lead–acid battery. http,//www.daviddarling.info/encyclopedia/L/AE_lead–acid_battery.html [accessed April 4, 2012]; Georgia State University. 2012. Lead–acid battery: hyperphysics (online). http,//hyperphysics.phy–astr.gsu.edu/hbase/electric/leadacid.html [accessed April 9, 2012]. With permission.)

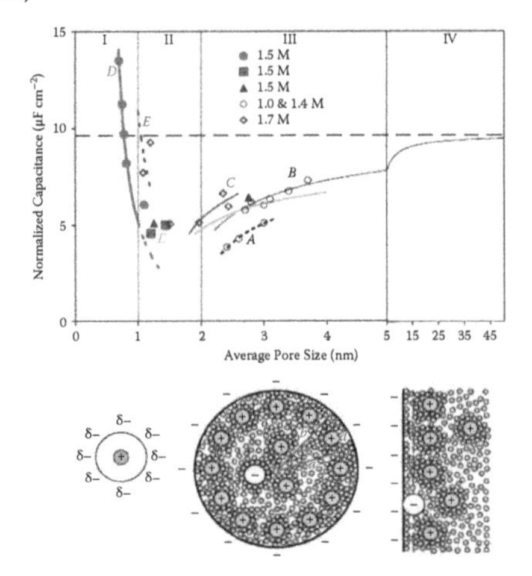

FIGURE 4.3
Capacitance tested with various ionics in acetonitrile (TEAMS: 1.7M, TEABF4: 1,1.4,1.5M) for various carbon structures. Templated mesoporous carbon (A, B), activated carbon (C), microporous carbide derived carbon (D, F), and microporous activated carbon (E). The bottom images from right to left illustrate model of planar EDLC with negligible curvature, EDLC with pores of non-negligible curvature, and model single ion wire within cylindrical pore. The models can accurately estimate capacitance in their pore regions. (Source: Simon, P. and Y. Gogotsi. 2008. *Nature: Materials*, 7, 845–854. With permission.)

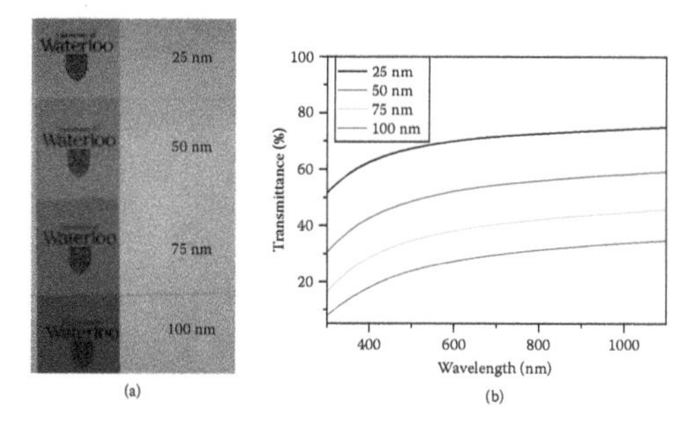

FIGURE 4.16
(a) Capacitive performance of chemically reduced graphene over large range of current density for different handling methodologies. (b) Ragone plot illustrating strong energy performance and average power density in ionic liquid electrolyte. (*Source*: Yang, X. et al. 2011. *Advanced Materials*, 23, 2833–2838. With permission.)

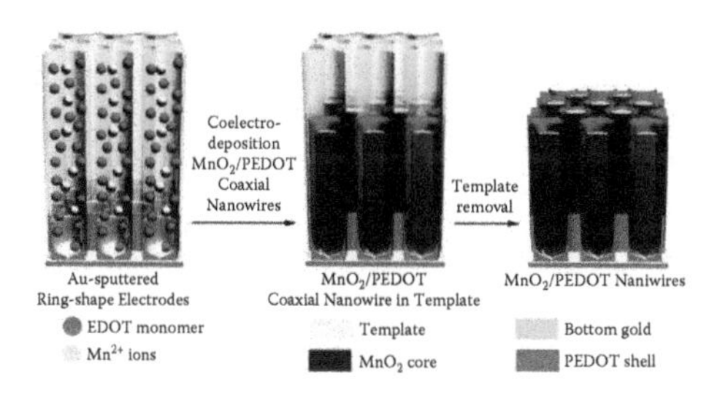

FIGURE 4.27
(a) Graphene composite film with polypyrrole deposited for 120 sec. Inset shows SEM image at the observation area. White bar = 100 nm. (b) Cyclic voltammogram curves for pure graphene film. (c) Graphene with polypyrrole deposited for 120 sec in KCl solution between –0.4 and 0.6 V versus SCE at scan rates of 0.01, 0.02, 0.05, 0.1, and 0.2 V/sec. (*Source*: Davies, A. et al. 2011. *Journal of Physical Chemistry C*, 115, 17612–17620. With permission.)

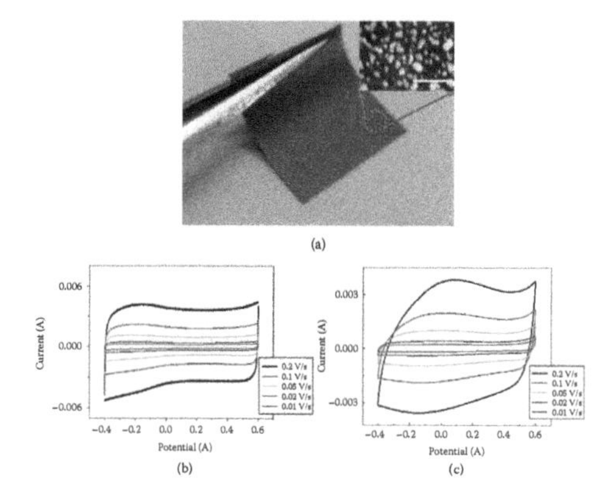

FIGURE 4.29
Thickness dependence of capacitance per area for CNT films comparing liquid (1 *M* H_2SO_4) and gel (PVA/H_3PO_4) electrolytes. (*Source*: Kaempgen, M. et al. 2009. *Journal of the American Chemical Society*, 9, 1872–1876. With permission.)

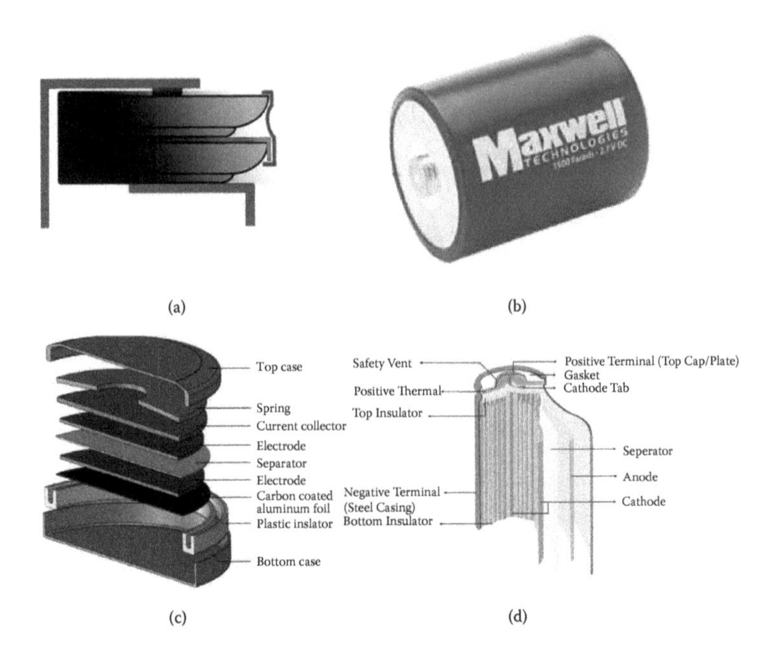

FIGURE 5.4
(a) Image of a manufactured two-coin cell stack covered by shrink wrap sleeve. (b) Manufactured rolled cell in metal can design. (c) Coin cell. (d) Rolled cell designs for ESs and batteries. See References 4 through 7.

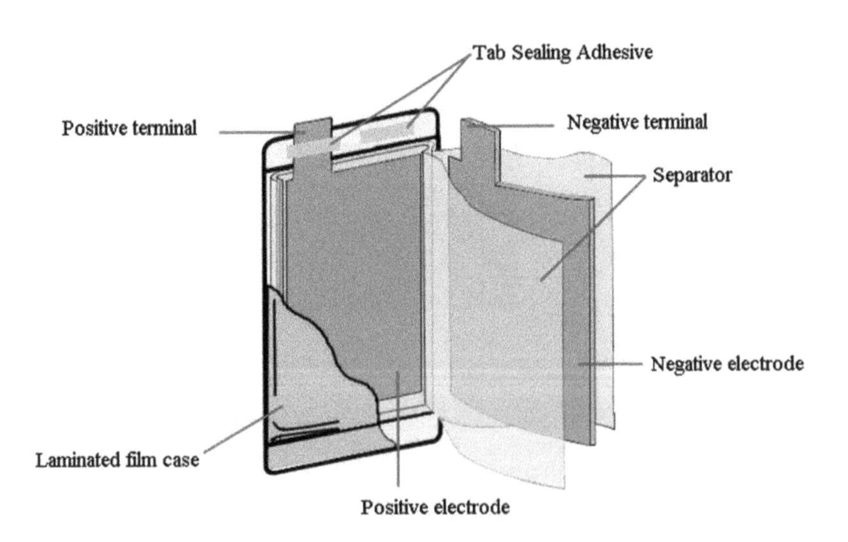

FIGURE 5.5
Pouch cell design for solid electrolyte lithium polymer battery. Design is easily adaptable to ECs made with organic electrolyte. (*Source: Electropedia: Cell Construction* (online). http://www.mpoweruk.com/cell_construction.htm [accessed April 5, 2012]. With permission.)

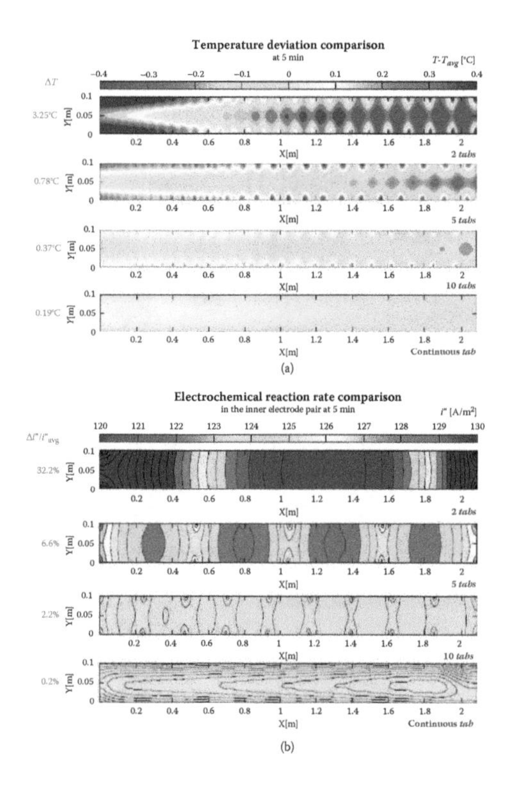

FIGURE 5.7
(a) Temperature and (b) electrode utilization profiles over area of a battery electrode for increasing number of solder tabs within roll of wound cell. (*Source*: Lee, K. J., G. H. Kim, and K. Smith. 2010. In *Proceedings of 218th ECS Meeting*, NREL/PR-5400-49795. With permission.)

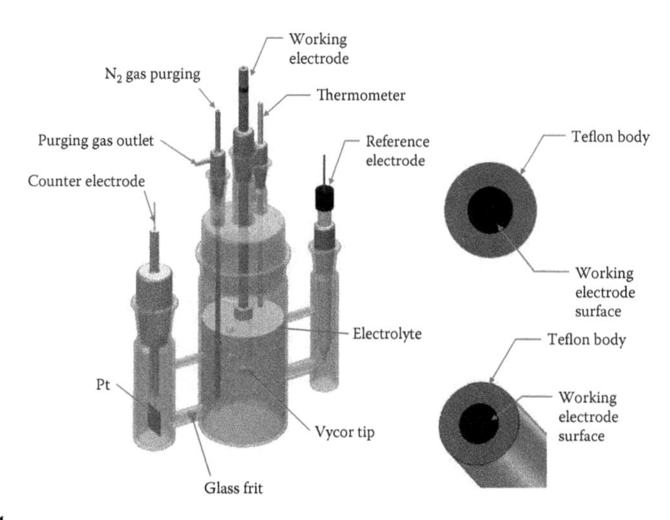

FIGURE 7.1
Conventional three-electrode cell. (*Source:* Tsay, K. C., L. Zhang, and J. Zhang. 2012. *Electrochimica Acta*, 60, 428–436. With permission.)

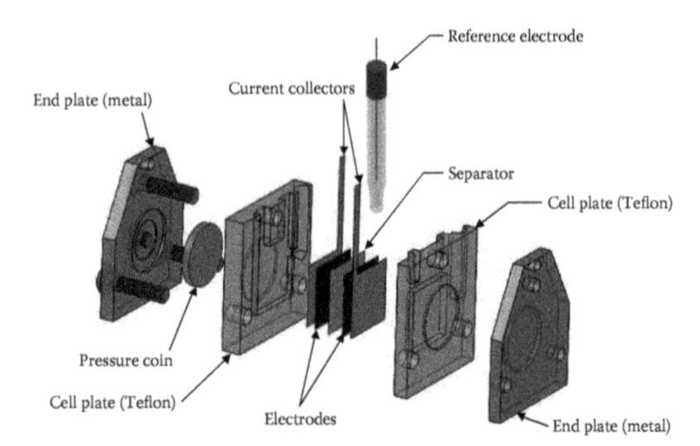

FIGURE 7.2
Two-electrode test cell for supercapacitor testing. (*Source:* Tsay, K. C., L. Zhang, and J. Zhang. 2012. *Electrochimica Acta,* 60, 428–436. With permission.)

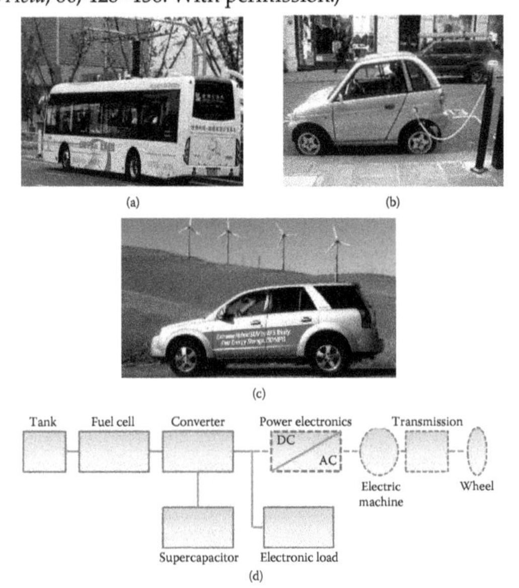

FIGURE 8.7
(a) EC powered Sinautec buses during charging; (b) and (c). Hybrid REVA (Image courtesy of F. Hebbert) and AFS trinity vehicles using ES. (d) Electric drive train combining fuel cells with ESs.

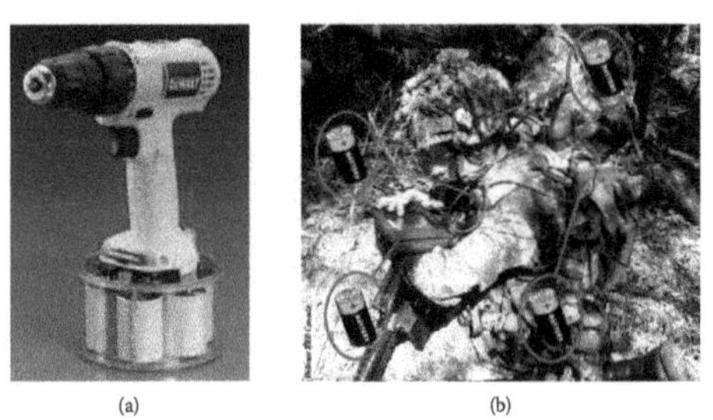

FIGURE 8.8
(a) NASA cordless drill powered by supercapacitor. (b) U.S. Army electromagnetic pulse weapons powered by ESs.

6

Coupling with Batteries and Fuel Cells

6.1 Introduction

In industry, uninterrupted power supply systems and power trains supported by battery technology have not proven to be complete solutions mainly because of restrictions on chemical power sources such as relatively low power density and insufficient cycle lives. Thus, high power and long cycle life ES devices have been identified as viable alternatives. In recent years, progressive research intended to develop ES devices and accelerate their applications has affirmed their status, and subsequent funding by the U.S. Department of Energy (DoE) indicates they are equivalent to batteries for addressing future energy needs. The increasing number of investigations focusing on ES development and integration into applications proposed several benefits arising from the ability of such devices to address rapid, short (<1 sec), high-power demands.

In this chapter, considerable attention is given to the relevance of ESs hybridized with batteries and fuel cells to integrate their respective strengths. Several available methods permit their integration, yet each design possesses key parameters for system operation and often requires further optimization based on its degree of controllability. The benefits of a hybridized energy system rest on efficiently maintaining the synergy between the individual devices. The difficulties of this task arise when component control and efficiency must be balanced by the cost and weight of the system. Ultimately, the specific purpose or application will decide which parameters must be optimized in a defined priority.

6.2 Coupling ES Systems with Other Energy Devices

The current systems used in commercially available low-energy products such as electronics are activated carbon-based double-layer ES devices. As a result, the advanced development and manufacturing of materials focused on enhanced energy storage via increases in capacitance or operating voltage. Both the energy and power densities of modern ES systems can be improved significantly.

In comparison with typical primary energy sources such as batteries, fuel cells (FCs), and internal combustion engines (ICEs), ESs are particularly useful for addressing typical or periodic high pulse power demands made by a system load to an otherwise slow-to-respond energy supply. Coupling ES with another energy device can form a hybrid system. The purpose of coupling is to increase overall energy efficiency of a system and extend its useful life.

For example, a promising near-term alternative is hybrid electric vehicle (HEV) technology that seeks to combine the best characteristics of fuel-driven engines, electric motor drives, and energy storage components to address energy efficiency issues. Energy efficiency, as a critical design factor for these alternative electric power storage systems, serves to benefit from supercapacitor systems to handle peak power demands for acceleration and capture regenerative braking energy.

The primary (fuel cells) and secondary (lithium ion batteries) energy sources used in HEV technology (Figure 6.1) are either inefficient or incapable of serving in this capacity without coupling to supercapacitors [1]. Without ES coupling, the strain on these chemistry-dependent battery–fuel cell systems caused by providing bursts of power over a short time can be costly. Moreover, operating conditions (e.g., temperature and high load demands) can significantly affect reliability. After coupling an ES into a system, a synergistic effect can be achieved in terms of high power demands and energy efficiency.

6.3 Hybrid Systems

Hybrid energy-storage systems (HESS) contain at minimum two dissimilar energy storage systems. The primary objective of their development is an improvement of electrical energy storage. The performance from their mutual integration is expected to be superior to that of either source individually. The ES in such an HESS generally consists of multiple cells in series with each other (stack) or in parallel (bank) to match the operating voltage

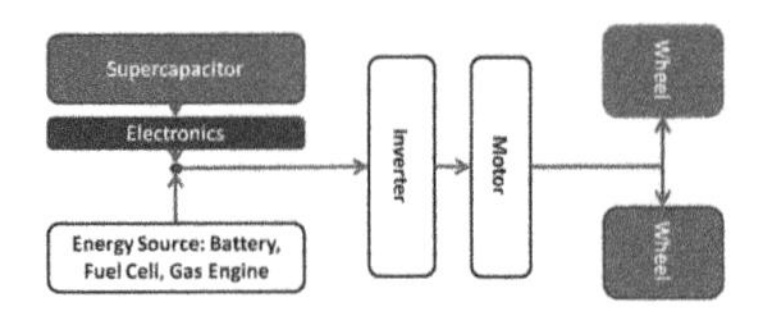

FIGURE 6.1
Schematic diagram describing a general HEV powertrain integrated with a supercapacitor.

range of the circuit and power requirements. Commercial module banks manufactured by companies such as Maxwell Technologies, Nesscap, and Power Systems are thus accessible and often used in evaluations of simulation designs, prototypes, and commercial HESS to assess their performance under various environmental and operating conditions required by an application.

Efforts in hybrid electric vehicle design initially considered the performance requirements of a hybrid power system from a standard set of operating parameters, along with specific requirements related to usable peak power (kW) and energy storage capacity (kWh). After identifying adequate energy storage units that could address the defined needs of an integrated engine, these parameters were used to evaluate the effects of power management and control strategies necessary to ensure operation of the engine and electric drive train at optimal levels over various driving cycles.

The energy required by various hybrid vehicles ranges greatly, depending on type, as shown in Table 6.1. Investigations centered on the practical relevance of ESs in various types of hybrid vehicle systems, but a direct comparison among them is difficult because load and system requirements, control strategies, and hybridization factors often vary. Therefore, a brief review of supercapacitor module power and energy characteristics for use in HESS focuses primarily on the benefits of their integration.

To analyze HESS use in hybrid vehicle designs, laboratory simulations are extensively explored at the Institute of Transportation Studies at the University of California–Davis, where an effective comparison of supercapacitor performance can be shown for several types of HEVs. The three main types are micro, charge-sustaining, and plug-in designs. Employment of the Advanced Vehicle Simulator (ADVISOR) software developed by the National Renewable Energy Laboratory for a single-shift parallel industry hybrid power-train (Honda) model demonstrated that a reasonable comparison for supercapacitor performance in

TABLE 6.1

Energy Storage Requirements for Various Types of Hybrid Electric Vehicles [2]

Type of Hybrid Driveline	System Voltage (V)	Usable Energy Storage	Maximum Pulse Power at 90–95% Efficiency (KW)	Cycle Life (Number of Cycles)	Usable Depth-of-Charge
Plug-in	300–400	6–12 kWh battery; 100–150 Wh Supercapacitor	50–70	2500–3500	Deep 60–80%
Charge sustaining	150–200	100–150 Wh Supercapacitor	25–35	300–500K	Shallow 5–10%
Micro-hybrid	45	30–50 Wh Supercapacitor	5–10	300–500K	Shallow 5–10%

these hybrids is achievable. A sawtooth control strategy implemented in testing was also used in two modes: charge-depleting inherently-efficient electric mode (engine off), and recharging high-power engine mode. Furthermore, regenerative braking was implemented for supercapacitor recharging.

The micro-hybrid electric vehicles are the simplest versions that improve fuel efficiency. Burke et al. [2] used a small electric motor and a module of 18 commercial carbon–carbon double-layer-based ESs. They demonstrated a 40% improvement in fuel economy with the Federal Urban Driving Schedule (FUDS) and an increase in engine efficiency of 30% in comparison to an ICE of 19%.

Charge-sustaining hybrid vehicles that are more reliant on high-power, high-energy-density HESS devices, used an 80 cell module to power a 35 kW electric motor. The result was a slight improvement to FUDS fuel economy (~45%). However, the change is not significant relative to micro-hybrids. Finally, a hybrid plug-in with a power of 70 kW (45 kW provided by the supercapacitor module) yielded a similar improvement to fuel economy through simulation.

Hybrid carbon ESs (Chapter 2) with larger energy densities of 8 to 12 Wh/kg were used in simulation testing for comparison. An unbalanced increase in power density (i.e., less than double) and reduced power efficiency impacted their suitability for micro-hybrids that rely on high-efficiency power. Charge-sustaining and plug-in hybrids appear to be more suitable applications.

The use of ES in HESS has thus far demonstrated dependability for power-reliant vehicle operations. Successful integration of ESs to improve fuel efficiency of hybrid buses and micro-hybrid passenger vehicles also suggests further benefits for their assimilation into hybrid transport applications.

6.4 Supercapacitor Integration with Batteries

Batteries and ESs possess complementary characteristics that lean toward a highly synergistic hybridization of these two energy storage systems. Generally the overlapping of a battery's high energy density with an ES's high power density produces a straightforward benefit over either individual system by taking advantage of each characteristic. The resulting performance is in actual fact highly related to the interconnections and controls implemented in the system to exploit their strengths and avoid their weaknesses. The power flow coordination and control management for improved energy efficiency is critical for any design, and must be weighed for a desired application in terms of computational and economic costs. An overview of solutions proposed for integration is provided for consideration in performance and system design.

6.4.1 ES–Battery Direct Coupling: Passive Control

The direct coupling of a supercapacitor and battery energy source in parallel is shown in Figure 6.2. The advantages provided by this simple and robust integrated system relative to a battery-only system include a capability to elevate peak power, greater efficiency, and extended battery life.

Simulation studies through a directly coupled pairing of a supercapacitor (BCAP0310 P250) and lithium ion (Li-ion) battery (MP 176065) have also shown that under certain load profiles the supercapacitor demonstrated superior performance to the state-of-the-art Li-ion battery. Despite these inherent benefits, a theoretical analysis of a representative passive ES–battery model [3] outlined the limitations arising from their direct coupling.

First, the load and supercapacitor voltages both float based on the battery voltage that is affected by its state of charge, and therefore limit exploitation of the power capability of the supercapacitor. In addition, the requirements of the supercapacitor module or cell voltage are defined by this same issue, in that the upper limit of the module voltage must match that of the battery. As a result, control over the module bank size becomes restricted.

Second, the augmented power provided by the hybrid energy storage system is largely governed by the equivalent series resistance of both coupled energy devices. The fixed partitioning of current supply shared by battery and supercapacitor can thus experience rippling during a pulse demand, particularly in the battery where a magnitude peak is endured at the end. This is a concern for Li-ion batteries that commonly possess intrinsic protection circuits to shut off the battery against such an occurrence.

Third, the terminal voltage of the HESS follows that of the battery rather than being properly regulated; thus the voltage difference between complete charge to discharge of a battery stack can have a significant effect on the power provided to the load. These issues led to the addition of circuit controlling elements to produce indirect coupling topologies [4].

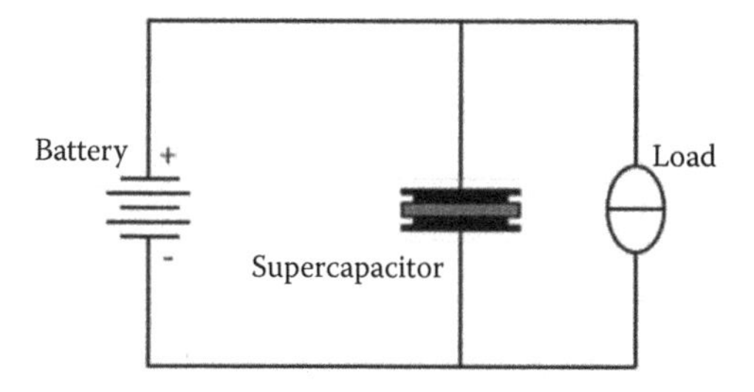

FIGURE 6.2
Passively controlled battery–supercapacitor hybrid system. (*Source:* Gao, L., R. A. Dougal, Member et al. 2005. 20, 236–243. With permission.)

6.4.2 ES–Battery Indirect Coupling: Active Control

Indirect coupling of a supercapacitor and battery via the addition of a DC–DC power converter affords a means of stepping the voltage up or down. This can provide HESS systems with supplementary degrees of freedom for operation and rectify problems and constraints surrounding the passive direct coupling described earlier. These advantages are:

1. The supercapacitor and battery voltages can now differ from one another, providing design flexibility for both types of arrays.
2. The weight of the power source to meet peak power requirements is now readily reduced compared with passive direct coupling.
3. A greater power capacity is realizable while circumventing a battery current that may surpass the safety limit.
4. A constant terminal voltage (with small variation) can be maintained for the HESS.
5. Regulated recharging of the battery can be achieved through a DC–DC converter without a need to introduce a separate charger.

Several topologies exist for implementing a power converter into the HESS circuit in conjunction with the benefits and challenges arising from their use. By reviewing a system that requires the HESS and power converter through a top-down approach, high level interfaces where energy and data flow are exchanged can be defined, and an early estimation of the "intelligence" required by the unit and subsequent control system can be made.

Topologies often vary based on control scheme, switch rating, component count, and circuit complexity. While various topologies have been compared and presented in the literature, the robust buck–boost (half-bridge) DC–DC converter remains the most popular for use in hybrid battery–supercapacitor systems as a result of its simplicity and efficiency.

Utilizing a DC–DC converter with an ES requires that it be bidirectional to permit the governance of both discharge and charge cycles. Depending on the configuration chosen, either the supercapacitor or the battery is connected to the DC bus (load). This is beneficial for controlling the battery current supplied to the DC bus, but the bus voltage will fluctuate according to the state of charge (SOC) of the supercapacitor. Furthermore, several ES cells are required to match the DC bus voltage.

Through introduction of a DC–DC converter, active control can be implemented by a microcontroller. The active control design, control settings, and rules vary, depending on the applications. For the additional topology shown in Figure 6.3, the average smaller current (relative to the peak) demanded by the load is met by the feedback-controlled buck DC–DC converter to discharge the battery at a steady rate, independent of the battery voltage variations that may occur.

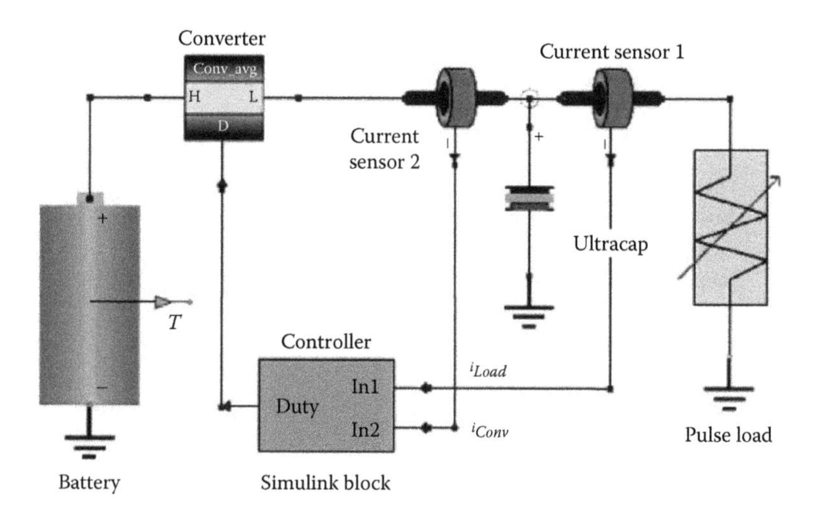

FIGURE 6.3
Representation of actively controlled hybrid battery/supercapacitor system. (*Source:* Gao, L., R. A. Dougal, Member et al. 2005. 20, 236–243. With permission.)

With this method, the safety limit of the battery is not exceeded. A parallel charging of the supercapacitor can occur at this time as well. At high load currents, both the battery and supercapacitor work simultaneously to supply power; however, the controller maintains the steady discharge of the battery while the supercapacitor supplements the remaining high current.

A comparison of passive and active control using this topology was conducted by Gao, Dougal, and Liu [4] with investigation of the power capacity and discharge cycle time of each HESS. Their results (Table 6.2) were gathered from a simulation performed in a virtual test bed environment with pre-validated models of a Sony US 18650 Li-ion battery and a Maxwell PC100 supercapacitor, and further validated with a hardware prototype [4].

This coupling of battery and supercapacitor demonstrated some successful improvements in the power capacity of a passive hybrid that measured 2.3 times that of a standalone battery [4]. A further 4 times increase in the power capacity was realizable by incorporation of a controlled boost DC–DC converter. However, the benefits of implementing active control for enhanced power capacity and reduction of battery loss are offset by the cost. Furthermore, the discharge time could also be reduced (~14.4 min) for a battery to go from complete charge to discharge in comparison to passive control. Therefore, reduced operating time and added energy loss are both attributed to the power loss through operation of the converter, and supercapacitor loss from its operation at high currents. Figure 6.4 describes an evaluation of the energy loss as a function of duty cycle for both passive and active hybrid ESS circuits.

TABLE 6.2

Energy Distribution within Passive and Active Hybridized Battery–Supercapacitor Systems

	Passive Hybrid	Active Hybrid
Discharge cycle time (sec)	8969	8105
Battery delivered energy (kJ)	38.11	38.69
Heat generation in battery (kJ)	3.63	2.07
Battery final temperature (K)	311	305
Energy loss in supercapacitor (kJ)	0.62	1.97
Energy received by load (kJ)	37.49	33.87
Energy loss in converter (kJ)	n/a	2.85
Power source efficiency	89.80%	83.10%

Source: Dougal, R. A., S. Liu, and R. White. 2002. *IEEE Transactions on Power Electronics*, 25, 120–131. With permission.

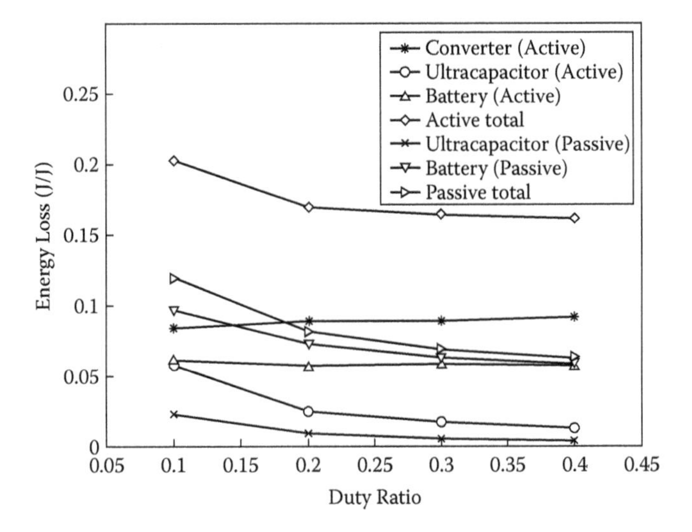

FIGURE 6.4
Energy losses distributed in both passive and active hybrid battery/supercapacitor systems in relation to duty ratio. (*Source:* Gao, L., R. A. Dougal et al. 2005. 20, 236–243. With permission.)

6.4.3 Control Strategies

The objective of a control strategy for supercapacitors implemented in HESS is to source power from the supercapacitor at high load demands, and subsequently allow it to receive pulse power such as regenerative braking for energy recovery. A pulse width modulation (PWM) power converter controlled through application of small signal modeling has been shown to be an effective control strategy in some work. In this approach, a proportional integral (PI) controller tuned with a transfer function for the supercapacitor current to the system duty cycle or bus voltage is used to demonstrate a reliable means of matching the system load power demands [4].

Another approach is to use algorithms derived from heuristics or through a well-trained neural network implemented in the control of the HESS to provide load power. The latter is used by optimizing the set of currents required of the HESS in various driving conditions while maintaining defined border conditions such as the supercapacitor SOC and operation of the battery within safety limits.

Control strategies using buck–boost DC–DC converters are used for a range of diverse topologies typically involving defined-by-system requirements. While systems may focus on a specific choice of power converter and module, further optimization and/or compromise may be required to satisfy system dynamics, cost and weight issues, and other factors. The advantages of an indirect coupling and control strategy include operation of the battery and supercapacitor at independent voltages, improved utilization of supercapacitor power capacity, and control of the battery current. The disadvantages of converter power loss and increasing component count must, however, be taken into account. In general, the improvements to a battery–ES HESS cannot optimize both the energy efficiency and battery life simultaneously; rather one property must be prioritized over the other.

6.5 Supercapacitor Integration with Fuel Cells

The fuel cell (FC) presents a clean power source alternative to current internal combustion engines. Among the many types of FCs characterized by their electrolytes, the polymer electrolyte membrane fuel cell (PEMFC) is lightweight and small, has a reasonably facile membrane fabrication, and shows great promise. However, one key weak point that continues to draw attention is the slow dynamic limitation demonstrated by PEMFCs during experimental use and the negative consequences that can result.

Sizing a PEMFC to match the average current demanded by a system load does not address high current pulses. Atypical or infrequent high currents are thus met by a delayed response of the fuel supply system and a subsequent voltage drop to the region defining limited reactant transport operation (concentration polarization) occurs. Furthermore, the rate at which the voltage drop occurs is increased by a reaction breakdown initiated by an undersupply of reactants. Consequently, ensuing reactions will fail and reactant starvation takes place. Irreversible damage can then arise from the continuing current demand on the PEMFC.

Adding to the slow dynamics inherent to FCs, the hydrogen and oxygen delivery systems (pumps, valves, hydrogen reformer, etc.) endure mechanical stress under high power demands, leading to mechanical failures. To avoid these phenomena and increase the expected lifetime of a PEMFC, control limitations must be set to the current supplied by the system. As a main energy source similar to batteries for high power applications, an FC can provide continual energy to power operations for extended periods but there remains an obvious need for a fast auxiliary power source necessary to improve the overall system performance.

HESS coupling of ES to FC is done to exploit the ability of the former to respond to rapid increases in power demand. The extended lifetime of a supercapacitor is also useful in meeting a PEMFC objective to minimize required maintenance. Once more, various architectures are possible with this hybrid system, with each possessing benefits and detriments that largely depend on the resulting applications.

Figure 6.5 shows the direct passive control coupling of an FC and ES connected in parallel directly to a DC bus or load. The benefits of this architecture are analogous to those described previously for battery coupling. Having fewer components yields a very robust system with reduced cost, weight, and size in contrast to one employing a DC–DC converter. However, several properties make the FC–ES combination largely impractical in applications. Without any active control to govern its operating dynamics, this hybrid design could result in a dynamic DC bus voltage and uncontrolled contribution of the combined FC and ES currents. Normally, typical applications require that power to the load be provided with a constant DC bus voltage, thus requiring one or more DC–DC converters to be included.

Incorporating a single power converter into the Figure 6.5 system can be done in three ways to provide additional degrees of freedom (Figure 6.6). An additional DC–DC converter is necessary to ensure complete control over the two power sources, with one acting to control the DC bus voltage and the other serving to filter the power demand of the system.

Architectures C and D shown in Figure 6.6 make use of one DC–DC converter fixed to a supercapacitor and fuel cell, respectively. An underlying problem encountered by these configurations is the need to oversize the supercapacitor or fuel cell in maintaining DC–bus voltage stability.

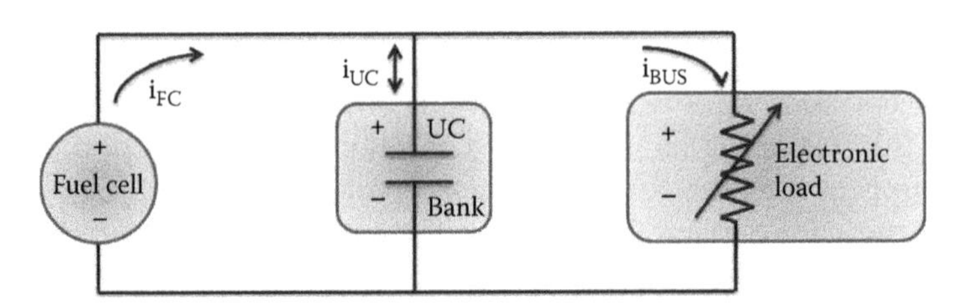

FIGURE 6.5
Passively controlled fuel cell–supercapacitor system.

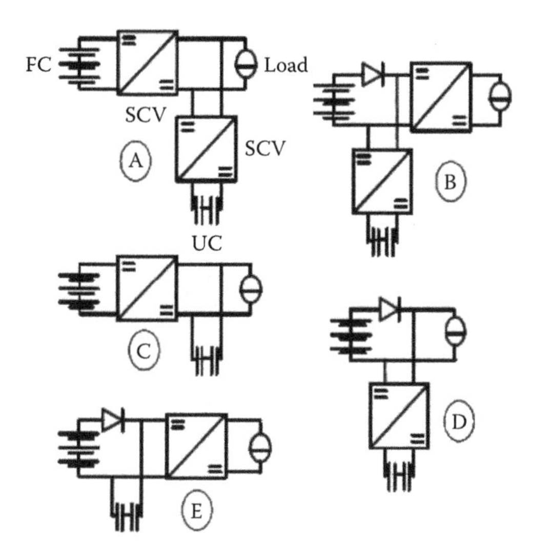

FIGURE 6.6
Various architectures for hybridized fuel cell–supercapacitor systems. (*Source:* Turpin, C. and S. Astier. 2007. *IEEE Transactions on Power Electronics*, 33, 474–479. With permission.)

Configuration E of Figure 6.6 was suggested by Garcia-Arregui, Turpin and Astier [5] to circumvent this problem while providing a reliable system with reduced weight. However a lack of power filtering leaves the energy management to the natural coupling of the FC and supercapacitor. Power profiles of the two power sources in Figure 6.7 appear acceptable. However, the internal resistance of the supercapacitor causes perturbations to the peak power demand on the fuel cell. Through a sizing strategy, the operating voltage criterion imposed upon the fuel cell and supercapacitor was also successful, as seen in Figure 6.8.

However, the implementation of these strategies with power converters brings associated issues of increased weight, economic cost, complexity of the implemented energy management strategy, and chances for component failure.

6.6 System Modeling and Optimization

Theoretical modeling has played an important role in HESS performance validation and optimization. Several methods and models exist to connect electrochemical supercapacitors with primary power sources, as discussed earlier in reference to FCs and battery hybrid systems. In consideration of the vast quantity of topologies, including active and passive design strategies, a

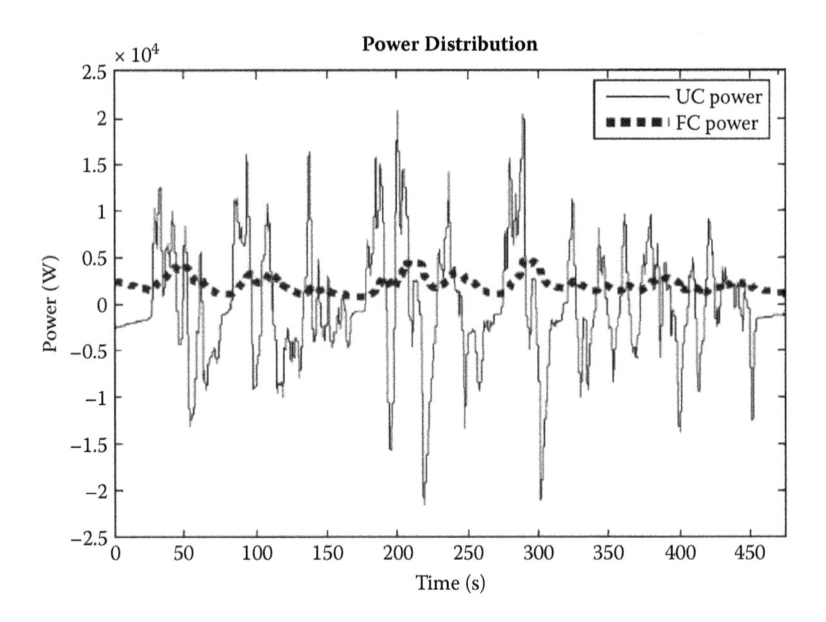

FIGURE 6.7
Power sharing profile of directly connected fuel cell–supercapacitor hybrid system. (*Source:* Turpin, C. and S. Astier. 2007. *IEEE Transactions on Power Electronics*, 33, 474–479. With permission.)

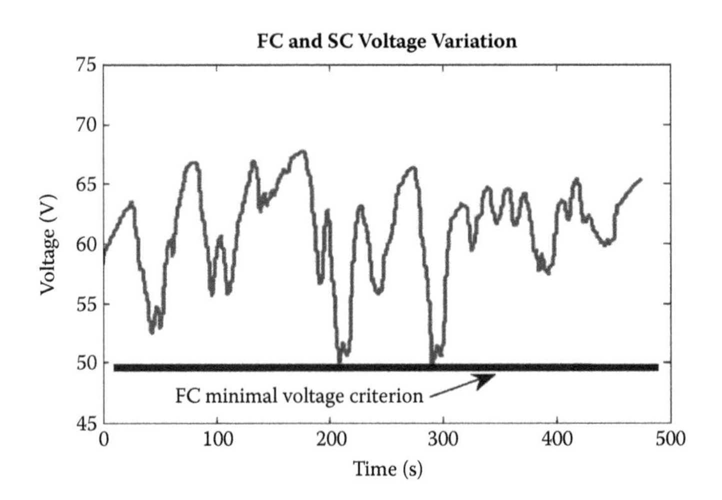

FIGURE 6.8
Directly connected fuel cell–supercapacitor system voltage profile. (*Source:* Turpin, C. and S. Astier. 2007. *IEEE Transactions on Power Electronics*, 33, 474–479. With permission.)

discussion of models pertaining to general design schemes is presented in this section.

The modeling of integrated hybrid energy storage systems for applications often requires an electrochemical model for each component to properly perform accurate simulation and feasibility studies, particularly with the use of FCs. In addition, power system studies benefit from simulation studies; this avoids the high costs and technical operating difficulties associated with prototype testing. Thus, the accuracy gained from model development is critical for a proper assessment of overall system performance.

A number of models describing supercapacitor resistor and capacitor behaviors used to mimic their performances in power systems have been reported and include classical equivalent, ladder circuit, and lumped or distributed parameter electrical and Debye polarization cell models [6]. An established design of a dynamic model of the often-used polymer electrolyte membrane fuel cell (PEMFC) is included in MATLAB® and Simulink software to simulate performance under varying conditions specific to applications.

6.6.1 Supercapacitor Modeling

Predictive operating dynamics of a complex supercapacitor device can be achieved through the implementation of one of several model circuit analogies. Distributed parameter models are often used; the most common is the classical equivalent model described in Figure 6.9. The ladder circuit models represent extended distributions of capacitances and resistances in reference to the classic equivalent and can be expanded to include several resistance and capacitance elements in parallel for the consideration of non-uniform pore charging in highly porous materials.

Designation of a ladder circuit model using L_n (n is an integer greater than 0) determines the model to contain R_n resistors, C_n capacitors, and the RL element in parallel. The primary difference between the classic and ladder circuit models is the means by which these parameters are determined. In general, alternating current (AC) impedance spectroscopy, as discussed in Chapter 7, is very useful in characterizing energy storage devices (i.e.,

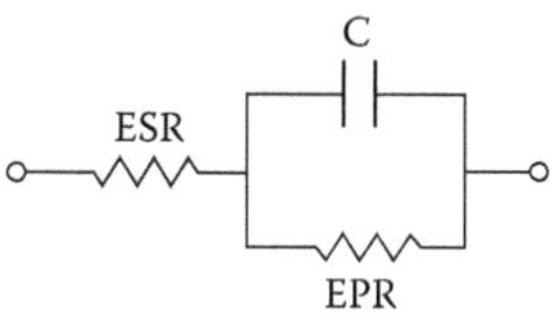

C
ESR
EPR

FIGURE 6.9
Classic equivalent model of supercapacitor circuit. ESR = equivalent serial resistor. EPR = equivalent parallel resistor.

batteries, FCs, etc.) and determining the parameters used in a number of supercapacitor models that will be treated discretely.

A simplified portrait of this technique describes the use of a small amplitude sinusoidal voltage (often <10 mV) applied to produce a resulting sinusoidal current. Measurement of this current permits the calculation of impedance and phase angle through which the double-layer capacitance can then be assessed. Further characterization of capacitance can be achieved through sweep analysis of the capacitance at various voltages and temperatures to gain practical assessment of device performance.

6.6.1.1 Classic and Advanced Equivalent Series Models

The classic equivalent series model parameters, as discussed in Chapters 2, 3, and 7, are relatively simple to determine and thus have less accuracy in predicting device behavior in comparison to more complex circuit models. The parameters of the classic model are obtained through a Nyquist plot (Figure 6.10) of the imaginary and real impedances measured over a range of frequencies (e.g., $\omega = 10^6$ to 10^{-2} Hz), where the double-layer capacitance can be derived, as will be discussed in Chapter 7.

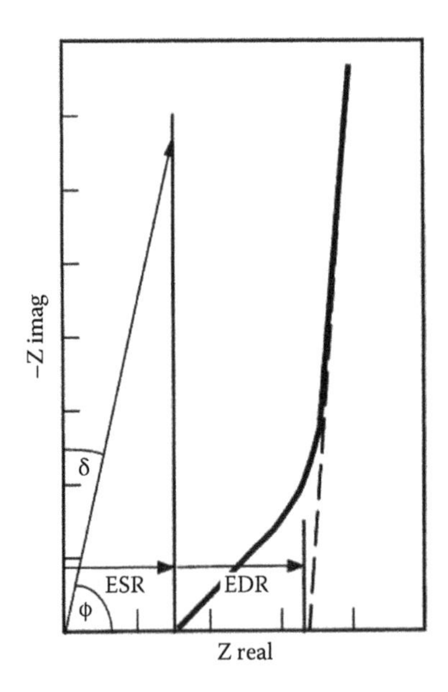

FIGURE 6.10
Nyquist impedance plot comparing ideal vertical impedance of capacitor (thin line) and that of supercapacitor (thick line). Equivalent series resistance (ESR) is derived from the intercept of the real impedance axis followed by the equivalent distributed resistance (EDR) of a porous electrode. (*Source:* Kötz, R. 2000. *Electrochimica Acta*, 45, 2483–2498. With permission.)

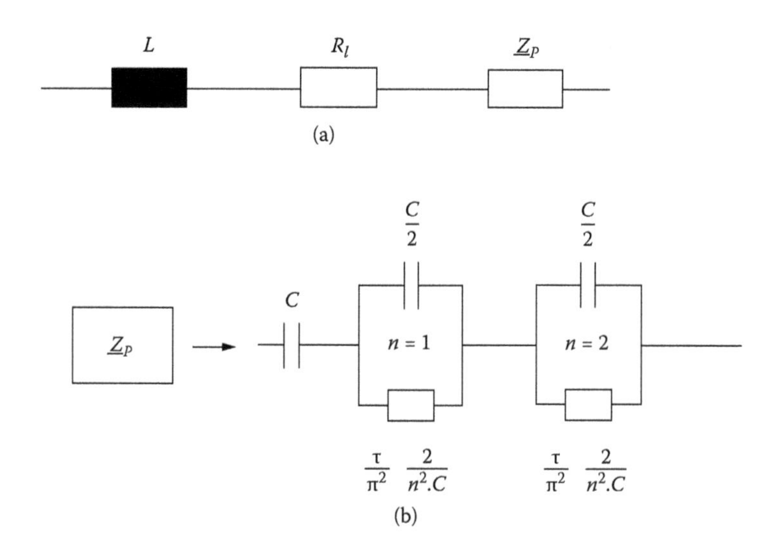

FIGURE 6.11
Advanced equivalent circuits describing supercapacitor. *L* is an inductor, R_i is internal resistance, and Z_p is a complex pore impedance element. (*Source:* Buller, S., Member, E. Karden et al. 2002. *IEEE Transactions on Power Electronics*, 38, 1622–1626. With permission.)

More advanced modeling of the dynamic behavior of ESs employs an advanced equivalent series model (Figure 6.11) using inductor *L*, internal resistance R_i, and complex pore impedance Z_p elements [9–11]. For more detailed modeling process, please see Reference 11.

6.6.1.2 Ladder Circuit Model

Ladder circuits have been used to model double-layer capacitive behavior in pulse load and slow discharge applications and have demonstrated success in modeling nickel–carbon fiber electrodes. Through the use of software employing various statistical techniques, the parameters for several ladder circuits (L_1 to L_n) can be assessed.

An evaluation of ladder circuits performed by Nelms, Cahela, and Tatarchuk [7,12] utilized a nonlinear least squares fitting technique to determine circuit parameters of L_1 to L_5 circuits. They attempted to match the analysis of an ELNA 50 F, 2.5 V double-layer supercapacitor through a comparison of developed models. In their approach, five time constant parameters of R_1-C_1 to R_5-C_5 were determined using software analysis of AC impedance data. The measurement of the leakage current was done by application of a measured current over time to maintain double-layer charging. Model results concluded that ladder circuits of L_3 or greater were necessary to provide accurate fitting to real data.

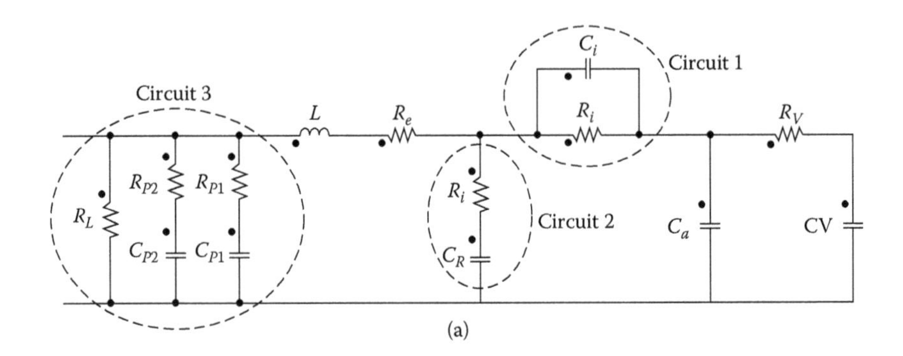

FIGURE 6.12
(a) Electric model of equivalent supercapacitor. (b) Real part of impedance plotted to frequency with 2.5 V bias voltage at 20°C. (c) Imaginary part of impedance plotted to frequency with 2.5 V bias voltage at 20°C. (*Source:* Rafik, F., H. Gualous, R. Gallay et al. 2007. *Journal of Power Sources*, 165, 928–934. With permission.) (continued)

6.6.1.3 Multifactor Electrical Model

A model by Rafik et al. [9] proposed using a limited number of variables to account for the dependencies of capacitance on frequency, voltage, and temperature, avoiding the complex *RC* element determinations necessary for ladder circuits. The model (Figure 6.12a) shows three incorporated circuits, two of which are similar to those described in the previous two model discussions.

Initial analysis of this model reviews the real part of impedance as a function of frequency in Figure 6.12b for a hypothetical series *RLC* circuit in parallel with a leakage resistance. A division of the real impedance plot into four distinct regions provides a means of quantifying separate resistance elements detailed as follows.

Zone I at low frequency range (1 to 10 mHz) incorporates both series and parallel resistances, where the latter is related to separator leakage current, self discharge, and charge redistribution along the pore length. Of the two resistances, the contribution of parallel resistance is more significant; however this could be expected to decrease with extended periods of polarization.

Zone II (10 mHz to 10 Hz) provides quantitative assessment of the series electronic resistance R_e of the conductors and the ionic electrolyte resistance $R_i(T)$. In this range, the equivalent series resistance is composed of $R_{esr} = R_e + R_i(T)$ resistances and varies according to the dependence of R_i on cell temperature. The ionic resistance is more prevalent at low frequencies as a result of better ion penetration into the pores of the electrode material.

Zone III shows a predominantly electronic resistance between the frequencies of 10 Hz to 1 kHz attributed to contact resistances of the electrodes, measurement connections, and electrolyte.

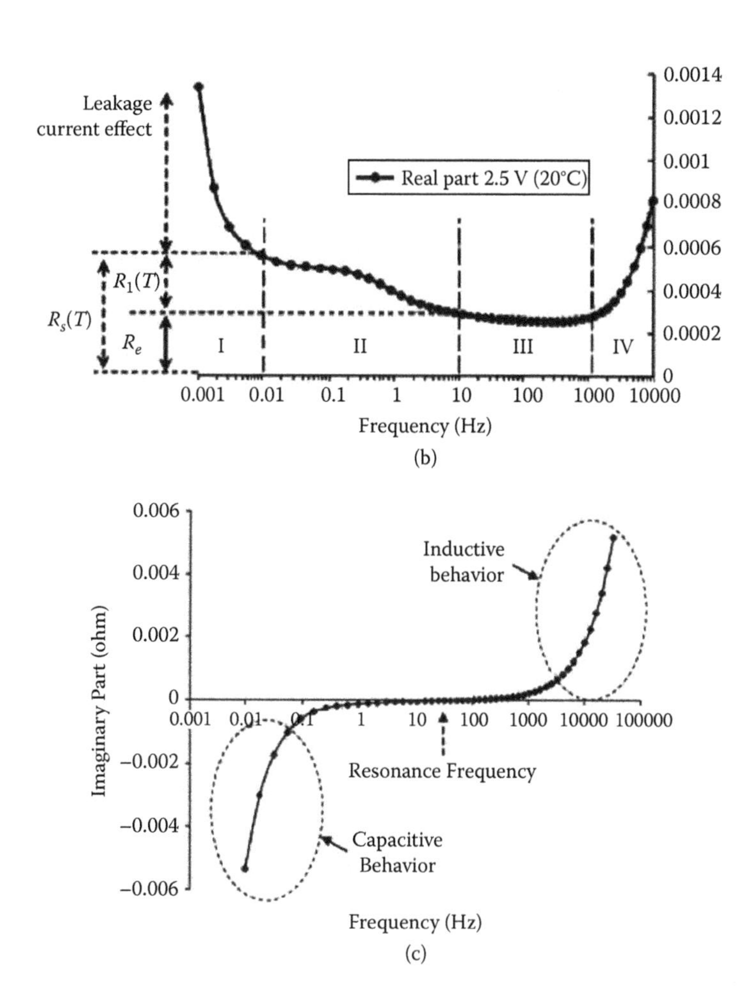

FIGURE 6.12 (CONTINUED)
(a) Electric model of equivalent supercapacitor. (b) Real part of impedance plotted to frequency with 2.5 V bias voltage at 20°C. (c) Imaginary part of impedance plotted to frequency with 2.5 V bias voltage at 20°C. (*Source:* Rafik, F., H. Gualous, R. Gallay et al. 2007. *Journal of Power Sources*, 165, 928–934. With permission.)

Zone IV depicts inductances present between 1 to 10 kHz resulting from the supercapacitor device and contact wires.

A plot of the imaginary impedance with respect to frequency in Figure 6.12c can provide additional information pertinent to the model. First, the inductance of the inductor is determined in the high frequency range. Furthermore, the resonance frequency (see Chapter 1) can be determined from the intercept of the plot where the inductive and capacitive impedances negate each other.

6.6.2 Polymer Electrolyte Membrane Fuel Cell Modeling

Modeling of a PEMFC is presented here to review the important parameters of operational control. The operation of a PEMFC relies on electrochemical oxidation and reduction reactions that take place at the anode and cathode of the cell, respectively. The anodic oxidation reaction catalytically splits hydrogen gas supplied to the anode into protons and electrons. The produced current travels through an external load circuit to provide electrical power, while protons permeate through a proton-conducting polymer membrane. Both products arrive at the cathode in addition to a supplied oxygen gas stream in which a second catalytically driven reduction reaction takes place at the cathode, reducing oxygen and hydrogen protons to water.

The dynamic model of a PEMFC can be realized in MATLAB and Simulink software for implementation in power systems [10]. Beginning with hydrogen flow, the three significant factors are input, output, and reaction flows during operation [11]. The thermodynamic potential of the chemical energy that can be converted into electrical energy is derived from Nernst's law and is dependent on the partial pressures of the reactants and temperature. For reaction kinetic consideration, overpotentials at both anode and cathode essentially constitute the energy required to drive a reaction beyond the state of thermodynamic reversibility.

Activation overpotential, as the preceding notation implicitly states, is the energy necessary to overcome the energy barrier for electron transfer from the electrode to the analyte or vice versa. Similarly, the diffusion overpotential describes the potential difference required to overcome the concentration gradient of the reactants or charge carriers between the bulk solution and the surface of the electrode. This is generated when the electrochemical reaction occurs at a sufficient pace to reduce the concentration of the reactants at the surface. Each of these overpotentials can be described in relation to cell current, voltage, energy, and power parameters used for whole HESS modeling [10].

6.6.3 Power Systems Modeling

Power system designs are required in every application and are achieved by the implementation of a chosen topology. These system integrations often include one or multiple power electronic converters through a parallel or series connection of supercapacitor banks with FCs or batteries as mentioned earlier. Each direct or indirect connection imparts both positive and negative aspects related to the intended use of the power system and more specifically to the robustness of control, system efficiency, and cost. As with design, power system modeling and control can be achieved by a variety of methods. A simplified method is presented here. Approaches that rely heavily on an electronic or computer engineering basis can be further explored in the literature.

The simplest manner of HESS integration is through direct connection. The direct integration of a supercapacitor with primary power sources benefits from a power requirement less than that necessary for a DC–DC converter operation. As an alternative to converter use, a diode can be applied as a replacement to assist in power control while achieving high efficiency. However, the terminal voltage of the FC will then determine the UC bank voltage and therefore limit full utilization of its power capabilities.

For simple design purposes, the total resistance between these two energy systems connected in such a manner determines their respective roles in power sharing. By following this notion, a primary control strategy often used is summarized here.

1. At periods of low power demand (<5 kW), the FC supports the load up to its limit. Any excess power is directed to charge the supercapacitor, while the terminal load voltage requirements determine supercapacitor charging or discharging.
2. At periods of high power demand (>50 kW), the FC supplies its rated power in addition to discharging the supercapacitor to deliver the necessary supplementary power beyond the FC capability.
3. Interrupted power for small time intervals is supplied only by the supercapacitor bank.
4. Design of the UC requires avoiding conditions that permit overcharging or undercharging of the bank.
5. Control of the operating voltage of supercapacitor V_{initial} to $\frac{1}{2}V_{\text{initial}}$ permits ~75% of the stored energy to be used.

Achieving these control strategies requires the implementation of PI controllers, ideal switching elements, and sensors to monitor currents and voltages; all of which can be evaluated via integration into a simulation model.

The load requirements of an application may call for either DC or AC electrical energy. HESSs produce DC energy; back-up power and residential applications operate under AC conditions, and therefore necessitate the inclusion of a DC–AC inverter. In the model presented for a residential power supply using a power conditioning unit (PCU), a simplified inverter is presented. In addition, although a DC–DC converter is included in real-world systems, it is not included in the following model due to its negligible effects on the dynamic response and the complications and numerical instabilities associated with its integration.

6.6.4 Optimization of Models

The integration of a battery–supercapacitor ESS with FCs continues to develop along with topologies that can be used to integrate these systems and their applications. Four common topologies are reviewed along with

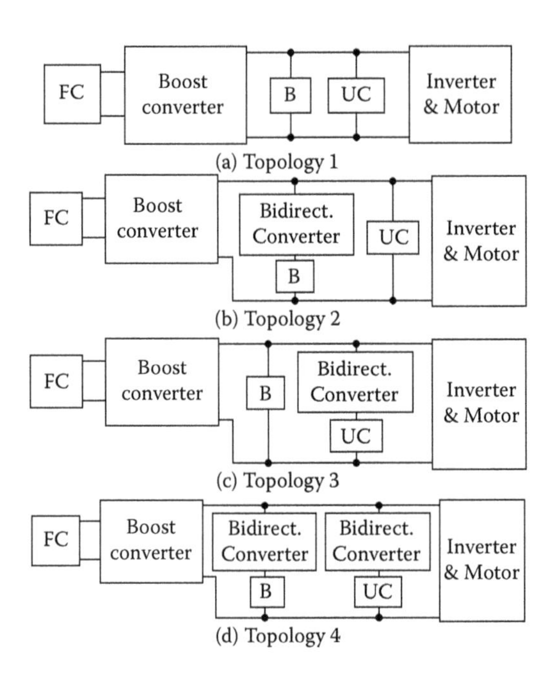

FIGURE 6.13
Various fuel cell, battery, and supercapacitor topology architectures. (*Source:* Bauman, J. and M. Kazerani. 2009. *IEEE Transactions on Power Electronics,* 1, 1438–1488. With permission.)

discussions of their advantages and disadvantages and the optimization of one topology for use in hybrid vehicle power trains.

The topologies presented in Figure 6.13 all utilize FC, battery, and supercapacitor hybridized ESSs to provide power to an inverter and motor. In addition, these systems all use boost converters to control FC power and step up or step down the voltage to match the range of operating voltage of the inverter. A DC–DC boost converter is used over a buck converter to take advantage of the higher efficiency offered by a high voltage–low current inverter in contrast to one that operates at high current and low voltage with greater I^2R losses. This also allows the use of a smaller FC with a lower operating voltage to reduce cost and weight.

Because the battery and supercapacitor are directly connected in parallel in Topology 1, the additional financial, mass, and efficiency costs associated with the use of an additional high power converter are avoided. The disadvantage of this system arises from the combined current flow shared by the battery and supercapacitor in which each system provides a current controlled only by its respective impedance. This defeats the motivation to extend battery life by not allowing the supercapacitor to provide most of the high power demanded by the load on the system.

Topology 2 provides a correction to Topology 1 by including a bidirectional DC–DC converter between the battery and the voltage bus leading to the inverter or motor. The converter enables a means of controlling the power supplied by the battery. Indirect control of the supercapacitor power is subsequently provided as the difference in power required by the inverter and that provided by the battery and FC. Disadvantages to this topology include the additional costs associated with a second converter and the inefficient means of recharging the battery.

High power converters are inefficient at low load operations, but high current recharging of the battery induces I^2R losses related to the internal resistance of the battery. Furthermore, the power to recharge the battery derived from the FC must travel through two DC–DC converters, causing further efficiency losses. Therefore, while control is implemented in Topology 2, it provides an inefficient means of recharging the battery.

Topology 3 is similar in principle to Topology 2 in which the supercapacitor rather than the battery is connected to the high voltage bus line via a bidirectional DC–DC converter. This strategy imparts similar benefits and disadvantages as Topology 2 regarding full power control of each system in exchange for higher costs, additional power losses, and increased system mass. This system presents the case reiterated throughout the chapter regarding a choice between improved control and efficiency. The additional DC–DC converter may be optimized in both Topology 2 and Topology 3, depending on the sizes of the battery and supercapacitor. However, the strength of the former lies in using the battery less often to extend its lifetime, while the latter provides better control of the supercapacitor for its intended continuous use, thus improving the charging efficiency of the battery.

In Topology 4, a bidirectional DC–DC converter is connected between both the battery and supercapacitor to further increase the costs and mass of the system. This topology is generally undesirable in comparison to the previous three, with the additional converter proving no significant advantage.

A new design (Figure 6.14, Topology 5) by Bauman and Kazerani [17] presents a means of overcoming the disadvantages of Topology 2 and Topology 3 with no detrimental consequences. This topology incorporates one high power unidirectional DC–DC converter to boost the FC voltage, while a more

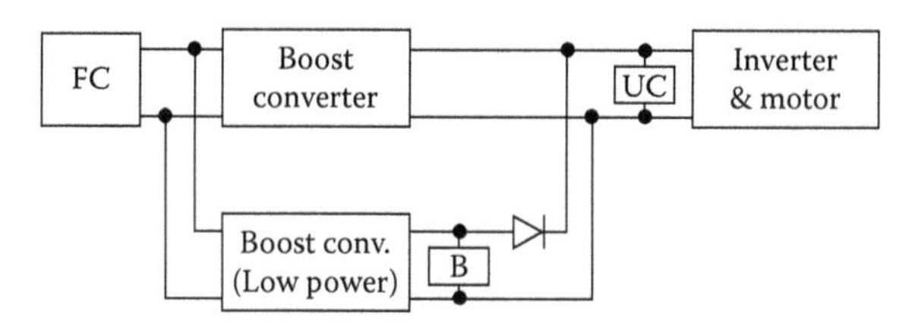

FIGURE 6.14
Topology 5: Improved architecture for fuel cell–battery–supercapacitor hybridization. (*Source:* Bauman, J. and M. Kazerani. 2008. *IEEE Transactions on Power Electronics*, 58, 3186–3197. With permission.)

reliable diode replaces the DC–DC converter implemented in Topology 2 to connect the battery; both lead to the high voltage bus line.

With a supercapacitor directly connected to the high voltage bus, efficient power is provided on demand, discharging to the extent that it matches the battery–diode terminal voltage. Following this event, the battery and supercapacitor share the power demands according to their respective impedances (Topology 1). The location of the supercapacitor also suits its ability to capture regenerative brake energy for recharging well beyond the voltage of the given battery.

This configuration better utilizes the power capabilities of a supercapacitor to efficiently discharge and charge the inverter and motor of the power train, while conserving the use of the battery for infrequent demands beyond the power capacity or voltage limit of the supercapacitor. The charging of the battery takes place through a low power unidirectional boost converter for operation in a high efficiency region and minimizes I^2R power losses within the battery [17].

6.6.5 Control and Optimization of ESS

Of the supercapacitor-battery-fuel cell (ES-B-FC) systems discussed, Topology 5 was proposed as a superior means of enhancing the utilization of each system through efficient power supply and distribution to complement their inherent operational discharging and recharging mechanics. The control strategy, improvements, and optimizations are considered in relation to their application in power train systems.

A unique aspect of Topology 5 is how power sharing is governed for the simultaneous operation of the battery and supercapacitor. The size of either system can be manipulated to affect power performance. A greater capacitance for a supercapacitor will translate to reducing the rate of voltage decline and subsequently extend the point at which the battery begins to provide supplementary power. In an alternative manner, a higher voltage battery begins to supply power earlier than a low voltage battery. Thus, these two design parameters can be used to control the power split between the battery and supercapacitor. Simulations involving various battery and supercapacitor sizes were evaluated to determine optimal performance.

Figure 6.15 is a simplified diagram of the control strategy for Topology 5. The two variables for input control are (1) the fuel cell power command provided to the high power DC–DC converter; and (2) the command from the motor inverter to the inverter. In this manner, the power requested by the load is partitioned into separate requests to the FC and the ESS (battery–supercapacitor). To effectively split the power requirement, a low -pass filter with time constant τ is used as a variable. In this manner the filter evaluates the power request made on the FC. Through its limiters, it blocks any demands below a prescribed low efficiency level, rerouting them to the ESS.

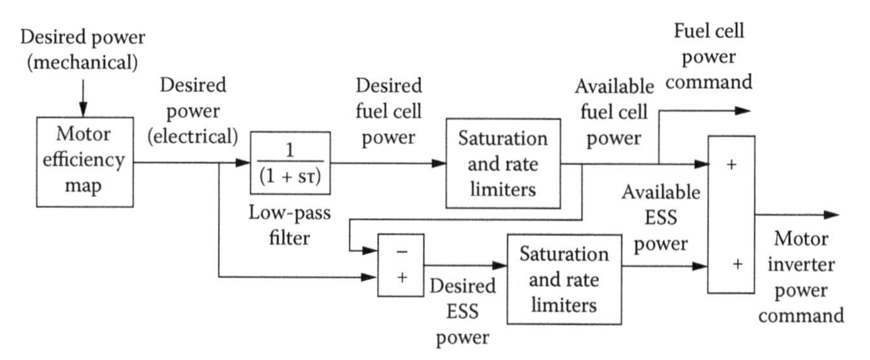

FIGURE 6.15
Block diagram of topology control strategy. (*Source:* Bauman, J. and M. Kazerani. 2008. *IEEE Transactions on Power Electronics*, 1, 1483–1488. With permission.)

When a power request is negative, the power is then accepted by the supercapacitor as regenerative braking energy up to a point where the bank voltage matches the maximum bus voltage, after which a mechanical braking mechanism is applied. Throughout system operations, the battery is charged by the FC at a constant 1 kW up to a 99% state of charge, provided that the battery is not assisting the supercapacitor in delivering power and the FC is operating within a high efficiency region. Analogously, the battery is used to charge the supercapacitor on the condition that it is not providing power to the inverter and motor.

One design limitation of Topology 5 is that only the supercapacitor is able to accept regenerative brake energy. Thus, two design factors can be implemented to accommodate this limitation: (1) the supercapacitor is designed to be large enough to avoid these potential losses, and (2) adding an antiparallel switch across the battery diode to enable battery charging through regenerative braking. The latter option allows a recharging current to travel across the diode, provided that it is within an acceptable range.

When the supercapacitor voltage declines to the point where the battery contributes and the power request from the inverter switches from positive to negative, the antiparallel switch can remain closed to pass regenerative power. At this stage, the low power provided is divided between the supercapacitor and battery. The anticipated increase in charging current is limited to the battery limiting current, at which point the antiparallel switch is opened and the remaining current is accepted by the supercapacitor alone. This option can possibly increase fuel economy if a relatively small supercapacitor is used. However, it requires additional cost, adds mass, and increases control complexity in addition to being less efficient for storing energy due to higher internal resistances. The increased use of a battery also limits its lifetime. Therefore, these options must be weighed by a designer in accordance with applied constraints.

Optimizing an ESS involves minimizing the costs of the system while maximizing its efficiency. To quantify this objective, Kazerani [16] used a function (f_{obj}):

$$f_{obj} = a \cdot x \cdot Eff - b \cdot Cost - y \cdot c \cdot Mass \tag{6.1}$$

where *a*, *b*, and *c* are weight coefficients representative of the importance of improving efficiency and reducing the costs and mass, and *x* and *y* are the parameters used to scale the compared quantities, respectively. Each parameter is related such that $a + b + c = 1$. The *Eff* term representing efficiency is derived by combining the efficiency calculated from the acceleration and regeneration phases of operation. Maximizing the objective function can be achieved by implementation of a nonlinear method using appropriate numerical solution software.

Through the use of the ADVISOR program noted earlier, the overall power requirements of a specific motor and drive cycle for a vehicle can be determined. From implementation of the model in the program, the ESS power component that generally includes the peak power and higher frequency power requests can then be isolated and optimized. Figure 6.16 depicts the electrical circuits describing the ESSs of Topologies 5a and 5b; the latter contains the antiparallel switch. The parasitic resistances of the battery and supercapacitor from which the efficiency calculations are derived are noted as R_b and R_c, respectively.

During operation of the ESS according to Figure 6.16, it is assumed that the brief stages for acceleration and regenerative braking do not significantly vary the internal voltages of the battery cells, and thus the internal battery voltage V_b remains constant. As a result of the internal resistance of the battery, the terminal voltage will vary with the battery current i_b. The voltage

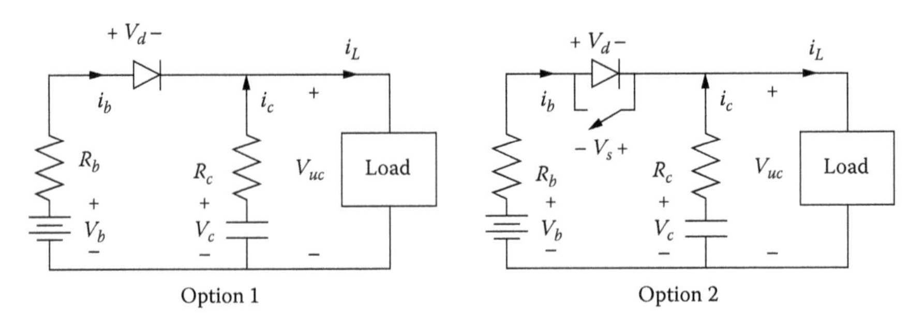

FIGURE 6.16
Electrical circuits for Options 1 and 2 of Topologies 5a and 5b, respectively. (*Source:* Bauman, J. and M. Kazerani. 2009. *IEEE Transactions on Power Electronics*, 58, 3186–3197. With permission.)

of the supercapacitor V_c will vary based on its operation, and likewise the terminal voltage V_{uc} varies with the input or output current i_c.

A general optimization method is presented in relation to the system and component parameters. Because both the battery and supercapacitor are composed of several cells connected in a bank (series and parallel), a per-cell basis is used for each component variable. Modeling each battery cell as a constant voltage with constant internal series impedance and the supercapacitor cell as a capacitor with a constant internal impedance using a linear model is preferred here for initial analysis. Other models can be used to account for more accurate changes in these parameters upon finalizing a design.

6.6.5.1 Sizing and Costs

Sizing a battery and supercapacitor components through series and parallel strings to assess the total number of cells required is done to control the voltage at which the battery begins to contribute (i.e., V_b to V_d, where V_d is the diode voltage drop). This involves simple calculations in which the subscripts s and p denote cells connected serially and in parallel. It is important to note that for a battery, the B_p (battery in parallel) only contributes to the size of the limiting current. Thus, for mass and cost controls, the B_s (battery in series) must be optimized. The following equation describes B_p [16]:

$$B_p = ceil\left(\frac{P}{V_{low} i_{bcell}}\right) \tag{6.2}$$

where the cell function returns the highest rounded integer, V_{low} is the low limit of the load inverter voltage, i_{bcell} is the maximum discharge current per battery cell, and P is maximum ESS positive power. Optimizing the B_s variable is then done using

$$V_b = V_{cellb} B_s \tag{6.3}$$

$$R_b = \frac{R_{cellb} B_s}{B_p} \tag{6.4}$$

where V_{cellb} is nominal battery voltage and R_{cellb} is battery internal impedance at DC. The supercapacitor sizing requires the necessary number of series-connected cells to match the maximum high voltage of the system V_{high} by the equation [16]:

$$C_s = ceil\left(\frac{V_{high}}{V_{cellc}}\right) \tag{6.5}$$

where C_s is the number of supercapacitor cells in series and V_{cellc} is the voltage of each capacitor cell. Optimizing the supercapacitor variable C_p is necessary, as it determines the bank capacitance of the supercapacitor module and its corresponding energy storage capability as well. The internal resistance of the supercapacitor bank R_c and its overall capacitance C [18,19] is obtained from

$$R_c = \frac{R_{cellc}C_s}{C_p} \tag{6.6}$$

$$C = \frac{C_{cell}C_p}{C_s} \tag{6.7}$$

By determining these component variables, the overall mass and costs associated with the system can now be derived simply by [19]

$$Mass = C_pC_s mass_{ccell} + B_pB_s mass_{bcell} + mass_{dperA} i_{bcell} B_p \tag{6.8}$$

The masses per battery and supercapacitor cell are denoted $mass_{bcell}$ and $mass_{ccell}$, respectively, while $mass_{dperA}$ represents the mass of the current-rated diode. A cost equation can be derived by replacing the $mass_i$ component with the respective $cost_i$ for each component i.

6.7 Improving Dynamic Response and Transient Stability

In consideration of the increasing number of studies of hybrid vehicle power train technology, an emphasis can be placed on the need for discussion of the dynamic stability of power systems in general. A rising trend for large power systems that can operate under an idle–stop–start function, including halting an engine at rest and instantly restarting upon demand for acceleration, can improve the energy efficiency of several dynamic applications.

For applications reliant on battery start-up as a primary energy source, this function can be expected to increase the number of engine ignitions approximately tenfold and would quickly decrease the life of a familiar lead–acid battery, disregarding the high power demands for quick acceleration. Furthermore, the rapid inversion of power expected to be recovered can be taxing and inefficient with a typical battery system. ESs can be used to

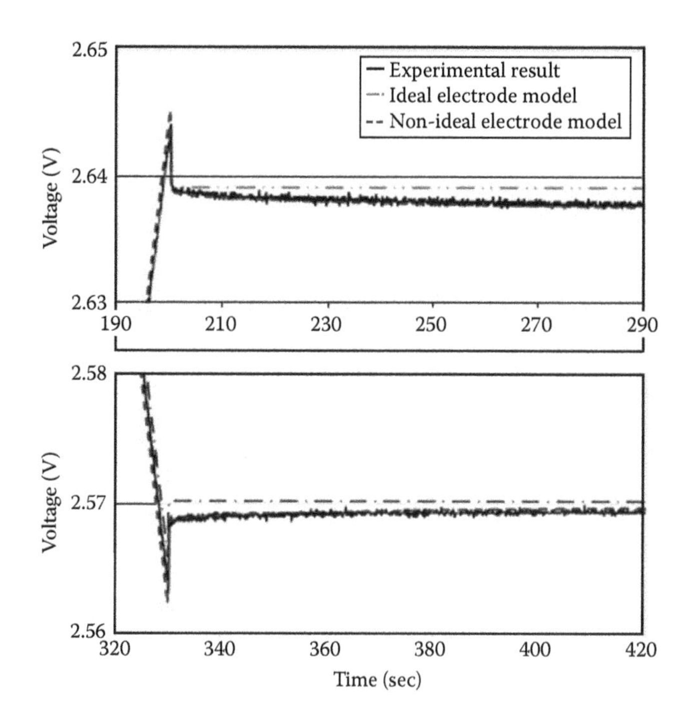

FIGURE 6.17
Simulation results comparing conventional and proposed models to experimental results with parameter variations. (*Source:* Kim, S. H., W. Choi, K. B. Lee et al. 2011. *IEEE Transactions on Power Electronics*, 26, 3377–3385. With permission.)

compensate the power demand, but the reliability of these highly dynamic systems requires models that can offer stable operation during irregular changes in system demands.

Power systems that are reliant on hybrid HESSs are inherently dependent on a good dynamic response to maintain operational stability. Functions implemented to minimize energy losses include intermittent shutdown when not in use, and this is where stability becomes critical for the rapid dynamics involved in these changes. Transient stability issues that arise from unexpected operating failures during start-up or at steady state can be predicted to an extent, and are important to consider during design.

The dynamic simulation models of supercapacitor systems that accurately account for dynamic variations in their parameters and self discharge are useful to improve stable ESS operations. A model designed by S. Choi proposed a constant phase element (CPE) to describe self discharge with nonlinear functions fitted to model parameters. Dispensing with its complex derivation, Figure 6.17 shows a comparison of their design model and a conventional model in predicting experimental results, and demonstrates the accuracy of the model for small variations overlooked by the conventional model [9].

Further expansion of dynamic modeling must consider a hybridized HESS, generally with a DC–DC converter. These systems must accommodate the operating effects on several highly relevant variables. Effects on current alone can derive from several sources such as the voltage current characteristics of each energy source, the power converter, and the filter and control parameters. Models that proposed control generally consider the individual currents and voltages of each energy source as well as the DC bus voltage, load current, and duty cycling of power converters. A division of these factors into state, control input, and external disturbance variables is subject to the model topology and strategy proposed by the designer.

Each topology design as a unique method of integrating ESS devices requires a separate investigation to develop dynamic stability models. The transient stability issues that are easily foreseen can also be addressed during this analysis.

6.8 Summary

Supercapacitor systems and their coupling with batteries and/or fuel cells are discussed in this chapter in terms of experimental tests and theoretical modeling. In particular, modeling these hybrid supercapacitor systems and their hybridization with primary power sources presented here demonstrates the growing trend toward exploiting the strengths of supercapacitors to improve isolated batteries and fuel cell systems. High power capabilities of supercapacitors can improve energy efficiency and lifetimes of costly primary power sources and are expected to improve the long term costs for some hybrid vehicle types. The increasing demand for implementation of ESSs in power systems and other applications arises from reduced cost and will provide greater economic advantages for their inclusion in the future.

References

1. Maher, B. 2009. *Ultracapacitor and the Hybrid Electric Vehicle*. Maxwell Technologies White Paper. [Acessed March 2012] http://lvrimn.com/products/ultracapacitors/docs/200904_whitepaper_ultracapacitorsandhevs.pdf
2. Burke, A., H. Zhao, and E. V. Gelder. 2009. Simulated performance of alternative hybrid electric power trains in vehicles during various driving cycles. *EVS24 International Battery, Hybrid and Fuel Cell Electric Vehicle Symposium*, Norway. *Fuel Cells*, 1–16.

3. Dougal, R. A., S. Liu, and R. White. 2002. Power and life extension of battery–ultracapacitor hybrids. *IEEE Transactions on Power Electronics*, 25, 120–131.
4. Gao, L., R. A. Dougal, et al. 2005. Power enhancement of an actively controlled battery–ultracapacitor hybrid. *IEEE Transactions on Power Electronics*, 20, 236–243.
5. Turpin, C. and S. Astier. 2007. Direct connection between a fuel cell and ultracapacitors. *IEEE Transactions on Power Electronics*, 33, 474–479.
6. Uzunoglu, M. and M. S. Alam. 2006. Dynamic modeling, design, and simulation of a combined PEM fuel cell and ultracapacitor system. *IEEE Transactions on Energy Conversion*, 21, 767–775.
7. Tatarchuk, B. J. 2003. Modeling double-layer ladder circuits. *IEEE Transactions on Power Electronics*, 39, 430–438.
8. Kötz, R. 2000. Principles and applications of electrochemical capacitors, *Electrochimica Acta*, 45, 2483–2498.
9. Kim, S. H., W. Choi, K. B. Lee et al. 2011. Advanced dynamic simulation of supercapacitors considering parameter variation and self-discharge. *IEEE Transactions on Power Electronics*, 26, 3377–3385.
10. Karden, E., S. Buller, and R. W. De Doncker. 2002. A frequency domain approach to dynamical modeling of electrochemical power sources. *Electrochimica Acta*, 47, 2347–2356.
11. Buller, S., E. Karden et al. 2002. Modeling the dynamic behavior of supercapacitors using impedance spectroscopy. *IEEE Transactions on Power Electronics*, 38, 1622–1626.
12. Uzunoglu, M. and M. S. Alam. 2008. Modeling and analysis of an FC/UC hybrid vehicular power system using a novel-wavelet-based load sharing algorithm. *IEEE Transactions on Energy Conversions*, 23, 263–272.
13. Rafik, F., H. Gualous, R. Gallay et al. 2007. Frequency, thermal and voltage supercapacitor characterization and modeling. *Journal of Power Sources*, 165, 928–934.
14. Fontes, G., C. Turpin, S. Astier et al. 2007. Interactions between fuel cells and power converters: Influence of current harmonics on a fuel cell stack. *IEEE Transactions on Power Electronics*, 22, 670–678.
15. Uzunoglu, M. and M. S. Alam. 2007. Dynamic modeling, design and simulation of a PEM fuel cell–ultracapacitor hybrid system for vehicular applications. *Energy Conversion and Management*, 48, 1544–1553.
16. Bauman, J. and M. Kazerani. 2009. Analytical optimization method for improved fuel cell–battery–ultracapacitor power train. *IEEE Transactions on Power Electronics*, 58, 3186–3197.
17. Bauman, J. and M. Kazerani. 2008. Improved power train topology for fuel cell–battery–ultracapacitor vehicles. *IEEE Transactions on Power Electronics*, 1, 1483–1488.
18. Zhai, N. S., Y. Y. Yao, D. L. Zhang et al. 2006. Design and optimization for a supercapacitor. *IEEE Transactions on Power Electronics*, 00, 6–9.
19. Chen, H., T. N. Cong, W. Yang et al. 2009. Progress in electrical energy storage system: a critical review. *Progress in Natural Science*, 19, 291–312.

7

Characterization and Diagnosis Techniques for Electrochemical Supercapacitors

7.1 Introduction

In previous chapters, the structures and components of electrochemical supercapacitors (ESs) and their associated designs, cells, and stacks have been described. However, component characterization, performance testing, and diagnosis are vital steps for optimizing and validating the technology. Because a supercapacitor is a multi-component device, all the components are required to fully play their individual roles and function together synergistically. To investigate the individual function of each component and its synergistic effect, experimental characterization, testing, and diagnosis techniques present the most reliable ways to validate the designs of supercapacitor components and entire devices. Several important electrochemical techniques such as cyclic voltammetry (CV), charging–discharge curves (CDCs), and electrochemical impedance spectroscopy (EIS) using both conventional electrochemical and supercapacitor test cells have been used to characterize, test, and diagnose supercapacitors.

In developing and optimizing new ES materials and components (electrode materials, electrolytes, and current collectors) based on their structures, morphologies, and performance, physical characterization using sophisticated instrument methods serves as the necessary approach. These instrumental methods are scanning electron microscopy (SEM), transmission electron microscopy (TEM), X-ray diffraction (XRD), energy-dispersive X-ray spectroscopy (EDX), X-ray photoelectron spectroscopy (XPS), Raman spectroscopy (RS), Fourier transform infrared spectroscopy (FTIR), and the Brunauer–Emmett–Teller (BET) technique.

This chapter will introduce these techniques with a focus on their application for ES characterizations.

7.2 Electrochemical Cell Design and Fabrication

For supercapacitor characterization, testing, and diagnosis purposes, the electrochemical cell designs are classified as the conventional three-electrode cell and the two-electrode test cell. The former is used for fast screening and characterization of electrode materials and their associated electrode layer structure and optimization. The latter is used to validate supercapacitor performance under real operating conditions.

7.2.1 Conventional Three-Electrode Cell Design and Fabrication

Figure 7.1 shows the design for a conventional three-electrode electrochemical cell. The three electrodes are: (1) the electrode material-coated working electrode made of a carbon material such as glassy carbon or a stable metal such as Au or Pt, (2) the counter electrode (Pt foil or net), and (3) a reference electrode such as a normal hydrogen electrode (NHE), reversible hydrogen electrode (RHE), or saturated calomel electrode (SCE). Note that the NHE uses large surface Pt black as the metal electrode and 1.0 *M* H^+ aqueous solution (such as 0.5 *M* H_2SO_4) as the electrolyte. Its electrode potential is defined as zero at 1.0 atm at any temperature.

The gas inlet and outlet shown in Figure 7.1 are used for gas purging. In particular, to avoid possible oxygen interference from dissolved air during surface CV measurements, N_2 gas is used to deaerate the electrolyte solution for 30 to 60 min [1]. In addition, a thermometer port monitors the temperature of the electrolyte solution. For controlling the temperature, the whole cell is emerged into a thermal bath that allows the temperature of the liquid to be adjusted to the desired level.

To prepare the working electrode layer, the electrode active material such as carbon or pseudocapacitive materials are mixed with conductive carbon and isopropanol for 30 to 60 min to form an ink. A desired amount of this ink (several microliters for a small electrode surface area such as 0.2 to 0.5 cm^2) is pipetted gradually onto a prepolished electrode with a microsyringe. After drying, several microliters of diluted Nafion solution is used to cover the coated electrode to form a uniform electrode layer for electrochemical measurements.

7.2.2 Two-Electrode Test Cell Design and Assembly

Figure 7.2 shows the two-electrode test cell reported by Tsay at al. [1]. It has active electrode surfaces for both positive and negative electrodes. These two electrodes are placed between two metal plates within Teflon plates. The metal plates serve as the current collectors and at the same time as holders to tighten the electrode separator, containing the electrolyte–electrode assembly (ESEA) together. One metal plate is fitted with three screws in a three-point geometry that ensures a better pressure balance than a four-point

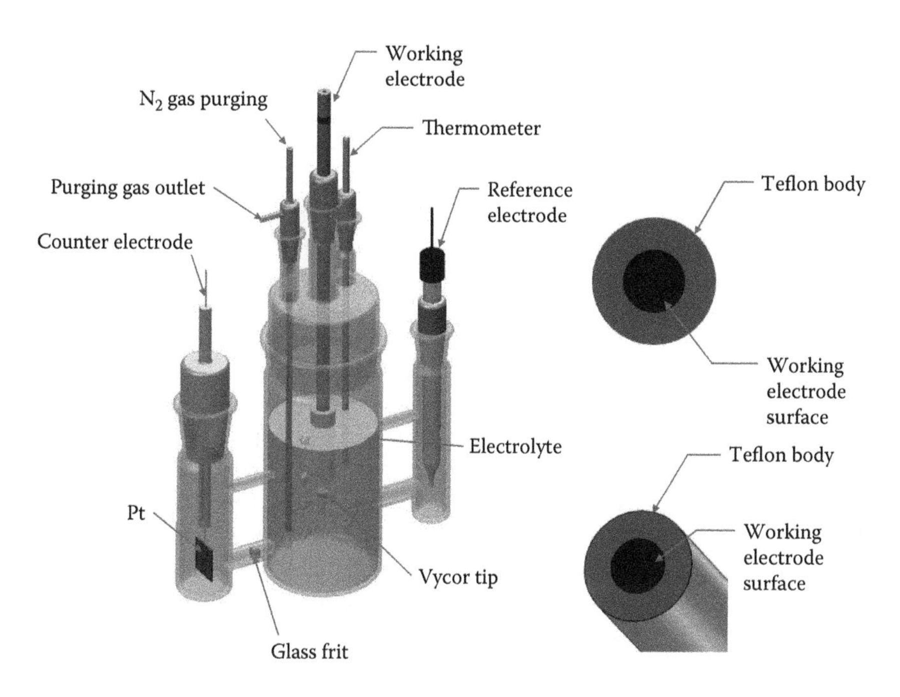

FIGURE 7.1
(See color insert.) Conventional three-electrode cell. (*Source:* Tsay, K. C., L. Zhang, and J. Zhang. 2012. *Electrochimica Acta*, 60, 428–436. With permission.)

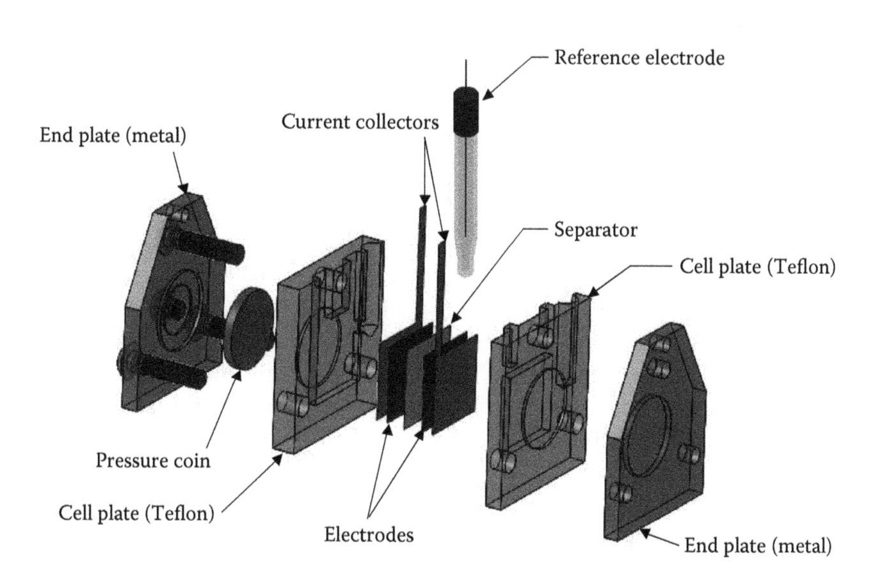

FIGURE 7.2
(See color insert.) Two-electrode test cell for supercapacitor testing. (*Source:* Tsay, K. C., L. Zhang, and J. Zhang. 2012. *Electrochimica Acta*, 60, 428–436. With permission.)

geometry. The pressure from the stainless steel plates can be directly transferred to the ESEA through stainless steel coins placed on the back of each Teflon plate of the internal cell

Before testing, the whole test cell is placed into a beaker filled with electrolyte solution for electrolyte intake, after which the cell is placed in a vacuum oven at 60°C for at least 30 min to remove trapped air inside the cell. Note that some test cells include a reference electrode to measure the individual electrode potentials, similar to the three-electrode measurement discussed above. However, locating this reference electrode inside the thin layer electrolyte (separator) is a challenge.

For the preparation of the electrode sheet, an active material is mixed with conducting carbon powders, formed into a paste using a solvent, and then manipulated by repeatedly folding, pressing, and rolling into a thin electrode sheet. For example, in the work by Tsay et al. [1], an electrode active carbon (BP 2000) and a conducting carbon were mixed for 30 min to form a uniform powder. The powder was then transferred to a beaker containing both PTFE binder and ethanol solution under constant stirring to form a powder suspension. By heating this suspension to remove most of the ethanol, a paste was formed. The paste was then manipulated by repeatedly folding and pressing using a spatula until sufficient mechanical strengthen was achieved. The next step was rolling the paste into a thin electrode sheet of the required thickness with a rolling press. Finally, the electrode sheet was placed into a vacuum oven at 90°C for at least 12 hr. The dry electrode sheet was then cut into two 2×2 cm^2 electrode layer squares that were sandwiched into the two-electrode test cell for examination.

7.2.3 Differences between Three- and Two-Electrode Cell Supercapacitor Characterizations

One primary difference must be understood in characterizing supercapacitors using the three- and two-electrode cells. The information obtained using the former technique comes solely from the target electrode process without interference from the other electrode. With the latter technique, the information obtained is the sum of contributions from both electrodes. Interference from the second electrode must be subtracted to determine the sole contribution of the target electrode. The information obtained from the three-cell technique is ex situ and does not necessarily reflect the real situation. However, it is commonly used for fast screening because it does not require an entire supercapacitor assembly The information obtained from the two-cell method is considered in situ or close to the real operating conditions.

For symmetric supercapacitor systems in which the two electrodes are identical, the information obtained using the two-electrode cell can be easily separated. For example, the obtained capacitance from the whole cell represents only half of the targeted electrode because the entire capacitance can be treated as two identical capacitances connected in series, as described in

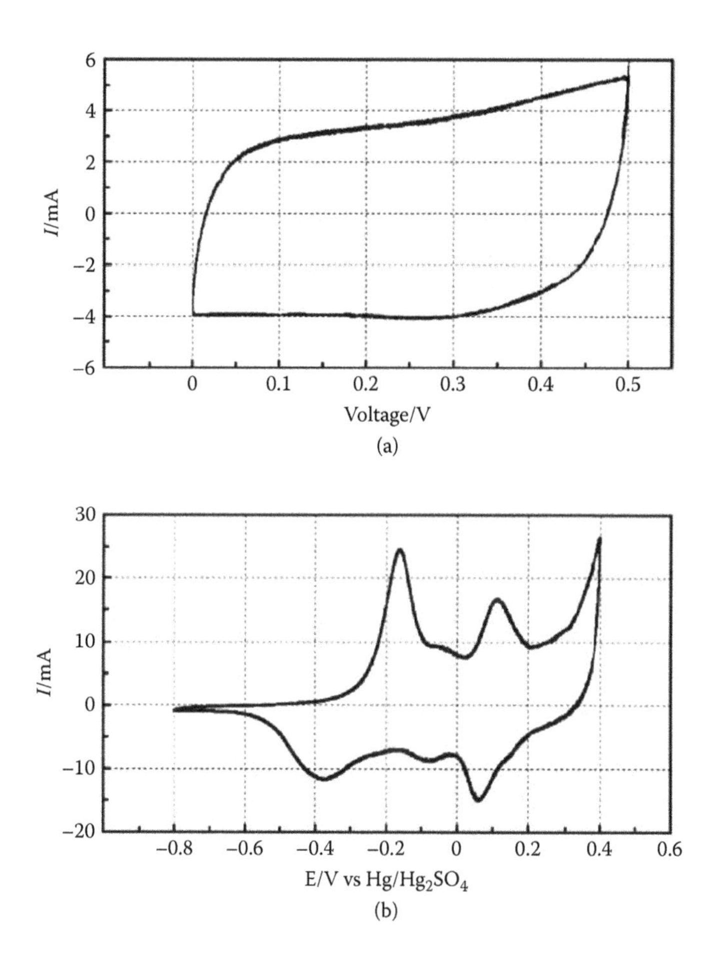

FIGURE 7.3
(a) CV of PANI/MWNT electrodes using three-electrode cell and (b) using two-electrode cell. (*Source:* Khomenko, V., E. Frackowiak, and F. Beguin. 2005. *Electrochimica Acta*, 50, 2497–2506. With permission.)

Chapters 1 and 2. However, for an asymmetric supercapacitor in which the two electrodes are not identical, it is not easy to separate the information from individual cells. In this case, the three-electrode cell may be used to distinguish the individual contributions from the two electrodes.

As an example, consider the pseudocapacitive study by Khomenko et al [2]. The three-cell image (Figure 7.3a) shows that the stable half cell operating voltage range is resolved as 0.2 V (limited by polymer degradation) down to −0.4 V (limited by conducting polymer, charge isolation state). The information gathered is visually very different from the more accurate cell performance data acquired during two-electrode testing (Figure 7.3b). The three-electrode spectrum illustrates two distinct faradic reactions. If capacitance is calculated from the three-electrode configuration, a value of 670 F/g is reported. For a symmetrical cell, the 0.6 V input voltage is split between the two electrodes charging from an internally defined open-circuit voltage (centered on the voltage window).

In the pseudocapacitive case, the separate reactions create a stable region around –0.2 V that likely defines the center voltage. CDC indicates that the positive electrode will have a lower capacitance of 410 $F.g^{-1}$, while the negative electrode will charge closer to 1110 $F.g^{-1}$. By considering the series capacitance as discussed later in this chapter, the full cell capacitance will be approximately 300 $F.g^{-1}$. This value is similar to the measured two-electrode configuration capacitance of 360 $F.g^{-1}$. The PPy composite evaluated in the same work exhibits a higher level of correlation with 196 $F.g^{-1}$ for three-electrode estimation of full cell capacitance and 200 $F.g^{-1}$ for the measured two-electrode configuration [2].

7.3 Cyclic Voltammetry (CV)

Cyclic voltammetry (CV), a widely used potential-dynamic electrochemical technique, can be employed to obtain qualitative and quantitative data about surface and solution electrochemical reactions including electrochemical kinetics, reaction reversibility, reaction mechanisms, electrocatalytical processes, and effects of electrode structures on these parameters. A potentiostat instrument such as the Solatron 1287 is normally used to control the electrode potential. The CV measurement is normally conducted in a three-electrode configuration or electrochemical cell containing a working electrode, counter electrode, and reference electrode, as illustrated in Figure 7.1. However, with alternative configurations, CV measurements can also be performed using a two-electrode test cell. The electrolyte in the three-electrode cell is normally an aqueous or non-aqueous liquid solution.

During CV measurement, the potential of the working or target electrode in the system is measured against the reference electrode via linear scanning back and forth between the specified upper and lower potential limits, as shown in Figure 7.4. The slope of the linear (forward and back) lines is called the potential (E) scan rate (v):

$$\nu = \frac{dE}{dt} \tag{7.1}$$

where E is expressed in V or mV units and v is in $V.s^{-1}$ or $mV.s^{-1}$. In Figure 7.4, the potential–time plot can be expressed as:

$$E = E_i + \nu t \text{ when } 0 \leq t \leq \lambda \tag{7.2}$$

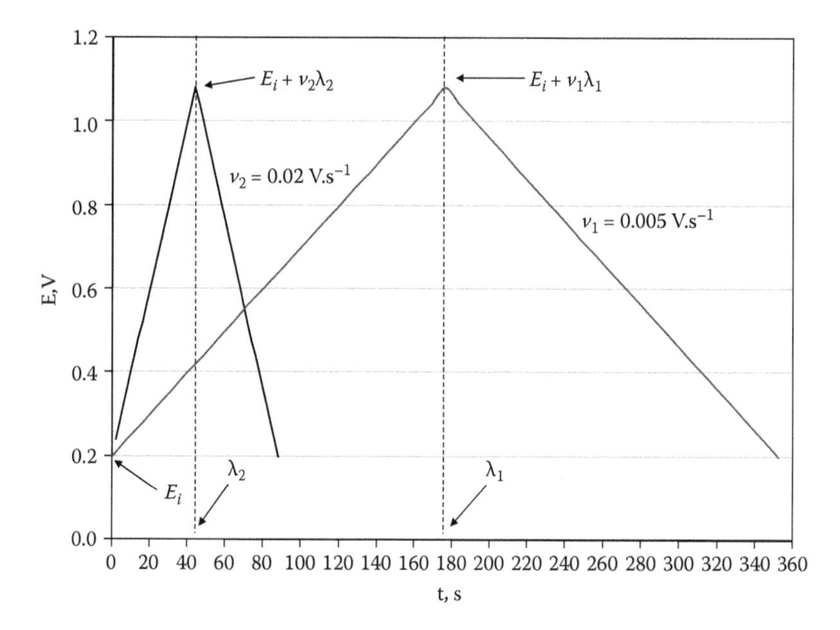

FIGURE 7.4
Potential–time curves in cyclic voltammetry at two potential scan rates, 0.005 and 0.02 V.s^{-1}, respectively.

$$E = E_i + 2\nu\lambda - \nu t \text{ when } \lambda \le t \le 2\lambda \tag{7.3}$$

where E_i is the initial potential and λ is the time at the maximum or up-limiting potential ($E_i + \nu\lambda$), indicating that the entire potential scanning from E_i to $E_i + \nu\lambda$ then back to E_i needs a time of 2λ. During experiments, the values of E_i and λ can be adjusted independently according to the desired potential scan range. Figure 7.4 shows two potential–time curves at two potential scan rates. The slower the potential scan rate, the more time needed to complete a CV cycle.

As discussed later, this potential scan rate can be used to study electrode kinetics. For example, if the potential scan rate is too fast, the electrochemical reactions on the electrode may not be able follow the electrode potential change, which will be reflected on the CVs (current-potential curves) recorded as a function of potential scan rate. From this potential scan rate dependence, the reaction kinetics can be deduced qualitatively and quantitatively.

During scanning of the electrode potential (difference between working electrode and reference electrode), the current passing between the working electrode and the counter electrode can be recorded. The current passing though the working electrode is then plotted as a function of electrode potential to yield a CV with a typical example plot shown in Figure 7.5 [3]. The CV was recorded on a glassy carbon electrode coated with an electrochemically active material (heat-treated Fe-N_x complex supported on carbon particles). The CV was recorded with an electrode material loading of 150 μg.cm^{-2} in

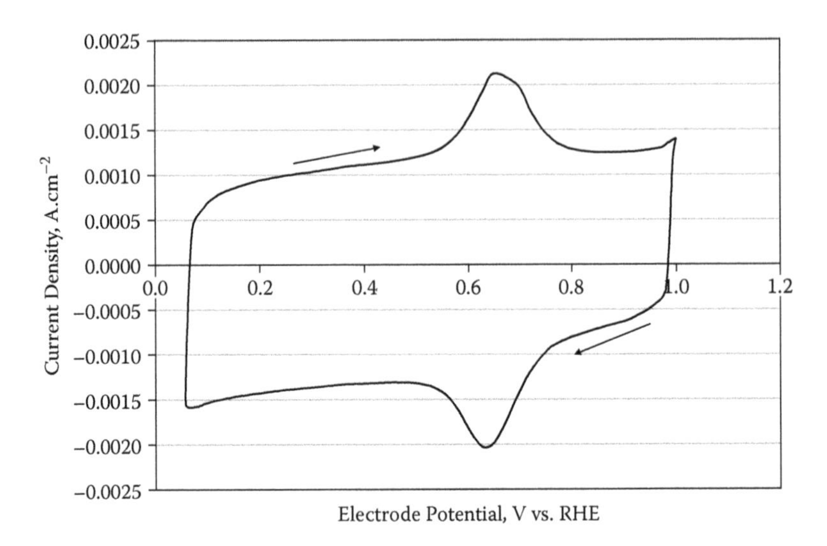

FIGURE 7.5
Cyclic voltammograms of 5 wt% Fe-N_x/C coated on glassy carbon electrode surface, recorded in N_2-purged 0.5 *M* H_2SO_4 solution. Fe-N_x/C loading = 150 μg·cm^{-2}. Potential scan rate = 50 mV/s. (*Source:* Zhang, L. and J. Zhang. 2011. NRC unpublished data. With permission.)

a N_2-purged 0.5 *M* H_2SO_4 solution. Both the background current (double-layer capacitance) and the redox current (pseudocapacitance) of Fe-N_x can be obtained at the same time. The anodic and cathodic peaks at 0.67 V (versus RHE) from the Fe(II)-N_x/Fe(III)-N_x center exhibit almost the same charge quantities.

Note that CV can be used for characterizing electrode surface processes without electron transfer from solution to solid phase and/or from solid to solution phase (Figure 7.5). It can also be used for cases involving electron transfer across an interface. However, because supercapacitors mainly employ the former case, in this chapter we will focus on surface CV. For solution CV and theory, please refer to the relevant literature [4].

7.3.1 Double-Layer Specific Capacitance Characterization Using Three-Electrode Cell

For characterizing pure double-layer supercapacitors, CV has been identified as the most reliable technique. Figure 7.6 shows the CVs recorded from a mesoporous carbon layer using a conventional three-electrode cell [5]. Note that the label on the left *Y*-axis is "specific current," which is the current (A) divided by the weight of carbon particles loaded on the electrode surface (g). The label on the right *Y*-axis is "specific capacitance," which is the specific current (A) divided by the potential scan rate (V.s^{-1}). As discussed in Chapter 2, specific capacitance can be expressed as

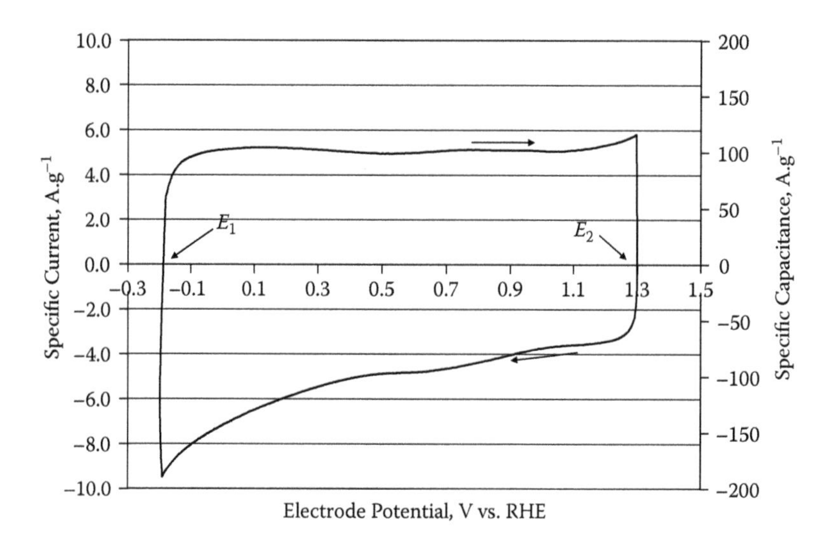

FIGURE 7.6
Cyclic voltammogram of mesoporous carbon coated glassy carbon electrode surface (0.2 cm^2), recorded in N_2-purged 0.5 *M* H_2SO_4 solution at 23°C and ambient pressure. Carbon loading = 100 $\mu g.cm^{-2}$. Potential scan rate = 50 $mV.s^{-1}$. (*Source:* Nicholson, R. S. and I. Shain. 1964. *Analytical Chemistry*, 36, 706–723. With permission.)

$$C_{sp} = \frac{C_m}{m} \tag{7.4}$$

where C_m is the measured capacitance (F) using the electrode layer constructed from the double-layer material and m is the mass of the electrode material (g). There are two ways to obtain C_m based on the measurement such as that shown in Figure 7.6. One way is to express C_m as a function of electrode potential:

$$C_m = \frac{dq(E)}{dE} = i(E)\frac{dt}{dE} \tag{7.5}$$

where $q(E)$ is the double-layer charge at electrode potential E and $i(E)$ is the current charging the double-layer at E. Combining with Equation (7.1), Equation (7.5) can be rewritten as

$$C_m = i(E)\frac{dt}{dE} = \frac{i(E)}{dE/dt} = \frac{i(E)}{\nu} \tag{7.6}$$

Combining Equations (7.6) and (7.4), the specific capacitance of the carbon-based double-layer material at the electrode potential E can be expressed as

$$C_{sp}(E) = \frac{i(E)}{m\nu} \tag{7.7}$$

Using Equation (7.7), the specific capacitance can be obtained at any potential point studied, as shown in Figure 7.6. However, this specific capacitance is taken only at a specified electrode potential. To determine the specific capacitance in the potential range of E_1 to E_2, integration is needed. An approximation method can be expressed as

$$C_{sp} = \frac{1}{m\nu} \sum_{j=1(E_1)}^{n(E_2)} \frac{i(E_i)}{n} \tag{7.8}$$

where n is the number of data points collected in the CV measurement. It can be seen that the larger the n, the more accurate the obtained C_{sp} should be.

The other way is to calculate C_m is using the measured charge quality Q, which is the total charge transferred during the forward or backward direction CV scanning in the electrode potential range of E_1 to E_2. If E_1 is the initial electrode potential, as shown in Figure 7.6, and E_2 is the end potential, C_m can be expressed as

$$C_m = \frac{Q}{|E_2 - E_1|} \tag{7.9}$$

Note that from the CV curve in Figure 7.6 the charge Q can be obtained through integration, as expressed by

$$Q = \int_{t=0(E_1)}^{t(E_2)} i(E)dt \tag{7.10}$$

Practically, this charge quantity can be obtained easily by measuring the areas under the CV trace scan or by using CV software. The specific capacitance can be obtained by

$$C_{sp} = \frac{1}{m|E_2 - E_1|} \int_{t=0(E_1)}^{t(E_2)} i(E)dt \tag{7.11}$$

Note that both Equations (7.8) and (7.11) are obtained from data collected during forward potential scanning. The same calculation can also be carried out using the data from backward potential scanning. Theoretically, the results from both potential scan directions should be the same. For example, the specific charge Q in Figure 7.6 in the positive scan direction can be measured to be 154 $C.g^{-1}$ in a potential range from –0.3 to 1.3 V ($E_2 - E_1 = 1.6$ V). Therefore, the specific capacitance (C_{sp}) of the electrode layer can be calculated to be 96.3 $F.g^{-1}$.

7.3.2 Double-Layer Specific Capacitance Characterization Using Two-Electrode Test Cell

Using a two-electrode cell, CVs can also be recorded by connecting both the reference and counter electrode probes of the potentiostat to one electrode of the test cell, while connecting the working electrode to the other electrode of the cell. In this case, the potential scan should be called the "cell voltage scan" because the voltage change over time is actually the cell voltage change.

The other difference is that the capacitance obtained using CVs recorded by the two-electrode test cell involves contributions from both electrodes, rather than only from one. If the two electrodes are not identical, the better approach is to use the three-electrode cell for measurement to obtain data about the individual electrodes. In the case of a symmetric cell where two electrodes are identical, the capacitance (C_T) measured by a CV can be expressed as the two capacitances connected in series:

$$\frac{1}{C_T} = \frac{1}{C_p} + \frac{1}{C_n} \tag{7.12}$$

where C_T is the total capacitance (F) of the two-electrode cell, C_p is the capacitance of the positive electrode layer, and C_n is the capacitance of the negative electrode layer. For a symmetric supercapacitor, C_n should be equal to C_p. Therefore, the target electrode capacitance (C_m) can be expressed by

$$C_m = C_p = C_n \tag{7.13}$$

Equation (7.12) can then be equated to Equation (7.14):

$$C_m = 2C_T \tag{7.14}$$

In the case of the symmetric two-electrode cell measurement, Equations (7.8) and (7.11) can be alternatively expressed as:

$$C_{sp} = \frac{2}{mv} \sum_{j=1(V_1)}^{n(V_2)} \frac{i(V_i)}{n} \quad (7.15)$$

$$C_{sp} = \frac{2}{m|V_2 - V_1|} \int_{t=0(V_1)}^{t(V_2)} i(V)dt \quad (7.16)$$

where V is the cell voltage, expressed as V_{cell} in Chapter 2. Again both Equations (7.15) and (7.16) are obtained from the data collected from the forward potential scanning and do not include data collected from backward potential scanning.

7.3.3 Potential Scan Rate Effect on Specific Capacitance

As mentioned above, the potential or voltage scan rate can affect measured capacitance. As an example, Figure 7.7a shows the CVs of a symmetrical supercapacitor cell recorded using a two-electrode test cell whose electrodes were composed of BP2000 carbon particles [1]. At low scan rates, the CVs display ideal capacitive behavior (rectangular shape). However, upon increasing the scan rate, this ideal behavior is distorted with a gradual loss in cell specific capacitance.

Figure 7.7b shows the dependency of specific capacitance on voltage scan rate. The observed decrease in specific capacitance with increasing scan rate was explained by the limited transfer of ions to the carbon particle surface, resulting in pore portions of the electrode layer that are inaccessible at high scan rates. This phenomenon is typical for all types of supercapacitors, reflecting the limited mass transfer kinetics within the porous electrode layer. As discussed in Chapter 2, this limitation can contribute to reduced capacitance values.

The magnitude of potential or voltage scan rate can be related to the charging and discharging rates of a supercapacitor. For example, a scan rate of 1.0 $V.s^{-1}$ (1.0 V voltage change per second) means that the supercapacitor can be charged or discharged from an initial cell voltage V_1 to $(V_1 + 1.0)$ V or $(V_1 - 1.0)$ in 1 sec. It is desirable for supercapacitors to have a high charging or discharging rate without capacitance loss. In practice, however, due to sluggish ion transport within the electrode matrix layer with rapid cell voltage or current changes, the apparent capacitance will be reduced. This is especially common in pseudocapacitors due to the slow electrochemical reaction and mass transfer within the electrode layer.

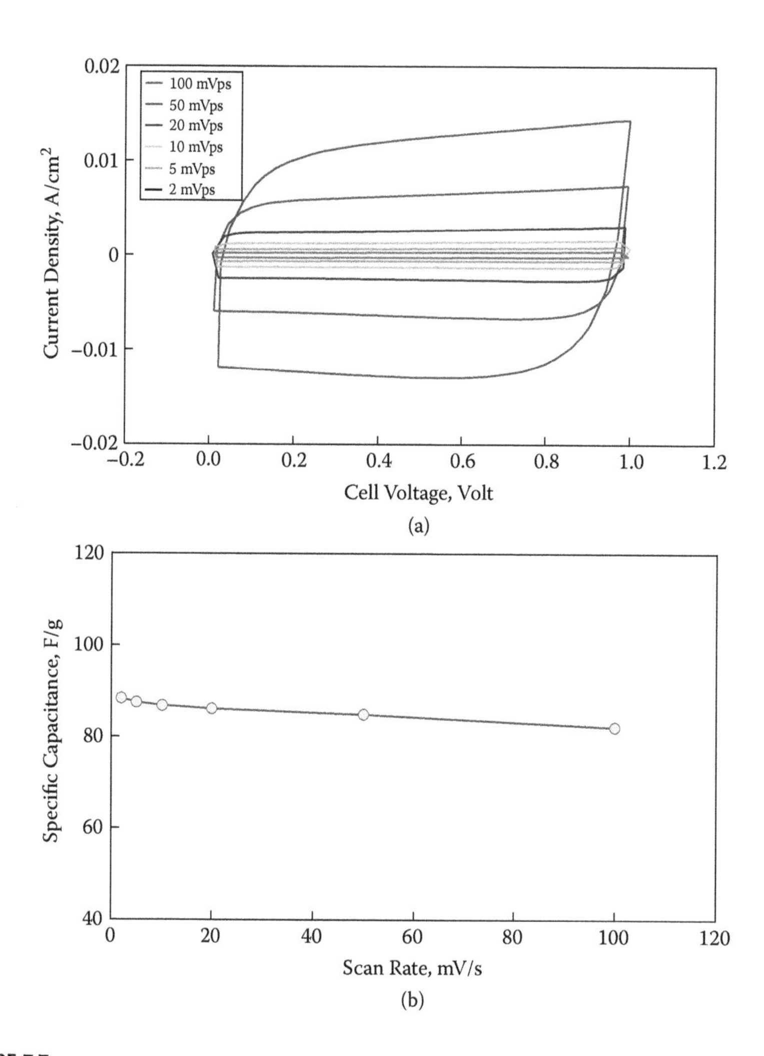

FIGURE 7.7
(a) Cyclic voltammograms recorded at various voltage scan rates using carbon BP2000-based supercapacitor with electrode composition of BP2000:Super C45:PTFE = 80:15:5 (wt%), electrode thickness of 100 μm, and active carbon loading of 3.0 mg.cm^2. (b) Specific capacitance as a function of voltage scan rate. (*Source:* Tsay, K. C., L. Zhang, and J. Zhang. 2012. *Electrochimica Acta*, 60, 428–436. With permission.)

7.3.4 Pseudosupercapacitor Characterization by Cyclic Voltammetry

As discussed in Chapter 3, two parallel processes within the electrode layer of a pseudocapacitor contribute to the overall capacitance. The first is double-layer charging and discharging, and the second is from faradic electrochemical reactions. This kind of device is called a pseudosupercapacitor, because the electrode charging–discharging process involves electrochemical reactions giving rise to pseudocapacitance. For an ideal electrode layer inside a pseudosupercapacitor, the electrode capacitance (C_m) can be treated as the sum of two capacitances connected in parallel:

$$C_m = C_{dl} + C_{pc} \tag{7.17}$$

According to the definition of capacitance and Equation (7.1), Equation (7.17) can be alternatively expressed as

$$C_m = i_{dl}(E)\frac{dt}{dE} + i_{pc}(E)\frac{dt}{dE} = \frac{i_{dl}(E) + i_{pc}(E)}{\nu} \tag{7.18}$$

Equation (7.18) suggests that the momentary current recorded on the active layer-coated electrode of the pseudosupercapacitor is the sum of the double-layer charging or discharging current ($i_{dl}(E)$) and the electrochemical reaction current ($i_{pc}(E)$). Note that both currents are functions of the electrode potential. As shown in Figure 7.5, the background current is the double-layer charging or discharging current, and the redox wave current is the electrochemical reaction current of the Fe(II)/Fe(III) center assigned to Reaction (7.I) [6]:

$$\text{Fe(II)-N}_x\text{/C} \leftrightarrow \text{Fe(III)-N}_x\text{/C} + e^- \tag{7.I}$$

The charge quantity of the background current (Q_{dl}) during the positive potential CV (Figure 7.5) scan direction in the range of ($E_2 - E_1$) can be calculated by measuring the area under the CV trace. Then, for the redox peak observed near 0.67 V (versus RHE), the charge transfer Q_{pc} (measured by subtracting background current) can be determined and the entire or apparent electrode capacitance can be calculated according to

$$C_m = \frac{Q_{dl} + Q_{pc}}{|E_2 - E_1|} \tag{7.19}$$

For example, the Q_{dl} in Figure 7.5 can be measured to be 2.90×10^{-3} C and the Q_{pc} to be 4.40×10^{-4} C, respectively, in the potential scan range $|E_2 - E_1|$ of 0.95 V. According to Equation (7.19), the capacitance (C_m) of the electrode layer can be calculated at 3.44×10^{-3} F. Due to the total weight of the active material (m) of 3.0×10^{-5} g, according to Equation (7.4), the specific capacitance of the active electrode material (C_{sp}) can be calculated as 115 F.g^{-1}. Note that this value reflects only the apparent specific capacitance for the entire electrode layer rather than the individual material's specific capacitance.

Using the CV data in Figure 7.5, the individual specific capacitance of an electrode material can be estimated. For example, the electrode layer in Figure 7.5 is composed of two materials, the carbon support particles and the Fe(II)-N_x species attached to the particles. The specific capacitances for these two materials should be different because they provide double-layer and pseudocapacitance capabilities, respectively. For carbon particles, the double-layer charge is 2.90×10^{-3} C in the potential range of 0.05 to 1.0V, and the weight of carbon particles is 0.0000286 g according to the 95:5% weight percentage ratio of the carbon and Fe active center in the electrode layer. Therefore, its specific capacitance can be calculated to be 107 $F.g^{-1}$

For the Fe(II)-Nx/Fe(III)-Nx redox wave that is electrochemically active in the potential range of ~0.56 to 0.76 V ($E_2 - E_1 = 0.20$ V in this case), according to the definition of pseudocapacitance provided in Chapter 3, the specific capacitance of the electrochemical active material

$$(C_{pc})_{sp} = \frac{Q_{sp}}{m_{sp}\left|E_2 - E_1\right|},$$

where $Q_{sp} = 4.4 \times 10^{-4}$ C, $m_{sp} = 1.40 \times 10{-}6$ g, and $|E_2 - E_1| = 0.2$ V] can be calculated as 1570 $F.g^{-1}$, which is much higher than that of carbon particles (107 $F.g^{-1}$). However, due to both its low quantity in the electrode layer and its limited reaction potential range, its contribution to the apparent specific capacitance is still small.

Note that Equation (7.19) is for measurements using a three-electrode electrochemical cell. For a two-electrode test cell based on Equation (7.14), Equation (7.19) will become Equation (7.20) if the cell is symmetric:

$$C_m = 2\frac{Q_{dl} + Q_{pc}}{\left|V_2 - V_1\right|} \tag{7.20}$$

7.4 Charging–Discharging Curve

Characterization via the charging–discharging curve (CDC) is one of the most reliable approaches to determine capacitance energy density, power density, equivalent series resistance, and cycle life of a supercapacitor. In recording charging–discharging curves, the conventional three-electrode or two-electrode test cells can be employed, depending on availability, similar to the procedure described above for CV characterization. Again, both symmetric and asymmetric supercapacitors can be characterized by this CDC technique. However, for characterizing cycle life of a supercapacitor, two-electrode test cells are favorable because they more closely resemble practical operating conditions.

In normal situations, CDCs are performed by applying a constant cell current, during which the cell voltage is recorded as a function of charging or discharging time. Constant cell voltage mode can also be applied for characterization of supercapacitor properties. In this case, cell current is continuously measured as a function of charging and discharging time, but this technique is seldom used. This section will cover only supercapacitor CDCs using constant current mode.

7.4.1 Capacitance, Maximum Energy and Power Densities, and Equivalent Series Resistance Measurements

As discussed in Chapter 2, immediately after charging starts (charging time $t \geq 0$), the supercapacitor charging voltage (V_{cell}) can be expressed as Equation (7.21) if the two-electrode test cell and the constant current (I_{cell}) charging modes are employed:

$$V_{cell} = I_{cell}R_{esr} + I_{cell}R_p(1-\exp(-\frac{t}{R_pC_T})) \text{ (charging process)} \quad (7.21)$$

where R_{esr} is the equivalent series resistance, and R_p is the equivalent faradic leakage resistance, which is an electrode potential-dependent parameter. In Chapter 2, for simplifying the mathematics, this parameter was treated as a constant. Actually, with an electrochemical leakage reaction, a more complicated equation should be applied (please see our recent work [7]). If there is no parallel leakage reaction, $R_p \to \infty$. Thus, Equation (7.21) can be simplified as

$$V_{cell} = I_{cell}R_{esr} + I_{cell}\frac{t}{C_T} \text{ (charging process)} \quad (7.22)$$

Equation (7.22) suggests that cell voltage has a linear relationship with charging time. After the supercapacitor is charged to a cell voltage of $(V_{cell})_{max}$, a discharging process using a constant current (I_{cell}) can be started immediately. The cell voltage (V_{cell}) can be expressed as

$$V_{cell} = -I_{cell}R_{esr} + (V_{cell})_{max} - I_{cell}\frac{t}{C_T} \text{ (discharging process)} \quad (7.23)$$

Figure 7.8 shows the CDC using a two-electrode test cell in which both electrodes are identical (symmetric cells) [7]. According to Equations (7.22) and (7.23), these data can be simulated to obtain several parameters such as capacitance (C_T), maximum cell voltage ($(V_{cell})_{max}$), and equivalent series resistance (R_{esr}). In the case shown in the figure, the simulated capacitance

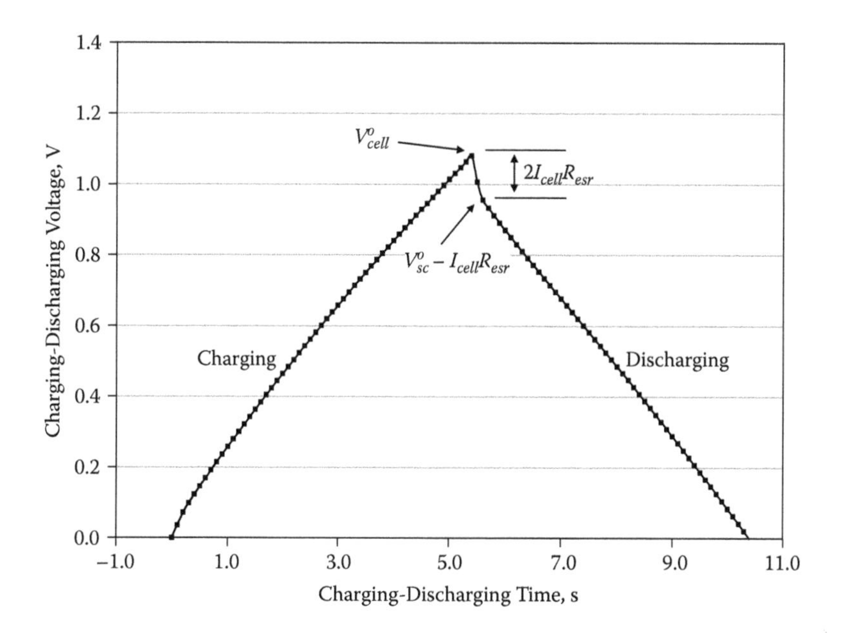

FIGURE 7.8
Charging–discharging curves recorded using symmetric supercapacitor cell with geometric area of 4.0 cm^2 for each electrode in 0.5 *M* Na_2SO_4 aqueous solution at ambient conditions. Charging and discharging current density = 0.025 $A.cm^{-2}$. Total BP 2000 carbon loading of electrode layers = 3.8 $mg.cm^{-2}$. Note that the R_{esr} in the figure should be the sum equivalent series resistances of both positive and negative electrodes. (*Source:* Ban, S. et al. 2012. *Electrochimica Acta*, in press. With permission.)

C_T is 0.127 $F.cm^{-2}$ or 0.509 F for the entire cell with an electrode area of 4 cm^2. The specific capacitance is 34 $F.g^{-1}$ for the carbon loading of 0.015 g, the maximum cell voltage $(V_{cell})_{max}$ is 0.991 V, and the equivalent series resistance R_{esr} = 0.806 $\Omega.cm^2$ or 0.202 Ω for the entire cell. Using these values, the maximum energy and power densities ($(E_m)_{max}$ and $(P_m)_{max}$, respectively) can be calculated according to the equations presented in Chapter 2:

$$(E_m)_{\max} = \frac{1}{2} C_T (V_{cell})^2_{\max},$$

and

$$(P_m)_{\max} = \frac{1}{4m} \frac{(V_{cell})^2_{\max}}{R_{esr}}.$$

The obtained $(E_m)_{max}$ and $(P_m)_{max}$ are 4.6 $Wh.kg^{-1}$, and 8.1×10^4 $W.kg^{-1}$, respectively.

For pseudosupercapacitors, charging–discharging curves can also be used to measure capacitance, maximum energy and power densities, and equivalent series resistance by simulating experimental charging and discharging curves. For more information, see Chapter 3.

Similar to the situation with CV, using different charging and discharging rates (current densities) can provide some information about the mass transfer kinetics of an electrode layer. Tsay et al. [1] demonstrated that the specific capacitance would decrease with increasing charging rate; similar to the result shown in Figure 7.7b. The decrease in specific capacitance with increasing charging rate is due to the limited transfer of ions to the carbon particle surface or pores, leading to inaccessible pore portions of the electrode layer at high charging rates.

7.4.2 Cycle Life Measurement Using Charging–Discharging Curves

The most reliable way to investigate the degradation of a supercapacitor is to charge and discharge it over many cycles (one cycle equals one charge plus one discharge) to observe the changes in both specific capacitance and equivalent series resistance. Normally, supercapacitors have much longer cycle lives than batteries. For example, a commercially available lithium ion battery has a normal life of 400 to 1200 cycles. A supercapacitor can have a life as long as 100,000 cycles. Fortunately, the fast charging and discharging rates of supercapacitors can reduce testing times dramatically. For example, if one charging–discharging cycle takes 10 seconds, 100,000 cycles will take only ~11.6 days.

Normally, with prolonged charging–discharging cycling, the capacitance of a supercapacitor will gradually reduce while the equivalent series resistance will increase, leading to decreases in both the energy and power densities of the device. If the degradation rate is defined as the energy or power density loss per cycle, the degradation rate can be calculated according to the number of cycles and the measured difference between the energy or power density before the cycle life test and after the test.

7.5 Electrochemical Impedance Spectroscopy (EIS)

EIS, also known as AC impedance spectroscopy, is a powerful technique for characterizing the properties of electrode–electrolyte interfaces related to metal corrosion and electrodeposition for batteries, fuel cells, and supercapacitors [5]. In characterizing supercapacitors, both capacitance and equivalent series resistance can be obtained from EIS [8].

EIS includes both ex situ and in situ measurements. The ex situ measurement is mainly used to characterize single-electrode materials and their associated electrode layers. This is helpful for the design, development, and screening of supercapacitor materials and components. In situ measurement is often used for the diagnosis of a single supercapacitor cell or stack under real operating conditions.

7.5.1 Measurement and Instrumentation

The theory and principle of EIS are beyond the scope of this chapter and explanations can be found in the literature [8]. This section focuses on the diagnostic applications of EIS in supercapacitors.

As a powerful diagnostic tool, one important advantage of EIS is the use of very small AC amplitude signals to analyze electrical characteristics without significantly disturbing the properties of the system measured. During EIS measurement, a small AC amplitude signal is applied to a supercapacitor cell over a frequency range from 0.001 to 3,600,000 Hz. Either voltage control (potentiostatic) or current control (galvanostatic) mode can be used. In the voltage control mode, an AC voltage single (usually 5 to 50mV) is applied to disturb the supercapacitor and the current response is measured to obtain the impedance of the system. In this mode, the frequency response analyzer (FRA, Solartron 1260A) and potentiostat (Solartron SI 1287) are employed. Similarly, in current control mode, an AC current single (usually 5 to 50 mA) is applied to disturb the electrochemical system, and the voltage response is measured to determine impedance.

Normally the EIS results measured by both control modes are consistent and show no significant differences. The impedance responses recorded by the EIS instrument are normally shown as Nyquist plots that illustrate the relationship between imaginary resistance or impedance and real resistance or impedance. For example, Figure 7.9 shows the Nyquist plots (dotted points) recorded from a two-electrode symmetric supercapacitor [9]. In a later section we will give a detailed analysis of these plots.

The three fundamental requirements of EIS measurement are: (1) the linearity between the perturbation signal and the system response, (2) the stability of the target system during measurement, (3) and the causality of the response. In particular, the perturbation must not cause the system to shift from its equilibrium state. To meet this requirement, it is better to measure the EIS of a supercapacitor at its open potential voltage (OCV) rather than under load. However, measurement at OCV does not necessarily reflect the situation under load conditions. Due to this concern, EIS results may not be as reliable as those obtained by CV and CDC.

7.5.2 Equivalent Circuits

Analysis of EIS data can be done by modeling or fitting of impedance spectra with an equivalent circuit (EC). This requires the construction of a physically meaningful EC containing several elements required by the studied

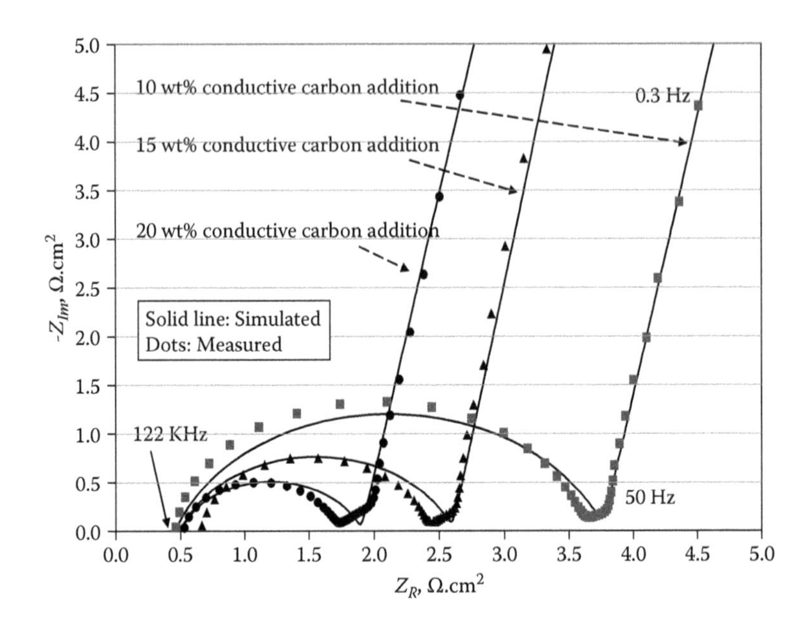

FIGURE 7.9
Nyquist plots recorded using two-electrode symmetric supercapacitor in 0.5 *M* Na_2SO_4 aqueous solution. The electrode layers with active surface area of 4.0 cm^2 are composed of BP2000 carbon particles as active material and stainless steel as current collector material. AC frequency range = 0.3Hz to 122KHz. (*Source:* Zhang, L. and J. Zhang. 2011. NRC unpublished data. With permission.)

system. Depending on the shape of the EIS spectrum, the EC model is usually composed of resistors (*R*), conductors (*L*), and capacitors (C) connected in series or in parallel. After an EC is designed, it can be used to fit the EIS spectra with a software program called Z-view. The quality of the fitting can be judged by how well the fitting curve overlaps the original spectrum at the same frequencies.

ECs were constructed for testing double-layer and pseudocapacitors discussed in Chapters 2 and 3, respectively. For a symmetric supercapacitor containing both double-layer and pseudocapacitances, an EC can be constructed (Figure 7.10a) in which R_{esr} is the equivalent series resistance, C_{dl} is the double-layer capacitance, R_{ct} is the charge transfer resistance of the electrochemical reaction producing pseudocapacitance, C_F is the pseudocapacitance, and R_p is the parallel resistance of the leakage reaction. Note that the pseudocapacitance is parallel to the double-layer capacitance.

In this way the entire capacitance of the electrode layer should be the sum of these two capacitances. If there is no parallel leakage reaction or its reaction kinetics are fairly slow, $R_p \to \infty$ and the EC in Figure 7.10a can be simplified as shown in Figure 7.10b. If the pseudocapacitance-generating electrochemical reaction kinetic activity is fairly slow, $R_{ct} \to \infty$ and $C_F \to 0$ and the EC in Figure 7.10a will be reduced to Figure 7.10c Moreover, if there are no parallel

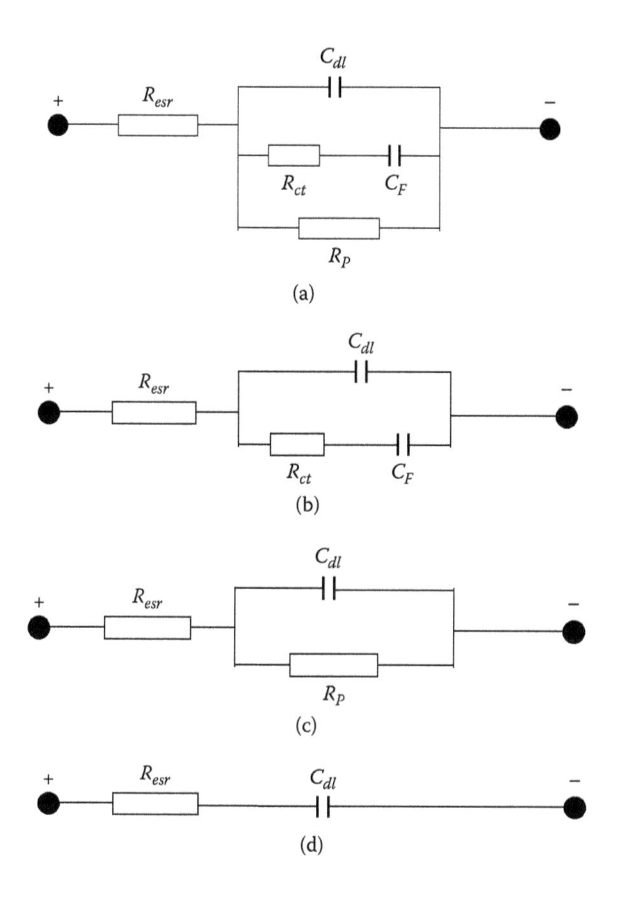

FIGURE 7.10
ECs for supercapacitor containing both double-layer and pseudocapacitances. (a) Complete model. (b) Model without parallel leaking reaction. (c) Model without pseudocapacitance generating reaction. (d) Model without both parallel leaking reaction and pseudocapacitance generating reaction.

leakage and pseudocapacitance generating reactions, the EC can be further reduced to Figure 7.10d.

For the EC shown in Figure 7.10a, the total AC impedance (Z) can be expressed as the sum of real (Z_R) and imaginary (Z_{Im}) components:

$$Z = Z_R + Z_{\text{Im}} = R_{esr} + \frac{R_p(a+(2\pi f)bC_F R_{ct})}{a^2+b^2} - j\frac{R_p(b-(2\pi f)aC_F R_{ct})}{a^2+b^2} \tag{7.24}$$

where $a = 1 - (2\pi f)^2 C_{dl} C_F R_{esr} R_{ct} R_p$ and $b = (2\pi f)(C_{dl}R_{ct} + C_F R_p + C_F R_{ct})$. This equation shows that both the real and imaginary components are functions of AC frequency (f). The magnitude of total impedance can be expressed as:

$$|Z| = \sqrt{\left(R_{esr} + \frac{R_p(a+(2\pi f)bC_F R_{ct})}{a^2+b^2}\right)^2 + \left(\frac{R_p(b-(2\pi f)aC_F R_{ct})}{a^2+b^2}\right)^2} \quad (7.25)$$

And the phase angle (θ) between the real and imaginary components can be expressed as

$$\theta = \tan^{-1}\frac{Z_{Im}}{Z_R} = \tan^{-1}\frac{(b-(2\pi f)aC_F R_{ct})}{(a+(2\pi f)bC_F R_{ct})} \quad (7.26)$$

If log(Z) and $log(\theta)$ are plotted as function of $log\ (2\pi f)$, the obtained curves are called the Bode magnitude plot and the Bode phase plot, respectively. More popularly, the EIS is expressed as a Nyquist plot—Z_{Im} versus Z_R, as shown in Figure 7.9. In general, Nyquist plots provide more visible and characterizing information about electrode processes than Bode plots. Therefore, we will focus our discussion on Nyquist plots.

Equation (7.24) indicates that both the real and imaginary components can be affected by the electrode parameters such as R_{esr}, C_{dl}, R_{ct}, C_F, and R_p. In theory, these five parameters can be simulated based on both the experimental impedance and proposed ECs similar to those shown in Figure 7.10. However, with too many parameters, a simulation will become difficult and the simulated values may become arbitrary. The effects of the magnitudes of these parameters can be seen in Figure 7.11.

Normally, the most reliable parameter an EIS can determine is $R_{esr}$when the AC frequency goes very high (>10 KHz). This can be seen from Equation (7.24), its associated plot in Figure 7.11a, and the ECs in Figure 7.10. For example, when $f \to \infty$, $Z_{Im} \to 0$, and $Z_R \to R_{esr}$, the Nyquist plot intercept on the Z_R-axis at the high frequency end in Figure 7.11a will be the value of R_{esr}. Figure 7.11a shows that the value of R_{esr} shifts from low to high Z_R values with an increase in the value of R_{esr}, while the shape of the plot is not changed as predicted from Equation (7.24).

Figure 7.11b shows the effect of double-layer capacitance (C_{dl}) on the Nyquist plot, demonstrating similar effects that pseudocapacitance (C_F) has on C_{dl}. For the effect of the parallel resistance of the leakage reaction (R_p), Figure 7.12 shows the example curves at different R_p values. It can be seen that with increasing R_p values, the R_p related semi-arc becomes larger, and when $R_p \to \infty$, as in the case without parallel leakage reactions, a vertical line will be obtained at the low end of the frequency. This can also be predicted from both Equation (7.24) and the EC in Figure 7.10b. In the case without parallel leakage reaction, the impedance corresponding to the EC in Figure 7.10b can be expressed as

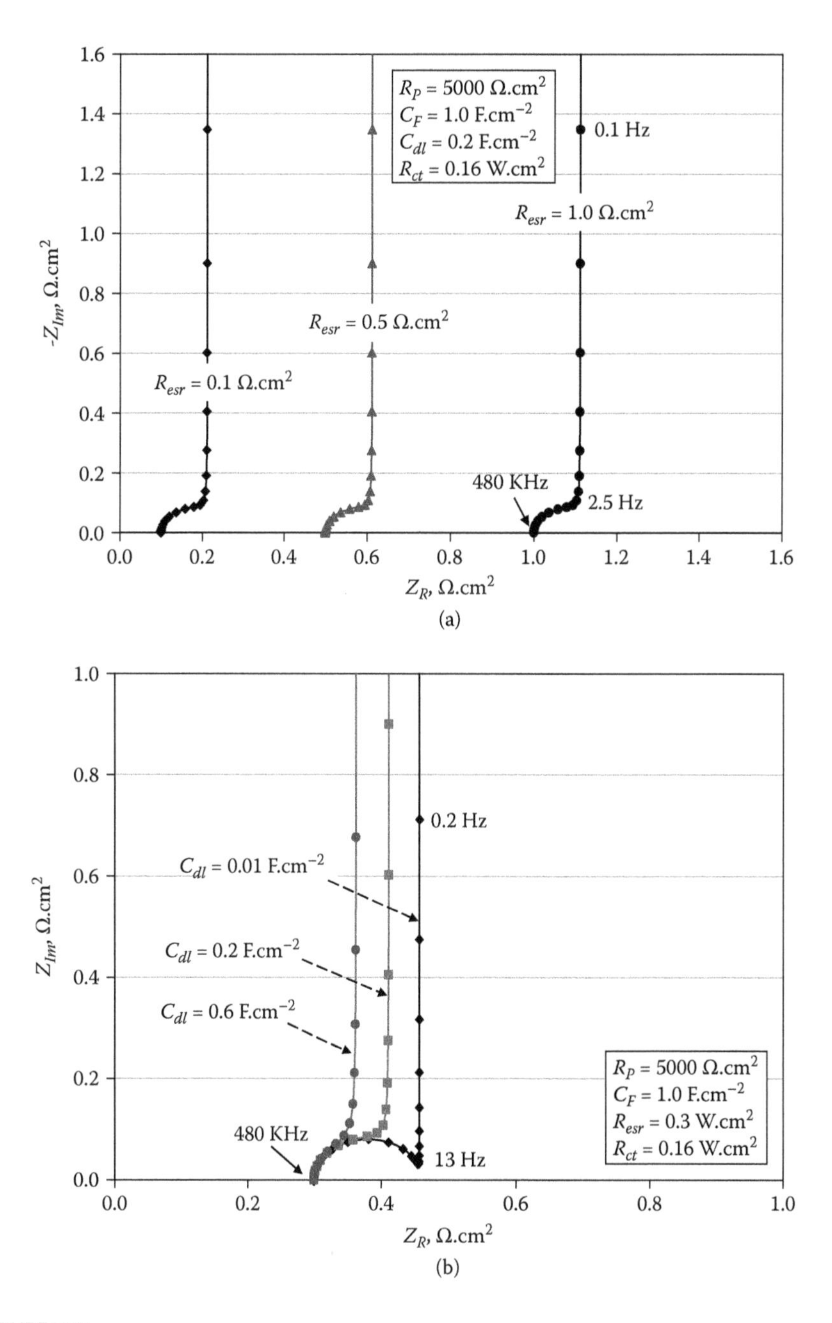

FIGURE 7.11
Calculated Nyquist plots according to Equation (7.28) to show effects of equivalent resistance (R_{esr}) (a) and double-layer capacitance (C_{dl}). The magnitudes of other parameters used to calculate these plots are also shown.

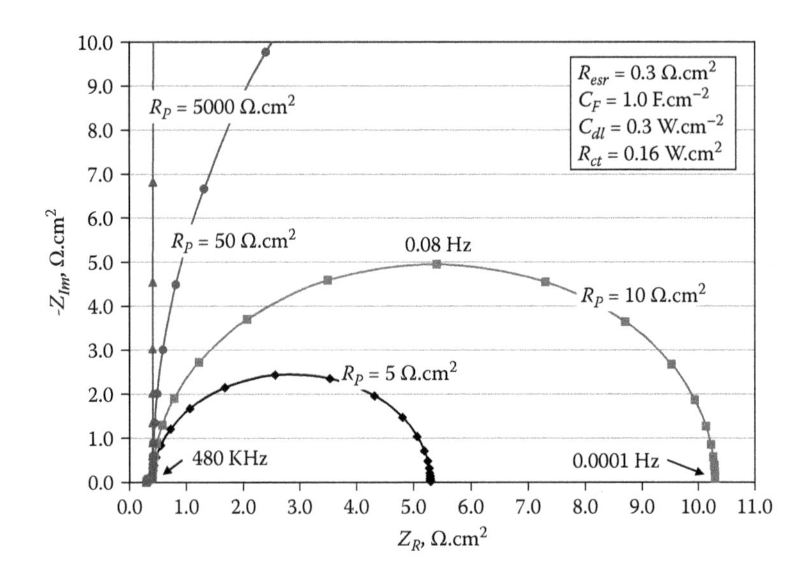

FIGURE 7.12
Calculated Nyquist plots according to Equation (7.28) to show effects of parallel leakage reaction resistance (R_p). The magnitudes of other parameters are also shown.

$$Z = R_{esr} + \frac{(2\pi f)^3 C_{dl} C_F^2 R_{Ct}^2}{(2\pi f)^4 (C_{dl} C_F R_{ct})^2 + (2\pi f)^2 (C_{dl} + C_F)^2} - j\frac{(2\pi f)^3 C_{dl} C_F^2 R_{ct}^2 + (2\pi f)(C_{dl} + C_F)}{(2\pi f)^4 (C_{dl} C_F R_{ct})^2 + (2\pi f)^2 (C_{dl} + C_F)^2} \tag{7.27}$$

In the case without pseudocapacitance generating reactions, the EC in Figure 7.10b can be reduced to the form of Figure 7.10c, and the corresponding impedance can be expressed as

$$Z = R_{esr} + \frac{R_p}{1 + (2\pi f)^2 R_p^2 C_{dl}^2} - j\frac{(2\pi f) R_p^2 C_{dl}}{1 + (2\pi f)^2 R_p^2 C_{dl}^2} \tag{7.28}$$

In the case without both parallel leakage and pseudocapacitance generating reactions, the EC can be further simplified into the form of Figure 7.10d, and the corresponding impedance can be expressed as

$$Z = R_{esr} - j\frac{1}{(2\pi f) C_{dl}} \tag{7.29}$$

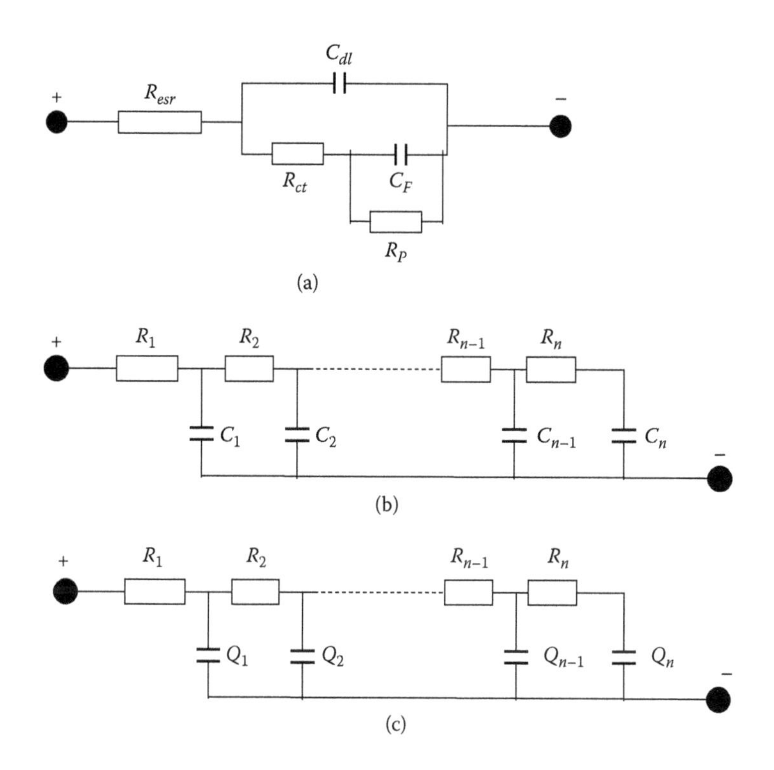

FIGURE 7.13
Equivalent circuits for supercapacitor containing both double-layer and pseudocapacitances. (a) Alternative model to Figure 7.10a. (b) Transmission line model. (c) Transmission line model with pure capacitance replaced by constant phase constant.

For supercapacitor development, special ECs have been proposed to fit the experimental results. For example, Figure 7.13 shows three proposed ECs cited in the literature. The first (a) is similar to that in Figure 7.10a, except that the parallel leakage resistance is connected in parallel to the pseudocapacitance C_F rather than to $R_{ct} - C_F$ [10]. The second EC (b) and the third (c) are transmission line models to take care of the porous electrode layer [8,10,11]. Note that the third EC model uses the constant phase elements (Q_i) rather than pure capacitances that mainly deal with the inclined Nyquist line at the low frequency range. In Figure 7.13b and c, the magnitudes of R_i, C_i, and Q_i can be the same or different, depending on the real situation. The constant phase element (Q_i) can be expressed as

$$Q_i = q_i^{-1}(2\pi f)^{-n}\cos\left(-\frac{\pi}{2}n\right) + jq_i^{-1}(2\pi f)^{-n}\sin\left(-\frac{\pi}{2}n\right) \tag{7.30}$$

where q_i is the factor of proportionality and n is the constant phase element exponent characterizing the phase shift. When n is in the value range 0.8 <

$n < 1.0$, q_i can be approximately treated as the capacitance in $\Omega^{-1}.s^n$ units. For a more detailed description of the constant phase element see Reference 8.

7.5.3 Supercapacitor Data Simulation to Obtain Parameter Values

As shown in Figure 7.9, the measured Nyquist plots contain a deformed semicircle at low frequency ranges and an inclined line at the high frequency range. Between these two frequencies is a narrow range with an inclined straight line. Because of the lack of a pseudocapacitance-generating reaction, this deformed semicircle should not be assigned to an electrochemical process. Note that, in this cell, a stainless steel plate was used as a current collector that provides less conductivity than other metals such as Al. Ni, etc. Therefore, it is reasonably believed that this semicircle comes from the interface between the low conductivity stainless steel plate and the electrode carbon layer. This was verified by further experimentation with the stainless steel plate replaced by a carbon plate (all other cell conditions remained constant). The deformed semicircle at the high frequency range disappeared although the other parts of the Nyquist plot could be still observed.

Based on the plots in Figure 7.9 and the experimental conditions, an equivalent circuit to fit the experimental data was proposed (Figure 7.14). The first part of this EC consists of R_1, R_2, and Q_1, which are the contact resistances and constant phase element, representing the deformed semicircle in Figure 7.9. R_3, Q_2, and Q_3 are used to deal with the porous electrode layer, representing the inclined straight lines in Figure 7.9.

For data fitting, an E-View program can be used. The simulated data are plotted as Nyquist plots for each curve in Figure 7.9 by solid lines. The model fits the experimental data well in a comparison of the dotted points with their corresponding solid lines. The simulated parameters are shown in Table 7.1.

Based on Table 7.1, when the impedance of Q_1 is much smaller, the sum of R_1 and R_2 may represent the equivalent series resistance of the supercapacitor. Due to the low conductivities of the stainless-steel current collectors and the carbon layer (10 wt% conductive carbon), this R_2 is very high. However, it can be significantly reduced by increasing the content of the conductive carbon. The EC in Figure 7.14 indicates that Q_2 should represent the

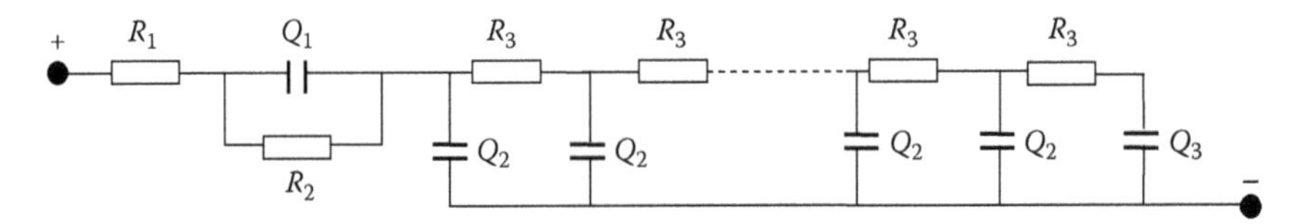

FIGURE 7.14
ECs for simulating data in Figure 7.10 for supercapacitor. R_1, R_2, and Q_1 represent contact resistances and constant phase element. R_3 and Q_2 describe the porous electrode layer. Q_3 represents the nontrivial boundary condition. (*Source:* Bisquert, J. 2000. *Physical Chemistry–Chemical Physics*, 2, 4185–4192. With permission.)

TABLE 7.1

Simulated Parameters Based on Data in Figure 7.9 for Changing Conductive Carbon (C45) Content in Electrode Layer

Carbon Content (wt%)	R_1 ($\Omega \cdot cm^2$)	R_2 ($\Omega \cdot cm^2$)	R_3 ($\Omega \cdot cm^2$)	q_1 ($\Omega^{-1} \cdot s^{n1}$)	n_1	q_2 ($\Omega^{-1} \cdot s^{n2}$)	n_2	q_3 ($\Omega^{-1} \cdot s^{n3}$)	n_3
10	4.10×10^{-1}	3.12×10^{0}	1.07×10^{0}	1.39×10^{-5}	0.916	1.05×10^{-1}	0.962	1.72×10^{-2}	0.934
15	6.03×10^{-1}	1.76×10^{0}	9.00×10^{-1}	1.21×10^{-5}	0.930	9.14×10^{-2}	0.961	1.37×10^{-2}	0.938
20	4.52×10^{-1}	1.22×10^{0}	1.08×10^{-1}	1.53×10^{-5}	0.903	7.70×10^{-2}	0.959	1.06×10^{-2}	0.930

Source: Bisquert, J. 2000. *Physical Chemistry–Chemical Physics*, 2, 4185–4192. With permission.

double-layer capacitance that has a value of ~0.1 $F.cm^{-2}$, consistent with the values obtained by both CV and CDC measurements of a similar electrode layer. Therefore, the EIS measurement and the proposed EC in Figure 7.14 should be reasonable.

7.6 Physical Characterization of Supercapacitor Materials

7.6.1 Scanning Electron Microscopy (SEM)

The SEM technique utilizes electron beams to scan the surface of a sample specimen, as shown in Figure 7.15. The specimen is irradiated by a focused electron beam and the signals produced create useful images describing the surface morphology of the specimen. Samples must be electronically conductive to prevent charging effects that can blur image quality at higher resolutions. To avoid this, some insulating samples are gold sputtered to provide a nanometer-thick conductive surface layer.

When the incident electron strikes the specimen surface, instead of bouncing off immediately, it penetrates for some distance before it collides with a surface atom and a region of primary excitation where signals are produced is created [12]. The most common signals used for imaging are secondary electrons, backscattered electrons, and characteristic X-rays. In normal conditions, the secondary electrons created from inelastic surface scattering can reach the detector in greater numbers, depending on incidence angle, and generate topographic information.

The backscattered electrons are higher energy electrons deflected elastically or scattered back to the detector. This backscattering provides specimen composition data because heavier elements produce greater backscattering intensity, resulting in brighter images than those produced by lighter elements [12,13]. Characteristic x-rays can reveal the distribution of chemical elements. Drawbacks of SEM include the requirement for a sample to be in a solid state and stable inside a vacuum. Normally, materials saturated with hydrocarbons, wet samples, and moisture-containing organic materials and clays are not compatible with SEM until they are lyophilized [14].

In the study of supercapacitors, SEM can provide important information about the material surface morphologies of cell components, specifically when analyzing separator membrane porosity and electrode morphology [15]. Images of the material surface can be collected before and after certain chemical or physical modifications or treatments to investigate their effects on material phases and morphologies. For example, SEM is used to examine the structural breakdown of single wall nanotube (SWNT) forests after the addition of liquid (stability within electrolyte), or investigating the layer interactions of graphene sheets in a transparent film [16]. SEM can also

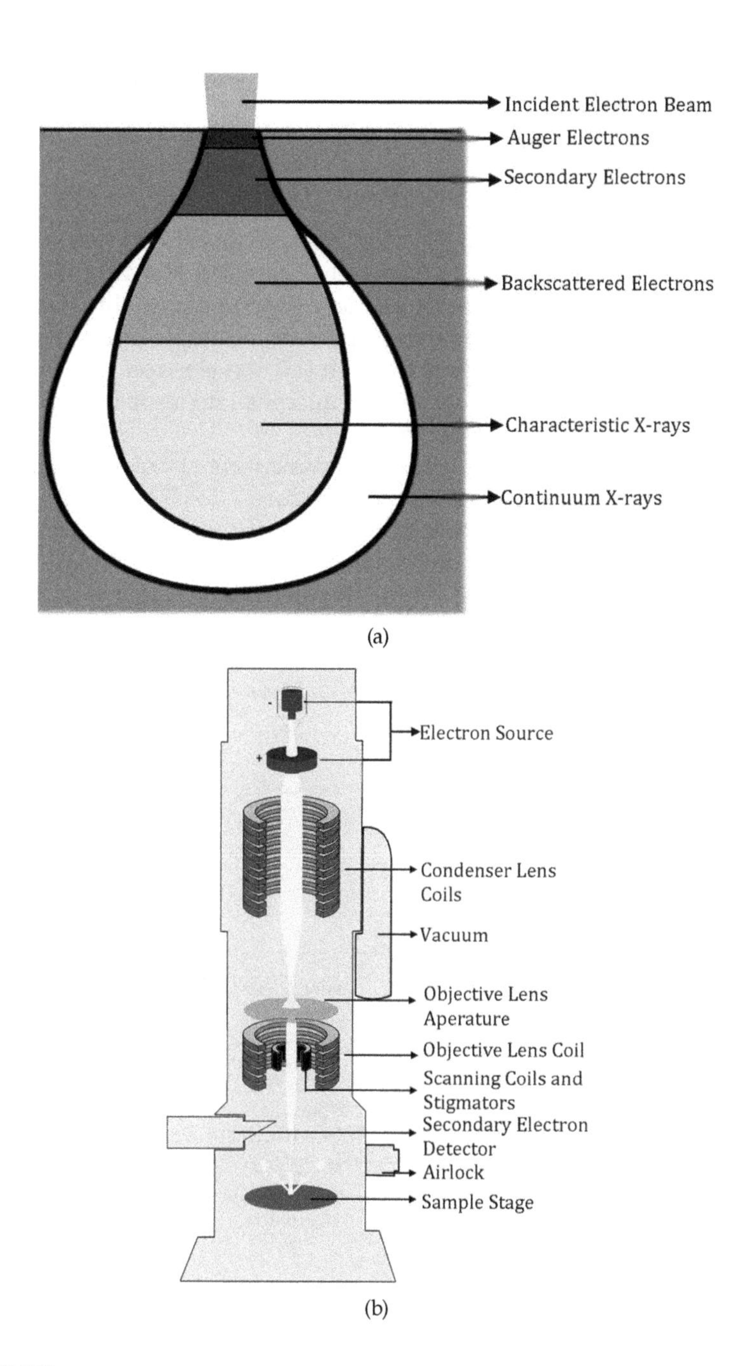

FIGURE 7.15
(a) Signals generated by interaction of electron beam and specimen and regions from which each signal type can be detected. (b) Scanning electron microscope. (*Source:* Zhou, W. and Z. L. Wang. 2006.)

be useful for identifying the positions of pseudocapacitive metal elements within composite carbon structures.

7.6.2 Transmission Electron Microscopy (TEM)

Similar to SEM, transmission electron microscopy (TEM) also utilizes a highly focused electron beam, as shown in Figure 7.16. However, TEM imaging requires a very thin specimen to achieve good image quality. This means that sample preparation is extremely important. A sample layer must be thin enough to allow electrons to pass through [17]. An electron gun emits high energy beams that can penetrate several microns into a solid. The electrons can penetrate through a thin specimen.

The essential components of a TEM instrument consist of an electron gun, lenses, detectors, and a specimen holder (Figure 7.16) [17]. A three- or four-stage condenser system will control variation of the illuminating aperture and the area of the specimen illuminated. The electrons that pass through the specimen then go through an imaging detection system and the image is displayed [18,19]. TEM is essential for material tomography or examining crystalline defects. However, TEM analysis is two-dimensional and it may be difficult to distinguish some features of an image. Further, the

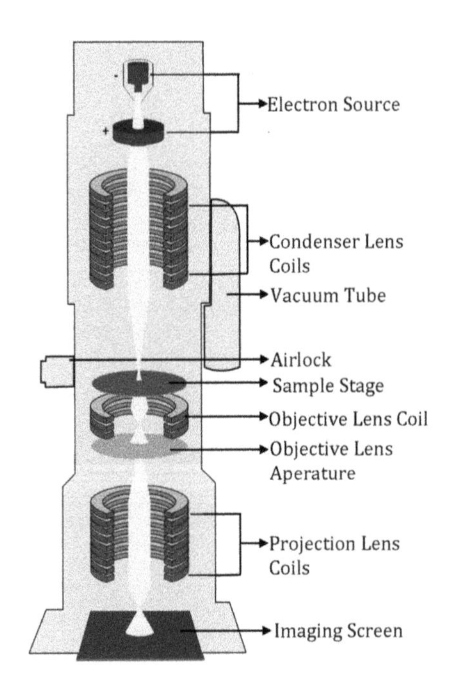

FIGURE 7.16
Typical TEM instrument and simplified schematic.

combination of ultrahigh voltage beam and high power electron sources makes the technique destructive for many sample types [17].

In electrochemical supercapacitors, TEM is used to collect and examine the microstructures of electrode materials, providing information about pore arrangements [15]. If a specimen is crystalline, the electrons will diffract or scatter off atomic planes inside the material, enabling the resolution of the crystal lattice structure down to atomic scale [20]. TEM can also estimate the level of carbon nanotube (CNT) aggregation or graphene layer thickness. Similar to SEM, TEM can also show changes in pore structures or arrangements after a material undergoes certain chemical or physical changes [21,22]. For example, morphological characterization of polymer-dispersed MWNTs and an image of a carbon template prepared using SBA-15 silica were characterized using TEM [23].

7.6.3 X-Ray Diffraction (XRD)

XRD is a non-destructive method of bombarding a sample with an x-ray beam to analyze the transmitted and diffracted beams [24]. As shown in Figure 7.17, the three basic components of an XRD instrument are the x-ray production unit arm, sample holder, and detector arm. The detector can rotate around the sample to measure the intensity of the diffracted x-rays at different angles. The angles, intensity, and peak widths of the resulting spectrum are keys for analyzing the sample against a materials database and calculating information about the sample. XRD performs best when a sample is homogeneous or single-phase. For nonisometric crystalline structures, indexing of patterns can be very difficult and amorphous materials cannot be identified [24–27].

The method is widely used for characterizing and identifying unknown crystalline materials, determining the structures and orientations of single crystals or grains, and measuring average spacings between layers or rows of atoms. XRD can also measure sample purity or texture [28]. In ES research, XRD is usually used to gather information about structure arrangements,

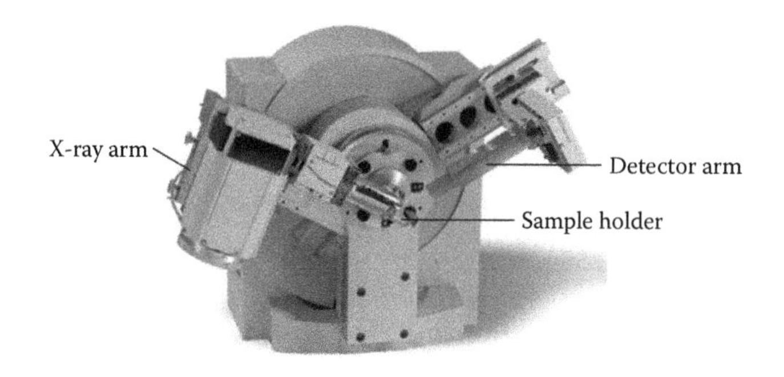

FIGURE 7.17
Modern X-ray diffractrometer. (Courtesy of Panalytics, XPert Powder [accessed March 28, 2012]. With permission.)

crystallite size, defects, and alterations of materials after they undergo certain processes [29–31].

For example, the relationship between pseudocapacitance and heat treatment temperature of nickel oxide was investigated using XRD, x-ray absorption spectroscopy, and CV. These techniques can provide important information about structure arrangements and the electrochemical properties of nickel oxide at various heat treatment temperatures. Similar studies were performed on composites (graphene–polyaniline, MnO_2–mesoporous carbon) [30,31]. XRD can reveal material structures revealing the relationship of electrochemical properties and the effects of certain chemical or physical alterations.

7.6.4 Energy-Dispersive X-Ray Spectroscopy (EDX)

EDX is commonly used as an addition to SEM, utilizing an electron gun and imaging equipment to locate the desired sample position. To perform EDX analysis of a sample, the electron imaging detector is replaced by an x-ray detector system. As previously noted, an electron beam that hits a sample produces a variety of signals including characteristic x-rays.

The x-rays are created when the incident electrons cause ejection of an electron in the inner shell. An outer shell electron fills the hole and releases the energy difference as an x-ray [32]. The x-ray energy is characteristic of the atomic structure and the difference between the electron shells. For example, the P Kα x-ray has an energy level of 2 KeV and is characteristic of a phosphorus atom undergoing a one-shell jump (α) from the L orbital to the K orbital. The detector collects the x-rays, converting them to voltage before they are sent for processing and analysis to generate an intensity spectrum.

EDX is a useful tool for ES research, for example, for determining the atomic dispersion of a sample surface. However, EDX loses measurement accuracy because of (1) overlapping peaks, (2) detector resolution, and (3) emission of x-rays in all directions—they must escape the sample before being reabsorbed to be detected. The third factor can mean lower energy x-rays are collected with lower intensity than is actually present and rough morphology can mask elements resulting in inaccurate atomic composition descriptions [33].

7.6.5 X-Ray Photoelectron Spectroscopy (XPS)

Also known as electron spectroscopy for chemical analysis (ESCA), XPS utilizes x-rays and photoelectric phenomena to study electronic structure, compound composition, electron and chemical states, electron bonding, and surface analysis [34–36]. In practice, a sample is bombarded by a monochromatic single wavelength x-ray beam. This causes core electrons from the sample to overcome their binding energy and escape to the sample surface where they are detected [34].

At different beam energies and wavelengths, specific characteristic bonds will interact with the beam. The whole process must be performed in an ultrahigh vacuum environment (UHV) and can reveal detailed information about the molecular composition of a surface. The technique is often used in industry to study catalysis, polymer surface modification, corrosion, adhesion, semiconductor and dielectric materials, electronics packaging, magnetic media, and thin film coatings [36,37].

For ES development, XPS is usually used to examine the oxidation states of different pseudocapacitive materials. A study of this use examined the oxidation states of ruthenium oxide powders with various water contents [38,39]. XPS is also used to study electrode functionalization through elemental analysis. For example, it can be used to investigate and improve the concentrations and types of nitrogen groups created by doping graphene and CNTs by various procedures. XPS also provides chemical characterization analysis for advanced electrolytes [40,41]. All these data help researchers determine the correlation between chemical structures and the capacitive characteristics of materials.

7.6.6 Raman Spectroscopy (RS)

RS is a spectroscopic technique based on inelastic or Raman scattering of a monochromatic light source on a sample. A typical Raman spectroscopy instrument should have an excitation source (laser generation), sample illumination and light collection optic system, spectrophotometer (for filtering and selecting wavelength), and a detector [42]. The technique measures weak inelastic scattering that occurs when photons interact with the electron cloud (Raman effect). The molecules absorb the photons, becoming excited, and then re-emit photons of different wavelengths, returning the molecule to a different rotational or vibrational state than its original ground state [42,43].

A filter removes all the elastically scattered light that retains the same wavelength as the incident photons. Energy shifting information collected over the spectrum of wavelengths is used to study vibrational, rotational, and other low-frequency transitions in molecules. RS gives results on most molecular samples and is flexible for testing solids, gases, and aqueous states. RS can also identify mixtures through characteristic peaks that fingerprint certain functional groups in a molecule. The technique is used both qualitatively and quantitatively in these composition analyses depending on clarity of the spectrum [43].

Raman is used as a complementary tool with TEM and XPS to examine structures and chemical composition changes of ES electrode materials that have undergone chemical or physical alterations, for example, characterizations of graphene, thin films, and electrode materials that will undergo pseudocapacitive redox reactions [44–46]. It has also been used successfully to study ion insertion into carbon materials for ES electrodes such as graphene–SnO2 nanocomposites in acidic solutions [47]. Raman investigation of CNTs after heat or chemical annealing has also been used to find optimum capacitive characteristics.

7.6.7 Fourier Transform Infrared Spectroscopy (FTIR)

FTIR is an analytical spectroscopy method that utilizes the infrared light spectrum to probe sample interactions (Figure 7.18). In principle, a sample is irradiated with infrared radiation (IR) and some of the IR light is absorbed by the material and some is transmitted through it [48,49]. The absorbed IR photons will excite molecules into a higher vibrational energy state and the wavelength absorbed is unique to the sample's molecular structure. The result is a unique profile of a material. FTIR fingerprints can be used quantitatively to determine concentration down to a few parts per million. Qualitatively, FTIR can be used to identify quality of a material [48–50].

For ESs, FTIR is usually used with TEM, SEM, XRD, BET and other techniques to characterize electrode materials. For example, FTIR is used to examine and chemically confirm the presence of uniform ultrathin polymer layers formed on carbon nanofiber electrodes [52]. FTIR, SEM, and XRD are also used to study surface morphologies of materials, for example, surface changes created during activation of polyacrylonitrile thin films deposited on carbon fibers. Other examples include analysis of CNT electrodes after polyaniline

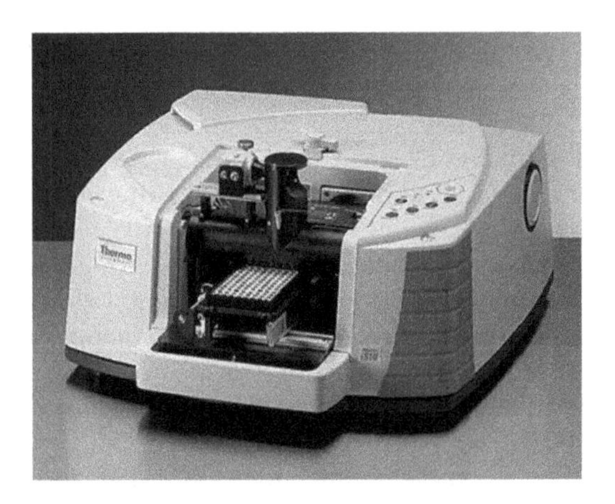

FIGURE 7.18
Modern FTIR instrument from Thermo Scientific. (*Source:* Thermo Scientific Instruments. *Direct Industry* (online). http://www.directindustry.com/prod/thermo-scientific-scientific-instruments/ft-ir-spectrometers-7217-56689.html [accessed March 30, 2012]. With permission.)

deposition [53] and porous structures of poly(ethyleneimine)-modified graphene sheets [54].

7.7 Brunauer–Emmett–Teller (BET) Method

The BET method is formulated on the thermodynamic principles for non-polar gas absorption on material surfaces, as shown in Figure 7.19 [55]. The method provides precise specific area measurements of a material by analyzing the absorption isotherm of nitrogen, argon, or carbon dioxide gas compared to a reference cell. Initially the sample is degassed fully to remove all the gas within the sample volume. During testing, a film of the test gas will form on the surface of the sample area, penetrating the pores. The gas desorption is also measured during a final degas as well. The analysis of the absorption and desorption isotherms provides total specific surface area (usually m^2/g). It also allows determination of pore size distribution [56,57].

BET is an important tool for ES characterization because surface area and pore size are both important parameters in determining material capacitance [59]. It is used to determine specific surface areas of electrode materials such as activated carbon and graphene [60]. Pore size measurements enable the estimation of electrochemical effectiveness of an active material when matched to a specific electrolyte [61–63]. Pore size and surface area are also good indicators of structural changes after chemical or heat treatment [64].

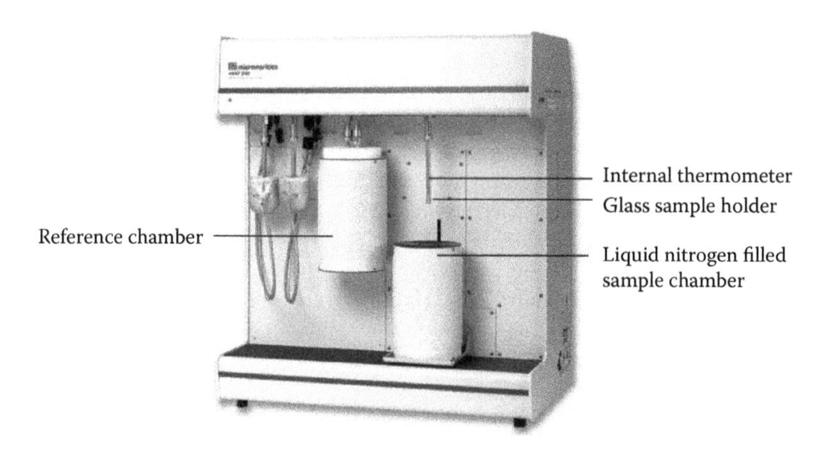

FIGURE 7.19
Modern BET instrument from Micrometrics. (*Source:* Micrometrics. *ASAP 2020* (online) http://www.micromeritics.com/ [accessed April 22, 2012]. With permission.)

7.8 Summary

In this chapter, both electrochemical and physical instrument characterizations for ES materials, components, and performance are discussed. Conventional three-electrode and two-electrode testing cells and their associated design and fabrication techniques for electrochemical characterization of supercapacitors in terms of equivalent series resistance, capacitance, and pseudocapacitance are presented.

Three common methods [cyclic voltammetry (CV), charging–discharging curve (CDC), and electrochemical impedance spectroscopy (EIS)] are briefly introduced. For fast screening of electrode materials, the conventional ex situ three-electrode cell is the choice test method. For in situ characterization of materials and supercapacitor performance, the two-electrode test cell can more closely represent the real conditions encountered during operation.

For equivalent series resistance and capacitance measurements, CV and CDC may be more simple or reliable than EIS in terms of data treatment. Although EIS may yield more information about the processes of supercapacitor operation, data simulation can induce arbitrary conclusions due to the equivalent circuit construction and the complexity of the simulation. Regarding physical instrument characterization of ES materials, components, development, and structure optimization, important methods such as SEM, TEM, XRD, EDX, XPS, RS, FTIR, and BET are briefly reviewed.

References

1. Tsay, K. C., L. Zhang, and J. Zhang. 2012. Effects of electrode layer composition and thickness and electrolyte concentration on both specific capacitance and energy density of supercapacitor. *Electrochimica Acta*, 60, 428–436.
2. Khomenko, V., E. Frackowiak, and F. Beguin. 2005. Determination of specific capacitance of conducting polymer/nanotubes composite electrodes using different cell configurations. *Electrochimica Acta*, 50, 2497–2506.
3. Zhang, L. and J. Zhang. 2011. Fe–N/C catalysts for PEM fuel cell. NRC unpublished data.
4. Nicholson, R. S. and I. Shain. 1964. Stationary electrode polarography. *Analytical Chemistry*, 36, 706–723.
5. Zhang, L. and J. Zhang. 2011. Mesoporous carbon as supercapacitor electrode material. NRC unpublished data.
6. Zhang, L. et al. 2009. Fe loading of a carbon-supported Fe–N electrocatalyst and its effect on the oxygen reduction reaction. *Electrochimica Acta*, 54, 6631–6636.
7. Ban, S. et al. 2012. Charging and discharging electrochemical supercapacitor in the presence of both parallel leakage process and electrochemical decomposition of solvent. *Electrochemica Acta*, in press.

8. Yuan, X. Z. et al. 2010. *Electrochemical Impedance Spectroscopy in PEM Fuel Cells*, New York: Springer.
9. Zhang, L. and J. Zhang. 2011. Supercapacitor cell structure optimization. NRC unpublished data.
10. Conway, B. E. 1999. Electrochemical Supercapacitors: *Scientific Fundamentals and Technological Applications*, New York: Kluwer.
11. Bisquert, J. 2000. Influence of the boundaries in the impedance of porous film electrodes. *Physical Chemistry–Chemical Physics*, 2, 4185–4192.
12. Zhou, W. and Z. L. Wang. 2006. *Scanning Microscopy for Nanotechnology: Techniques and Applications*, Berlin: Springer, 1–32.
13. Wells, O. C. 1974. *Scanning Electron Microscopy*, New York: McGraw Hill, 1–13.
14. *Scanning Electron Microscopy* (online). http://serc.carleton.edu/research_education/geochemsheets/techniques/SEM.html [accessed March 27, 2012].
15. Frackowiak, E. 2007. Carbon materials of supercapacitor application. *Physical Chemistry-Chemical Physics*, 9, 1714–1785.
16. Yu, A. et al. 2010. Ultrathin, transparent, and flexible graphene films for supercapacitor application. *Applied Physics Letters*, 96, 35.
17. Williams, D. B. and C. B. Carter. 2004. *Transmission Electron Microscopy: A Textbook for Materials Science*, New York: Springer, 141.
18. Reimer, L., and H. Kohl. 2008. *Transmission Electron Microscopy Physics of Image Formation*, New York: Springer. 1–15.
19. *The Transmitted Electron Microscope* (online). http://www.nobelprize.org/educational/physics/microscopes/tem/index.html [accessed March 17, 2012].
20. Egerton, R. F. *Physical Principles of Electron Microscopy: An Introduction to TEM, SEM and AEM*, New York: Springer, 11–16.
21. Yoon, S. et al. 2011. Development of high-performance supercapacitor electrodes using novel ordered mesoporous tungsten oxide materials with high electrical conductivity. *Proceedings of Royal Society of Chemistry*, 47, 1021–1023.
22. Miller, J. M. et al. 1998. Deposition of ruthenium nanoparticles on carbon aerogels for high energy density supercapacitor electrodes. *Journal of the Electrochemical Society*, 144, 309–311.
23. Reddy, A. L. M. et al. 2008. Asymmetric flexible supercapacitor stack. *Nanoscale Research Letters*, 3, 145–151.
24. *Introduction to X-ray Diffraction* (online). http://www.mrl.ucsb.edu/mrl/centralfacilities/xray/xray-basics/index.html [accessed March 20, 2012].
25. Clark, C. M. and B. L. Dutrow. *X-ray Powder Diffraction* (online). http://serc.carleton.edu/research_education/geochemsheets/techniques/XRD.html (accessed March 20, 2012].
26. Group, E. A. *X-ray Diffraction* (online). http://www.eaglabs.com/mc/x-ray-diffraction.html [accessed March 20, 2012].
27. Moeck, P. *X-ray Diffraction* (online). http://web.pdx.edu/~pmoeck/phy381/Topic5a-XRD.pdf [accessed March 20, 2012].
28. Vandier, L. *X-Ray Diffraction Laboratory* (online). http://www.lcc-toulouse.fr/lcc/spip.php?article120 [accessed March 28, 2012].
29. Nam, K. W., W. S. Yoon, and K. B. Kim. 2002. X-ray absorption spectroscopy studies of nickel oxide electrodes for supercapacitors. *Electrochimica Acta*, 47, 3201–3209.
30. Dong, X. et al. 2006. MnO_2-embedded-in-mesoporous-carbon-wall structure for use as electrochemical capacitors. *Journal of Physical Chemistry B*, 110, 6015–6019.

31. Zhang, K. et al. 2006. Graphene–polyaniline nanofiber composites as supercapacitor electrodes. *Chemistry of Materials*, 22, 1392–1401.
32. Bell, D. C. and A. J. Garratt-Reed. 2003 *Energy Dispersive X-ray Analysis in the Electron Microscope*. Microscopy Handbooks Series. New York: BIOS Scientific.
33. Goldstein, J. et al. 2002. *Scanning Electron Microscopy and X-ray Microanalysis*, 3rd ed., New York: Kluwer.
34. Vij, D. R. *Handbook of Applied Solid State Spectroscopy*. New York: Springer, 485.
35. Carver, J. C., G. K. Schweizer, and T. A. Carlson. 1972. Use of x-ray photoelectron spectroscopy to study bonding in Cr, Mn, Fe, and Co compounds. *Journal of Chemical Physics*, 57, 973–982.
36. Al., C. et al. 1996. X-ray photoelectron spectroscopy sulfur: Study of organic thiol and disulfide binding interactions with gold surfaces. *Journal of the American Chemical Society*, 12, 5083–5086.
37. Physical Electronics Division of ULCVAC-PHI. *X-ray Photoelectron Spectroscopy* (online). http://www.phi.com/surface-analysis-techniques/xps.html [accessed March 27, 2012].
38. Foelske, A. et al. 2006. X-ray photoelectron spectroscopy study of hydrous ruthenium oxide powders with various water contents for supercapacitors. *Electrochemical and Solid State Letters*, 9, 268–272.
39. Hu, C. C., and X. X. Lin. 2002. Ideally capacitive behavior and x-ray photoelectron spectroscopy characterization of polypyrrole. *Journal of the Electrochemical Society*, 149, 1049–1057.
40. Jiang, J. and A. Kucernak. 2002. Electrochemical supercapacitor material based on manganese oxide: preparation and characterization. *Electrochimica Acta*, 47, 2381–2386.
41. Toupin, M., T. Brousse, and D. Belanger. 2004. Charge storage mechanism of MnO_2 electrode used in aqueous electrochemical capacitor. *Journal of the American Chemical Society*, 16, 3184–3190.
42. Princeton Instruments. *Raman Spectroscopy Basics* (online). http://content.piacton.com/Uploads/Princeton/Documents/Library/UpdatedLibrary/Raman_Spectroscopy_Basics.pdf [accessed March 30, 2012].
43. Gardiner, D. J. 1989. *Practical Raman Spectroscopy*, Heidelberg: Springer.
44. Vivekchand, S. R. C. et al. 2008. Graphene-based electrochemical supercapacitors. *Journal of Chemical Sciences*, 120, 7–13.
45. Nam, K. W. and K. B. Kim. 2002. A study of the preparation of NiOx electrode via electrochemical route for supercapacitor applications and their charge storage mechanism. *Journal of the Electrochemical Society*, 149, 346–354.
46. Chmiola, J. et al. 2006. Anomalous increase in carbon capacitance at pore sizes less than one nanometer. *Science*, 313, 1760–1763.
47. Li, F. et al.. 2009. One-step synthesis of graphene–SnO_2 nanocomposites and its application in electrochemical supercapacitors. *Nanotechnology*, 20, 455–602.
48. *Handbook of Analytical Methods for Materials* (online). http://mee-inc.com/ftir.html [accessed March 27, 2012].
49. WCAS Group. *Fourier Transform Infrared Spectroscopy* (online). http://www.wcaslab.com/tech/tbftir.htm [accessed March 30, 2012].
50. Beckman Institute of California Institute of Technology. *Introduction to Fourier Transform Infrared Spectrometry* (online). http://mmrc.caltech.edu/FTIR/FTIRintro.pdf [accessed March 30, 2012].

51. Thermo Scientific Instruments. *Direct Industry* (online). http://www.directindustry.com/prod/thermo-scientific-scientific-instruments/ft-ir-spectrometers-7217-56689.html [accessed March 30, 2012].
52. Ghaemi, M. et al. 2008. Charge storage mechanism of sonochemically prepared MnO_2 as supercapacitor electrode: Effects of physisorbed water and proton conduction. *Electrochimica Acta*, 53, 4607–4614.
53. Deng, M., B. Yang, and Y. Hu. 2005. Polyaniline deposition to enhance the specific capacitance of carbon nanotubes for supercapacitors. *Journal of Materials Science*, 40, 5021–5023.
54. Yu, D. and L. Dai. 2010. Self-assembled graphene–carbon nanotube hybrid films for supercapacitors. *Journal of Physical Chemistry Letters*, 1, 467–470.
55. Kumar, K. R. and N. Balasubrahmanyam. 1986. Moisture sorption and the applicability of the Brunauer–Emmett–Teller equation for some dry food products. *Journal of Product Research*, 22, 205–209.
56. *NanoQAM Research Center on Nanomaterials and Energy* (online). http://www.nanoqam.uqam.ca/equipement.php?id = 9&lang = en [accessed April 2, 2012].
57. Materials Technology Company. *Ceram* (online). http://www.ceram.com/testing-analysis/techniques/brunauer-emmett-teller-surface-area-analysis-barrett-joyner-halenda-pore-size-and-volume-analysis/ [accessed April 2, 2012].
58. Micrometrics. *ASAP 2020* (online) http://www.micromeritics.com/ [accessed April 22, 2012].
59. Zhu, Y. et al. 2011. Carbon-based supercapacitors produced by activation of graphene. *Science*, 332, 1537–1541.
60. Brousse, T. et al. 2006. Crystalline MnO_2 as possible alternatives to amorphous compounds in electrochemical supercapacitors. *Journal of the Electrochemical Society*, 153, 2170–2180.
61. Xu, M. W. et al. 2007. Mesoporous amorphous MnO_2 as electrode material for supercapacitor. *Journal of Solid State Electrochemistry*, 11, 1101–1107.
62. Kim, C. and K. S. Yang. 2003. Electrochemical properties of carbon nanofiber web as an electrode for supercapacitor prepared by electrospinning. *Applied Physics Letters*, 83, 1216–1218.
63. Kim, C. et al. 2004. Characteristics of supercapacitor electrodes of PBI-based carbon nanofiber web prepared by electrospinning. *Electrochimica Acta*, 50, 877–881.
64. Ragupathy, P., Vansan H. N., and N. Munichandran. 2008. Synthesis and characterization of nano MnO_2 for electrochemical supercapacitor studies. *Journal of the Electrochemical Society*, 155, 34–40.

8

Applications of Electrochemical Supercapacitors

8.1 Introduction

The quest for alternative energy sources is one of the most important and exciting challenges facing science and technology in the twenty-first century. Environmentally friendly, efficient, and sustainable energy generation and usage have attracted large efforts around the world. For example, portable, stationary, and transportation applications are all areas that demand high levels of energy—areas in which electrochemical energy storage and conversion devices such as batteries, fuel cells, and electrochemical supercapacitors (ESs) are the necessary sources for delivering alternative energy.

Among the electrochemical energy storage and conversion advances, ESs are considered the most feasible complementary devices. They represent a new breed of technology among the other energy storage devices. They store greater amounts of energy than conventional capacitors and deliver more power than batteries. For example, Figure 2.20 (Chapter 2) is a Ragone plot showing various energy and power capabilities [1].

The Ragone plot also shows that although ES devices have lower energy densities than batteries and fuel cells, their power densities are 10 times greater than those of batteries and fuel cells. Several other advantages of the ES are worthy of mention:

1. Very short charge and discharge times, from fractions of a second to several minutes
2. Life cycles exceeding 100,000 charge–discharge cycles
3. Possible capacitance values ranging from 0.043 to 2700 F
4. Operating temperature ranging from –20°C to 55°C under various application conditions

ES application areas include power electronics, memory protection, battery enhancement, portable energy sources, power quality improvement, adjustable speed drives (ASDs), high power actuators, hybrid electric vehicles, renewable and off-peak energy storage, and military and aerospace applications. All these application areas will be discussed in this chapter.

8.2 Power Electronics

Modern portable electronics devices have been used in many industrial, commercial, residential, aerospace, and military applications. Recently, due to the high demand for energy conservation, power electronics are gaining great attention in the area of energy efficiency.

The systems of most power electronics need energy storage devices to provide energy back-up and decoupling between power conversion stages. Due to their low cost, the most common storage devices are conventional electrolytic capacitors and batteries, but they have drawbacks related to size, performance, and cycle life. Several new energy storage technologies now available appear to be promising options to traditional energy storage devices. Electrochemical supercapacitors (ESs) appear to be outstanding options for applications, particularly in areas requiring high power densities, fast transient responses, and reduced volume.

As discussed in previous chapters, ESs are new energy storage devices with excellent potential for portable applications such as power electronics systems due to their high capacitance values ranging from 1 to 2700 F, equivalent series resistance typically 10 times lower than resistances of conventional capacitors, and long cycle lives. The large capacitances of ESs make them suitable for backup energy storage in power electronics systems requiring reduced size. In the following sections, more applications of power electronics will be discussed.

8.3 Memory Protection

For many decades, several battery systems including lithium types have been used in memory protection or back-up power applications. However, batteries are not always the ideal solution. Some batteries need sophisticated recharging circuits; otherwise they can undergo temperature runaways and can explode. Due to limited cycle lives, they must be replaced relatively frequently and need to go through a series of conditioning steps before entering service. Finally, when you really need them, they may not last long enough to retain data during a long-term power outage. Although data centers worldwide still rely on batteries to protect critical data on their servers and storage controllers, long cycle lives and safety are still highly desired characteristics.

ESs are used throughout electronic circuit boards to regulate voltages that can vary due to component power usage, distances, and conductance across the devices. They work by storing a varying incoming electronic current while sending out a constant level. The larger the ES, the more power it can store. Like a battery, an ES can supply a circuit with enough energy to operate for a limited time. Actually, ESs have been used in this application for decades, for example, supplying enough power to keep a mobile computer alive while batteries are swapped.

Redundant arrays of independent disks (RAID) systems, are designed to preserve data under adverse circumstances. One example is a power failure that threatens data that is temporarily stored in volatile memory. To protect this data, many systems incorporate a battery-based power back-up that supplies short-term power—enough watt-seconds for the RAID controller to write volatile data to nonvolatile memory. With advances in flash memory performance such as dynamic random access memory (DRAM), density, lower power consumption, and faster write time, ESs with lower ESRs and higher capacitances per unit volume made it possible to replace the batteries in these systems with longer lasting and higher performance in a "greener" manner.

Servers and storage controllers use DRAM for many applications, but buffering or caching the output data is one of the most critical steps. If a system loses power, the cache is gone unless the DRAM has some kind of nonvolatile feature. If the cache is targeted for a rotating drive or RAID striped across multiple drives, data will be corrupted in the system and cannot be recovered. The nonvolatility of the DRAM is provided by various types of battery systems. The more recent applications involve the venerable lithium ion (Li-ion) battery and associated charge circuit.

However, by using an ES to provide power during operation, increased levels of protection (Figure 8.1) can also be achieved using an input voltage monitoring circuit to detect power failure and immediately write the contents of the DRAM to an onboard flash memory array [2]. For example, Viking Modular Solutions' ArxCis-NV™ [3] using an ES was designed and developed to back-up critical mission data in the event of a power failure. This data, which formerly resided in volatile DRAM historically relied on sub-optimal and unreliable batteries for protection. ArxCis-NV now offers a highly reliable, secure, maintenance-free, and disaster-free solution for critical RAID applications.

ArxCis-NV is also environmentally conscious. It is a safe and green product that requires no toxic battery disposal. Figure 8.2 shows a DDR3 ArxCis-NV module that combines supercapacitor DRAM and flash memory to deliver a persistent memory solution that saves critical data from power failures. The circuit in Figure 8.1 shows a supercapacitor-based power backup system using the LTC3625 charger. The system is also fitted with an automatic power

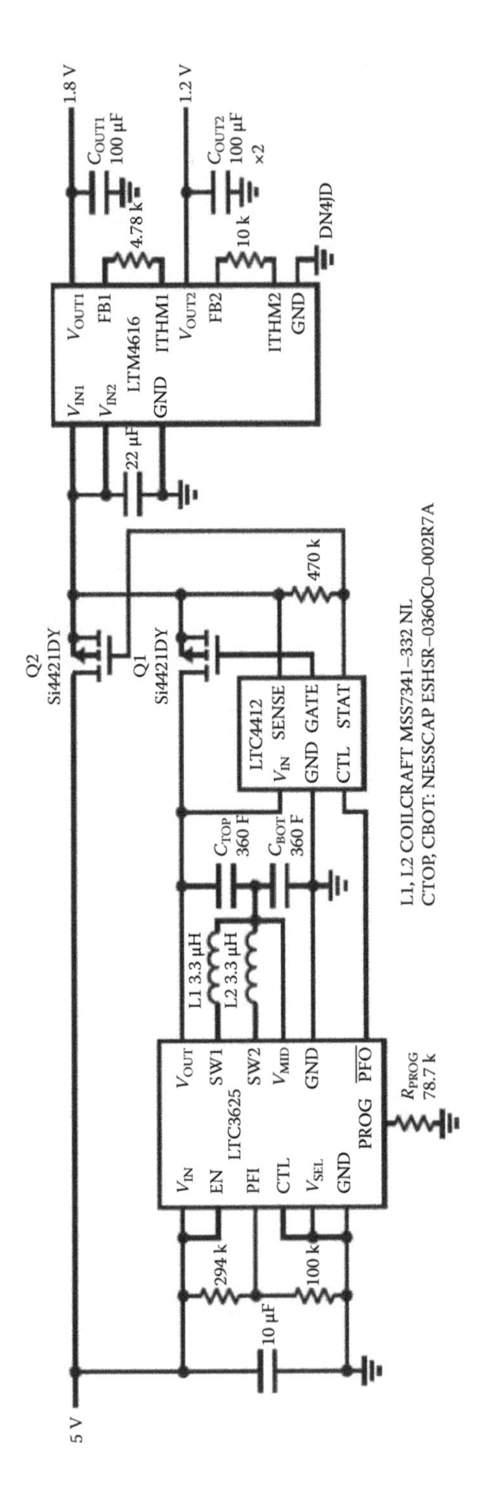

FIGURE 8.1
ES energy storage system for data backup. 1 (*Source:* Linear Technology. http://circuits.linear.com/all--supercapacitor_charging-385 [accessed March 1, 2012]. With permission.)

FIGURE 8.2
DDR3 ArxCis-NV module combines supercapacitor DRAM and flash memory to deliver a persistent memory solution that saves critical data from power failure events. (Courtesy of Viking Technology.)

crossover switch using the LTC4412 power path controller and an LTM4616 dual output μModule DC–DC converter.

Many electronic devices also include premature shutdown circuitry. These devices will power down upon detection of low voltage to prevent loss of data. A noisy supply voltage can sometimes trigger these shutdown circuits. An ES will help prevent premature shutdowns by reducing the severity of voltage transients [4].

8.4 Battery Enhancement

In today's world, the number of portable and mobile devices is increasing exponentially. Driven by the increasing demands for computing, communications, and transportation, these devices continue to become more advanced. All this progress requires batteries to deliver more performance in terms of power and energy. Actually, the power requirements of computer processors have increased steadily since 1986. Unfortunately battery manufacturers have not been able to catch up with the increasing needs of the new devices. Batteries are still limited by the dynamics and characteristics of the chemical reactions they create.

In battery-powered portable systems such as laptops and other mobile applications, the span of time during which a device can operate before the energy reserve is depleted is called the battery run-time. For example, current battery technologies have typical energy densities of 200 W/kg, depending on the type. For example a typical Li-ion laptop battery normally weighing 0.4 kg can only supply the computer load (25 Wh) for ~3 hours—insufficient for a full day of work or uninterrupted auonomous operation during a typical airplane flight. With the current state of technology, the most practical laptop power is a Li-ion battery with a rated capacity of 4 Ah. Despite considerable advances in Li-ion battery technology, the devices still cannot keep up with emerging notebook usage features providing digital entertainment, multimedia, and wireless connectivity. The expanded usage demands higher power and currents with higher slew rates.

To increase the run-time of a battery-powered device, the traditional method is to simply increase the number of cells constituting a battery pack. However, this solution only increases the size, weight, and maintenance needs of a portable system and is not acceptable because it reduces device portability and market acceptance.

ESs with low equivalent series resistances (ESRs) represent emerging technologies [5,6]. These types of capacitors do not possess the energy density of a conventional battery, but their power densities are more than 10 times higher. These features make ESs good complements for improving the performance of battery-based applications in which the battery supplies the energy and the ES delivers short-term power. Benefits of this combination include enhanced peak power performance, run-time extension, and reduction of internal losses. Current ESs are available with capacitance values ranging from 1 to 5000F, power densities of 4300 W/kg, and maximum energy storage of 8125 J. Their ESR values range from 0.3 to 130 mΩ—five times smaller than a conventional battery ESR. Thus, combining batteries and ESs is advantageous for supplying pulsating loads.

The coupling of a battery and an ES can be achieved in three ways. One is directly connecting the ES to the battery terminals. This requires the voltage rating of the ES to match the battery, so several ES cells connected in series are needed. The second method is to connect an inductor in series with the battery and an ES in parallel. This improves the current distribution between the capacitor and the battery. The third option is using a DC–DC converter along with an ES to interface with the battery. This scheme has the advantage of utilizing a lower voltage ES interfaced to a standard battery pack by a DC–DC converter.

In addition, the connection of an ES in parallel with a battery has the effect of reducing the peak value of the current supplied by the battery. This reduces internal losses of the system and the root mean square (rms) value of the current supplied by the battery. The power saved by this connection results in additional run-time for the system.

An ES can relieve a battery from the most severe load demands by meeting the peak power requirements and allowing the battery to supply average loads. The reduction of pulsed current drawn from the battery can result in an extended battery life. Figure 8.3 shows the expected increase in the run-time of a battery after combination with an ES [4]. The figure represents a 2 A GSM load on a 3.5 V 600 mAh Li-ion battery. Coupling an ES with a battery

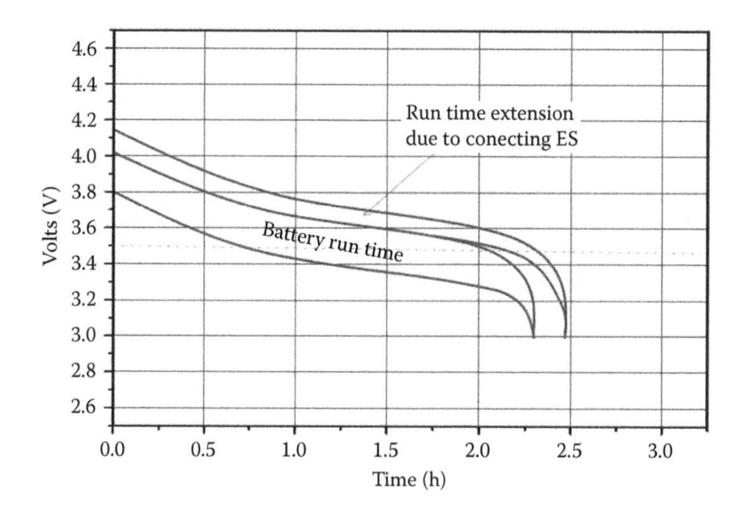

FIGURE 8.3
Run-time extension during use of ES in combination with 3.6 V 600 mAh lithium ion battery with 2 A GSM load. (*Source:* Smith, T. A., Mars, J. P., Turner, G. A. 2002. *Proceedings of 33rd Annual IEEE Power Electronics Specialists Conference*, 124–128. With permission.)

may extend the run-time of the battery almost 300% at 3.5 V (from ~0.75 to ~2.25 hr at 3.5 V), 33% at 3.3 V, and 14% at 3 V. The best extension in run-time is expected at high voltage.

8.5 Portable Energy Sources

For some portable electronic devices with moderate energy demands, ESs may act as rechargeable stand-alone power sources. Currently, batteries are the most convenient power supplies. However, they require long recharge times and need to be charged overnight. This is considered a limitation of the current technology. ESs offer opportunities to create devices that can be recharged quickly, perhaps in just a few seconds, and the repeated charging and discharging can proceed without significant losses in efficiency.

A typical application of ESs is to power light-emitting diodes (LEDs) that provide highly efficient and quickly rechargeable safety lights. Park et al. [7] used an ES to fabricate a no-battery power supply for the Mini-FDPM—a handheld, noninvasive breast cancer detector based on the principle of frequency domain photon migration (FDPM). The ES-based design imposed new challenges for choosing the voltage regulators, capacities, and output voltages. This new design provides power to the target application for the target duration.

A complete ES power supply package reported by Jordan and Spyker [8] incorporated a DC converter circuitry on a 50 F, 2.5 V ELNA Dynacap. By

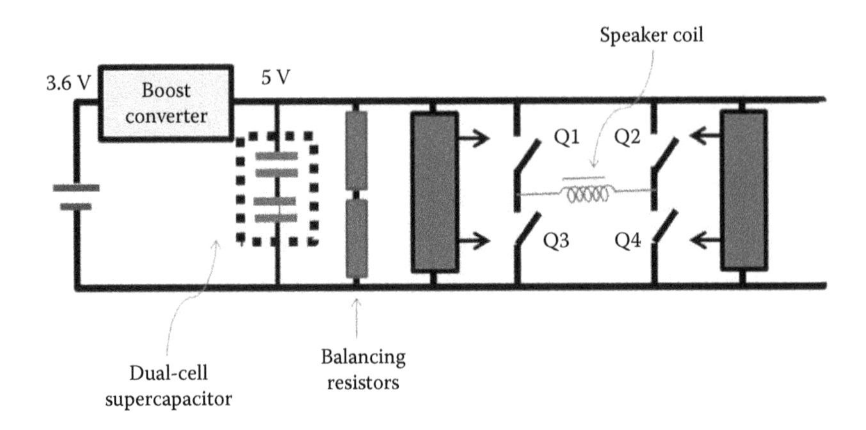

FIGURE 8.4
A 0.55 F 85 mΩ dual-cell ES delivers 5 W power bursts to drive peak power functions such as audio and LED flash. A battery covers a phone's average audio power needs of 0.5 to 1 W, recharging the ES between bursts.

using surface mount components, the entire package was kept to the size of a D-cell battery. The package was able to maintain a constant voltage for more than a minute with a 100 Ω load.

Compared to conversional capacitors for portable devices, ESs can overcome the power delivery constraints of batteries and the energy delivery limitations of conventional capacitors. They handle peak power events for wireless transmission, GPS, audio, LED flash, video, and battery hot swaps, and then are recharged from a battery at an average power rate.

For example, a typical cell phone design features a standard 3.6 V battery powering two Class D amplifiers to drive a pair of 8 Ω speakers. This typical design delivers a peak power of only 2.25 W, suggesting that it can power only thin-sounding music with a weak bass beat. Standard camera phones struggle to generate enough flash power to produce clear pictures in low light. High-current LEDs need up to 400% more power than a battery can provide to achieve full light intensity. An online report by Mars [9] showed that they were able to overcome power delivery limitations and boost flash and audio performances of portable devices using an ES. Figure 8.4 shows how a 0.55 F, 85 mΩ dual-cell ES can deliver 5 W power bursts to drive peak-power functions such as audio and LED flash.

8.6 Power Quality Improvement

An electrical power quality problem is defined as deviation and fluctuation of voltage that results in damage or mis-operation of electronic equipment or

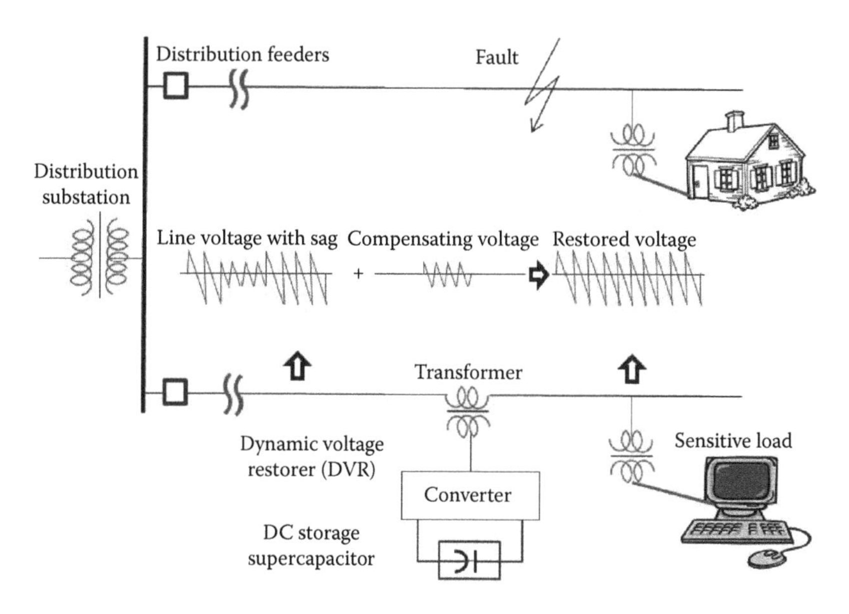

FIGURE 8.5
Sketch of improved power quality on distribution side by DVR.

other electrical devices. Power fluctuation is a serious matter that can severely affect sensitive devices. ESs can be used as energy storage devices for systems designed to improve the reliability and quality of power distribution.

For example, static condensers (statcons) and dynamic voltage restorers (DVRs) are systems intended to inject or absorb power from a distribution line to compensate for voltage fluctuations. They require DC energy storage devices that can store energy and allow it to be drawn. The length of voltage perturbation that can be effectively compensated for depends on the energy density of the DC storage device. Most voltage perturbations on a distribution bus are short-lived, most not lasting more than ten cycles [10]. For this application, the limited storage capability of an ES is not a problem. Furthermore, because an ES has the advantage of fast discharge time, it can respond quickly to voltage disturbances.

In general, batteries are not generally suitable for short duration and fast response applications such as statcons and DVRs because their lifetimes are shortened considerably when their charge is drained. Figure 8.5 shows improved power quality on the distribution side of a DVR.

It is worth noting that power quality applications require instantaneous reliable ride-through power. Using traditional lead–acid batteries for such rapid, deep-cycle electric power demands is harmful to batteries. Also due to the higher ESRs of traditional batteries, their use in this kind of application produces higher voltage drops on their terminal voltages and extra losses due to the pulsating characteristic of the load current. To ensure constant readiness when batteries are used, an application will require significant maintenance and incur high replacement costs. In this regard, the use of ESs offers significant advantages over the use of batteries. The initial cost per kilowatt hour of commercial ESs available today is higher than cost for conventional batteries. However, ESs are good alternatives because they offer much longer cycle lives and lower maintenance costs.

8.7 Adjustable Speed Drives (ASDs)

Adjustable speed drives (ASDs) are commonly used in industrial applications because of their efficiency, but they are often susceptible to power fluctuations and interruptions. Disruptions in industrial settings are highly undesirable, and down time of a machine in a continuously running process can cause significant monetary losses. Adjustable speed drives that can "ride-through" power supply disturbances are valuable tools for industry.

The major reasons for ASD shutdown are transients, interruptions, sags (under-voltages), swells (over-voltages), waveform distortions, voltage fluctuations, and frequency variations [11–13]. These problems can cause ADS symptoms such as premature trips or shuts down, resets and restarts, frequent repairs and replacements, erratic control of process parameters, unexplained fuse blowing and/or component failures, and frequent motor overheating trips and/or continuous operation of motor cooling systems.

For an ASD to ride-through disturbances at full power, an energy storage device must act as a backup power source. A number of options are available. Batteries and flywheel systems can provide ride-through up to an hour. The major disadvantages of batteries and flywheel systems are their sizes and maintenance requirements, although batteries are economic options. Fuel cells can store large amounts of energy but cannot respond quickly. Super-conducting magnetic energy storage (SMES) systems can provide reasonable ride-through capabilities but require sophisticated cooling systems.

In relation to these devices, ESs present unique advantages such as quick responses to voltage fluctuations, long lifetimes, and minimal maintenance. Furthermore, they can be monitored easily because their charge state depends on voltage [14]. However, the choice of energy storage option relies largely on power requirements and the desired ride-through time (Figure 8.6). Normally, the durations of most power fluctuations do not exceed 0.8 sec. An ES can provide ride-through times up to 5 sec to a rating of 100 kVA. Therefore, ESs are likely provide satisfactory ride-through in most applications.

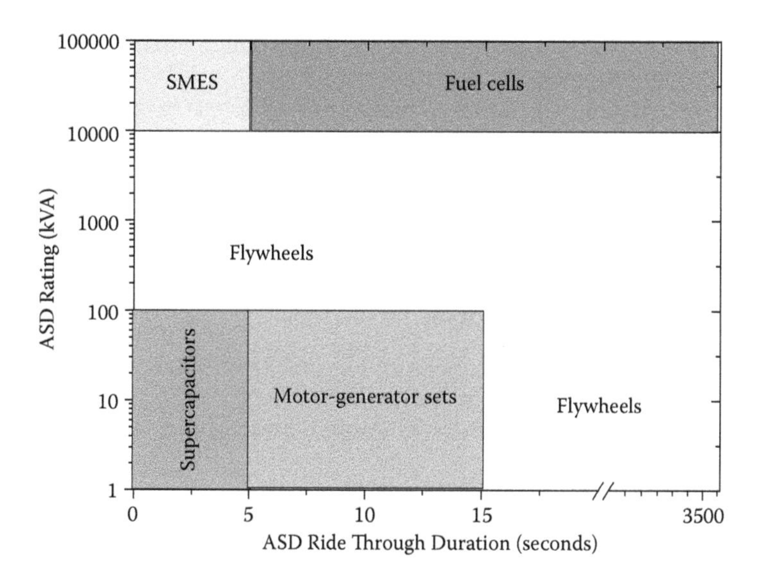

FIGURE 8.6
Energy storage options for different ASD power ratings.

8.7.1 Energy Storage Options for Different ASD Power Ratings

Voltage sag is a reduction of AC voltage at a given frequency for the duration of 0.5 cycles over 1 min. Common causes of voltage sags include large starting loads and remote fault clearing performed by utility equipment [15]. Voltage sags normally do not damage equipment but they can disrupt the operation of sensitive electronic loads such as the control boards of adjustable speed drives because sags make their protections trip.

Several topologies have been proposed to provide ride-through capabilities to adjustable speed drives for overcoming voltage sags. Gomez et al. [16,17] proposed a low cost approach to provide ride-through for voltage sags by adding three additional diodes and a boost converter to the DC links of adjustable speed drives. Corley et al. [18] proposed an approach to provide ride-through with an ES and a combination of boost and buck converters connected in parallel.

Deswal et al. [19] proposed a compensator scheme using an ES and an isolated boost converter. The operation of the sag compensator scheme was enabled when the magnitude of the voltage sag was larger than the threshold value set in the control block of the compensator, normally 10% sag, or during a short-term voltage interruption. Under this condition, the voltage provided by the ES was boosted by the DC–DC converter to the DC link voltage level of the ASD system. The time that the module could compensate for the voltage disruption depended on the amount of energy stored in the ES and the load supplied by the ASD. In this scheme, the ES could maintain the ASD DC bus voltage under voltage sag conditions, and the energy storage module was connected to support the DC link voltage during the power system faults. This topology was capable of providing full ride-through to an ASD by maintaining the DC link voltage level constant during the power quality (voltage sag) disturbance.

8.8 High Power Sensors and Actuators

An actuator is a type of motor for moving or controlling a system or mechanism. It uses electric energy for operation and converts it into some kind of motion. Mechanical actuators, for example, convert rotary motion into linear motion or vice versa.

Electromechanical actuators can perform thrust vector control for the launch of space vehicles and act as flood control actuators on submarines. Most actuation systems demand pulsed currents with high peak power requirements and fairly moderate average power requirements [20]. While an ES bank on its own is unlikely to be able to store enough energy, a battery combined with an ES can be designed to meet both average and peak load requirements. If a battery alone is used to meet both requirements, an oversized configuration is required. This is undesirable for space-limited applications in which weight must be kept to a minimum. A hybrid power source consisting of a battery and an ES bank can achieve a weight saving up to 60% when compared to use of a battery alone [21].

8.9 Hybrid Electric Vehicles

The prospective use of ESs in electric vehicles has drawn much attention to the technology. This technology appeals to the energy-conscious because of its energy efficiency and the possibility of recovering energy lost during braking. Many of the current power sources under consideration for installation in electric vehicles (EVs) do not meet the power requirements of acceleration. Fuel cells are promising due to their high energy densities, but their power specifications are still limited. Both the power and energy requirements of an EV can be satisfied with a combination of fuel cell and ES technology.

In general, a combined power source configuration allows a high energy density device such as a fuel cell to provide the average load requirements. Peak load requirements that result from accelerating or climbing up hills can be met by a high power device such as an ES bank. The utilization of ES also makes regenerative braking possible. Because the ES bank can be recharged,

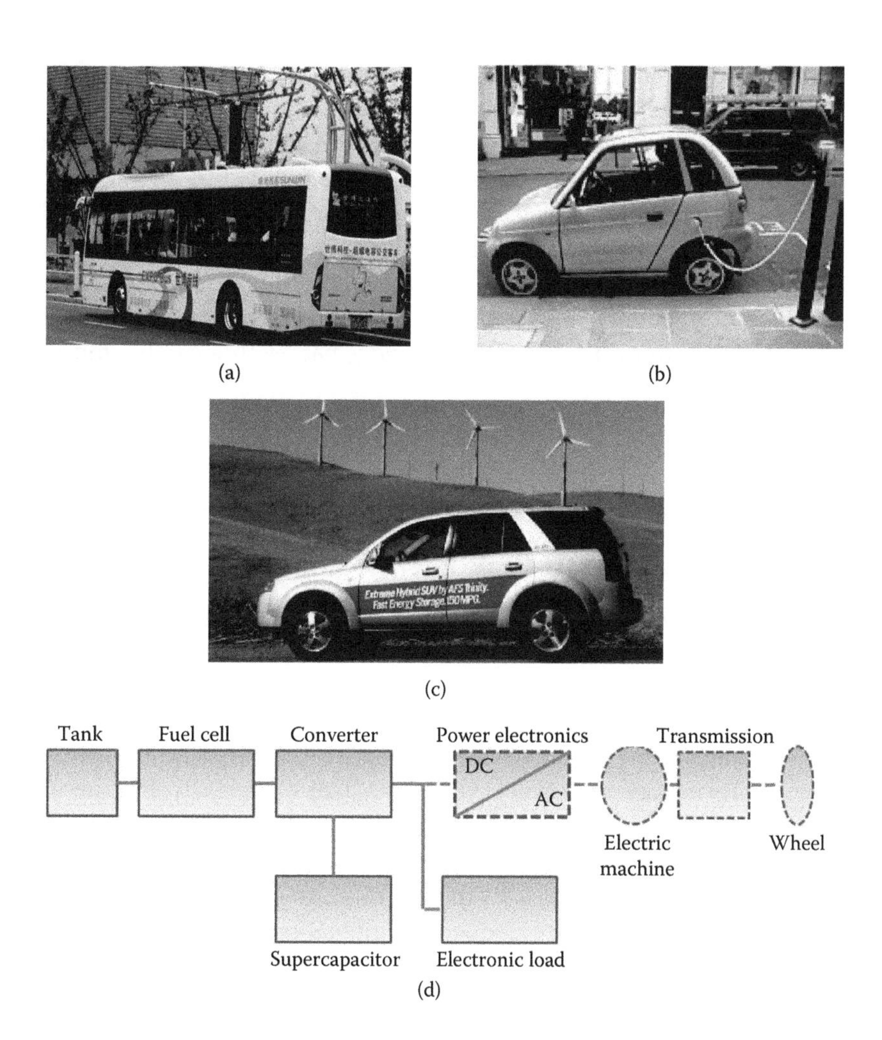

FIGURE 8.7
(See color insert.) (a) EC powered Sinautec buses during charging; (b) and (c). Hybrid REVA (Image courtesy of F. Hebbert) and AFS trinity vehicles using ES. (d) Electric drive train combining fuel cells with ESs..

it is possible to store some of the energy of an already moving vehicle, and therefore increase fuel efficiency as well. Such EVs have already been constructed as shown in Figure 8.7. A scheme of an electric drive train using a combined fuel cell–ES system is shown in Figure 8.7d [22].

In addition to their applications in EVs, ESs can also be used to maximize the efficiency of internal combustion engines (ICEs) in hybrid vehicles (HEVs). For example, 42 V electrical systems were proposed to meet the increased power demands of luxury vehicles, and alternatives to devices such as starter motors will become viable [23]. In a 42 V vehicle, one such an option will be the integrated starter alternator (ISA), an electrical part that can replace both the starter motor and the alternator.

The implementation of an ISA can provide greater generating ability and allow start–stop operation of the engine. The ISA can start an ICE quickly and easily, so when the vehicle must stop for an extended period, the ICE can be turned off rather than unnecessarily burning fuel. An ES bank in this configuration provides the power for engine cranking and is kept charged by a battery. The battery does not provide the power for starting the ICE; it is used only for charging the ES. In this way, battery life can be improved significantly.

8.10 Renewable and Off-Peak Energy Storage

Renewable and off-peak power supplies that derive their energy from intermittent sources such as wind or solar radiation must store energy to ensure that it is available at all times. Under such circumstances, ESs have a number of advantages over the common battery. For example, if a photovoltaic (PV) device generates power and stores it in a battery, the limited cycle life of the battery means replacement every 3 to 7 years [23]. Conversely, ESs can withstand large numbers of charge and discharge cycles without suffering significant losses in performance and thus need to be replaced every 20 years, which is also the lifetime of a PV panel [24]. Life-cycle costs are therefore reduced by the elimination of frequent maintenance.

Energy efficiency is always of primary concern in renewable power generation and ESs demonstrate higher charging efficiencies than batteries. A lead–acid battery, for example, can lose up to 30% of its energy during charging. ESs, on the other hand, may only lose 10% [24]. The ability to operate efficiently over a wide range of temperatures is also an advantage of ESs. Remote stations may be located in cold climates and if batteries are used for energy storage, the temperature must be maintained near room temperature by auxiliary systems, representing additional cost and energy consumption.

The major shortcoming of ES technology for application in intermittent renewable energy sources is its limited energy density. This results in excessive capital cost (equal to battery cost) to maintain energy storage. For this reason, ESs are rarely chosen for such applications. A study by Telstra Research Laboratories [25] emphasized the reduced life-cycle costs of network termination units powered by PV panels and ESs, and concluded that while present prices exclude their use, the capital costs can be expected to decrease significantly in the coming years [25].

8.11 Military and Aerospace Applications

With the development of ES technology, the military market segment has demonstrated increasing needs for ES devices to address a variety of power challenges based on the unique characteristics of ESs. They provide quick power delivery, superior low temperature operation, and the ability to handle up to a million cycles.

In military applications, the most frequent ES use is providing backup power for electronics in military vehicles, fire control systems in tanks and armored vehicles, airbag deployments and black boxes on helicopters, and backup power and memory devices for handheld emergency radios. ES-based modules are also appropriate in peak power applications to facilitate reliable communication transmission on land-based vehicles, active suspension for vehicle stabilization, GPS guided missiles and projectile systems, cold engine starts, and bus voltage hold-up during peak currents.

In marine applications, ES-based modules are used for high power discharge of unmanned boats in naval warfare, all-electric launches on aircraft carriers, power sources for magnetic drivers that power munitions elevators on aircraft carriers, and bridging power when switching from ground to onboard aircraft subsystems for data retention. ES components are also commonly employed in pulsed laser radar systems and unmanned aerial vehicles (UAVs). Figure 8.8 shows typical applications, including a NASA cordless drill powered by a supercapacitor and electromagnetic pulse weapons powered by ES and used by the U.S. Army.

(a)

(b)

FIGURE 8.8
(See color insert.) (a) NASA cordless drill powered by supercapacitor. (b) U.S. Army electromagnetic pulse weapons powered by ESs.

8.12 Summary

Electrochemical supercapacitors (ESs) present important advantages that qualify them for many applications. In this chapter, several important ES use areas are briefly reviewed. Examples are power electronics, memory protection, battery enhancements, portable energy sources, power quality improvement, adjustable speed drives (ASDs), high power actuators, hybrid electric vehicles, renewable and off-peak energy storage, and military and aerospace applications.

ESs have high power densities, long cycle lives, low ESRs, and reduced sizes and weights. When used with batteries, they improve battery cycle life significantly. For example, the use of a battery–ES hybrid connection has proven beneficial for run-time extensions in mobile applications. The extension is achieved by a reduction of internal battery losses generated by the lower impedance resulting from the battery–ES connection. Furthermore, the loss reduction is accompanied by an improvement in the power-delivering ability of the hybrid system, improving its performance for pulsating loads such as microprocessors. The addition of an inductor connected in series with the battery can increase the run-time of the system when compared to a parallel battery capacitor connection.

In a practical application intended to improve power quality, the use of an ES bank to provide ride-through in adjustable speed drives proved feasible in compensating for deep voltage sags and short voltage interruptions while maintaining a constant DC link voltage level during transient disturbances. It is worth noting that due to the current-dependent ES cell voltage, DC–DC converters are required to regulate cell voltage. This will cause some parasitic power loss and increase system cost.

References

1. Winter, M. and R.J. Brodd. 2004. What Are Batteries, Fuel Cells, and Supercapacitors? *Chemical Reviews*, 104, 4245–4270.
2. Drew, J. 2011. ES-based power back-up prevents data loss in RAID systems. *Linear Technology Design Notes*, DN487.
3. http://www.vikingmodular.com/products/arxcis/arxcis.html and http://www.ssg-now.com/viking-modular-eliminates-the-battery-in-ddr3-non-volatile-module/
4. Smith, T. A., J. P. Mars, and G. A. Turner. 2002. Using ES to improve battery performance. *Proceedings of 33rd Annual IEEE International Power Electronics Specialists Conference*, 124–128.

5. Mellor, P.H., N. Schofield, and D. Howe. 2000. Flywheel and ES peak power buffer technologies. *Proceedings of IEEE Conference on Electric, Hybrid and Fuel Cell Vehicles*, 8/1–8/5.
6. Ribeiro, P., B. Johnson, M. Crow et al. 2001. Energy storage systems for advanced power applications. *Proceedings of IEEE*, 89, 1744–1756.
7. Park, C., K. No, and P. H. Chou. 2011. *Turbocap: a batteryless, ES-based power supply for Mini-FDPM* (online). citeseerx.ist.psu.edu/viewdoc/summary?doi=10.1.1.143.2621 [accessed November 15, 2011].
8. Jordan, B. A. and R. L. Spyker. 2000. Integrated capacitor and converter package. *15th Annual IEEE Applied Power Electronics Conference and Exposition*, New Orleans.
9. Mars P. 2007. *ES for mobile phone power* (online). http://www.cap-xx.com and http://www2.electronicproducts.com/ES_for_mobile_phone_power-article-farr_capxx_dec2007-html.aspx [accessed December 29, 2011].
10. Hingorani, N. G. 1995. Introducing custom power. *IEEE Spectrum*, 32, 41–48.
11. Heydt, G. T. 1994. *Electric Power Quality*, 2nd ed. West Lafayette, IN: Stars in a Circle.
12. Hingorani, N. and L. Gyugyi. 1999. *Understanding FACTS*, Wiley/IEEE.
13. Bollen, M. H. J. 2000. *Understanding Power Quality Problems: Voltage Sags and Interruptions* , Power Engineering Series, New York: IEEE Press.
14. von Jouanne, A., P. N. Enjeti, and B. Banerjee. 1999. Assessment of ride-through alternatives for adjustable-speed drives. *IEEE Transactions on Industry Applications*, 35, 908–916.
15. IEEE. 1995. Recommended Practices on Monitoring Electric Power Quality, Standard 1159.
16. Gomez, J. L. D., P. N. Enjeti, and B. O. Woo. 1999. Effect of voltage sags on adjustable speed drives: a critical Evaluation and approach to improve performance. *IEEE Transactions on Industry Applications*, 35, 1440–1449.
17. Gomez, J. L. D., P. N. Enjeti, and A. von Jouanne. 2002. Approach to achieve ride-through of an adjustable speed drive with fly-back converter modules powered by supercapacitors. *IEEE Transactions on Industry Applications*, 38, 514–522.
18. Corley, M., J. Locker, J., Dutton et al. 1999. Ultracapacitor-based ride-through system for adjustable speed drives. *Proceedings of Annual IEEE Power Electronics Specialists Conference*, 26–31.
19. Deswal, S. S., R. Dahiya, and D. K. Jain. 2010. Improved performance of an adjustable speed drives during voltage sag condition. *International Journal of Engineering Science and Technology*, 2, 2445–2455.
20. Merryman, S. A. 1996. Chemical double-layer capacitor power sources for electrical actuation applications. *1st Intersociety Energy Conversion Engineering Conference.*
21. Merryman, S. A. and D. K. Hall. 1997. Characterization of CDL capacitor power sources for electrical actuation applications. *32nd Intersociety Energy Conversion Conference.*
22. Kotz R., M. Bartschi, B. Schnyder et al. 2001. ES for peak power demand in fuel cell-driven cars. *Proceedings of Electrochemical Society*, PV, 21.
23. Spillane, D., D. O'Sullivan, M. G. Egan et al. 2003. Supervisory control of a HV integrated starter–alternator with ultracapacitor support within the 42 V automotive electrical system. *8th Annual IEEE Applied Power Electronics Conference and Exposition.*

24. Barker, P. P. 2002. Ultracapacitors for use in power quality and distributed resource applications. *IEEE Power Engineering Society Summer Meeting*, Knoxville.
25. Robbins, T. and J. M. Hawkins. 1997. Powering telecommunications network interfaces using photovoltaic cells and ES. *Telecommunications Energy Conference*, Melbourne.

9

Perspectives and Challenges

9.1 Introduction

Research and development efforts have been plentiful in recent years in attempts to create new electrochemical supercapacitor (ES) component materials to improve the performance and commercial viability of this technology. With incremental progress in the field of material development, considerable technical advancements in ESs have been achieved in the past decade.

Two important material areas must be addressed: the ES electrode material, and the electrolyte material. Most of the research and development on electrode materials development focuses on high capacitive materials such as double-layer capacitive carbons, pseudocapacitive materials, and their composites. Increasing the capacitances of electrode active materials is desirable because it will allow higher energy storage capabilities. However, in developing these high capacitance materials, cost and cyclability challenges must be considered.

Electrolyte materials (solvents and ionic species) are also important components of ES devices. Traditionally, aqueous electrolyte solutions were used because they were nonhazardous and easy to handle. Aqueous electrolytes, however, offer limited electrochemically stable operating potential windows (up to ~1.0 V).

As the energy storage density is proportional to the square of the operating voltage, using alternative electrolyte systems with significantly wider operating voltage stability windows can significantly improve the energy storage capability of an ES device. On that front, organic and ionic liquid electrolytes are under investigation. Most commercially available ES systems utilize organic electrolytes. However, organic and ionic liquid electrolytes present several material handling and technical challenges that must be overcome.

Despite marked improvements in recent years in ES electrode and electrolyte material areas, several challenges remain. In this chapter, the market challenges of ES development efforts will be first discussed, followed by detailed discussions of the progress and technical challenges facing electrode and electrolyte material developments. In addition, the computational tools that can be utilized to supplement material development efforts will also be discussed. Finally, some perspectives and research and development directions will be explored.

9.2 Market Challenges

Although several ES devices have already reached the market, numerous issues still limit their attractiveness to global consumers: limited performance capabilities, long term durability issues, and high cost. ESs will undoubtedly be integral components of future energy systems, but strong research and development efforts are required to develop novel system components to overcome the challenges and increase the economic feasibility of this emerging technology.

To ensure the competitive advantages of ES devices over contemporary battery technologies, significant improvements in energy density are required. In recent years, steady progress has been made towards this objective through the development of unique active electrode materials with higher capacitance capabilities and the deployment of organic and alternative electrolytes with increased operational voltages.

Although these developments are extremely promising, closing the energy density gap between conventional batteries and ESs without sacrificing cyclability has proven a very difficult task. These long term durability issues stem from the electrode and electrolyte components. Specifically, during charge and discharge cycling, psuedocapacitive materials may undergo volumetric or compositional changes that affect the integrity of the electrode structures and cause performance degradation over long periods of operation [1]. Electrolyte decomposition is also a significant issue that is more prevalent at higher operating voltages and can occur on the surfaces of carbon-based electrode materials, causing pore blockages and increased electrode resistances [2]. Corrosion of aluminum current collectors is also a significant challenge that may be overcome by surface treatments for increasing stability.

Decreasing the cost of ES devices is critical for ensuring that they are attractive on a commercial scale. Currently, the cost of ESs exceeds $20 per watt hour—significantly higher than state-of-the-art lithium ion batteries at only $2 per watt hour [3]. This represents a very large gap. Despite several advantages of ESs over their battery counterparts, cost reduction is a necessity along with improving device performance and stability. Developing and applying uniquely designed electrolytes, electrode materials, and current collectors with improved performance at reduced cost will undoubtedly increase the global attractiveness of this emerging technology. The progress and future outlook of these efforts will be discussed in the following sections.

9.3 Electrode Material Challenges

9.3.1 Current Collectors

An ES electrode is traditionally composed of a conductive metallic (generally aluminum) current collector coated with an active electric double-layer carbon or psuedocapacitive component. The primary technical challenge in using a metallic foil current collector lies at the interface of the collector and the active material. This interface induces a charge transfer resistance that will inevitably increase the internal resistance of the entire system and reduce overall ES performance.

To address this issue, increasing the active material–current collector contact area and modifying the interface are feasible methods to decrease the charge transfer resistance. Several publications reported modifications of aluminum current collectors by a two-step procedure [4,5]. In the first step, a chemical or electrochemical etching procedure is carried out to induce surface roughness. This creates an increased interfacial area between the current collector and the active material, decreasing internal resistance and allowing for higher capacitance during operation. These etching procedures have been demonstrated to reduce the internal resistance of an ES cell from 50 to 5 $\Omega.cm^2$ after electrochemical etching of a commercial current collector [5].

After etching, a carbonaceous coating can be applied by chemical vapor deposition [6] or carbonaceous sol–gel deposit methods [7] deemed capable of further decreasing the internal resistances of ES cells as low as 0.4 $\Omega.cm^2$. Therefore, new and modified current collector surface treatment techniques can promise to further decrease internal cell resistance and improve the overall performance of modern ESs.

The development of novel current collector structures is also a feasible method for improving ES performance. For example, utilization of a carbon nanotube (CNT)-based current collector was reported to produce very high capacitance values and cyclability [8]. Using highly graphitic nanostructured materials such as CNTs can offer significant advantages including high surface areas, electronic conductivities, electrochemical and mechanical durability, and good compatibility with pseudocapacitor materials such as transition metal oxides [9–11] and conductive polymers [12,13]. These standalone current collector and active material composites show great promise and also have the potential to significantly decrease the overall weights of electrode structures. However, further research and development efforts are required to investigate their practical feasibility.

Electrochemical stability of the current collectors remains a significant technical challenge that must be considered throughout all aspects of research and development efforts. For example, over prolonged periods of operation, material corrosion may result in increased resistance, active material detachment, and inevitable significant performance loss. To address this challenge, investigating unique modification techniques for use in traditional current collectors and developing novel current collector materials are two promising approaches. The inevitable goal is to develop current collector structures that decrease internal resistance and cost without sacrificing operational durability.

9.3.2 Double-Layer Electrode Materials

Carbonaceous materials are almost exclusively utilized as active materials of double-layer electrodes due to their high conductivity, electrochemical stability, and porosity. Activated carbons still constitute the most practical active carbon-based electrode materials. They have high surface areas, are inexpensive to produce, and can be fabricated using a variety of readily available precursor materials.

It is desirable to develop active carbon materials with higher capacitances by increasing both surface areas and porosities. The common belief was that increasing pore volumes to sizes that are accessible to solvated ion species would directly increase capacitance. However, it was proposed recently that the presence of micropores (<2 nm) could increase the capacitances of these materials [14]. This indicates that some sort of ion desolvation mechanism occurring at the molecular scale allows ion transport and adsorption in these micropores.

A significant dependence on ion size was also observed. This suggests that designing active carbon materials with tunable pore sizes that could be optimized for specific types of ion species will be a challenge. With these methods, it will be possible to develop electrode materials with significantly improved capacitances through trial and error investigations supplemented by molecular scale computational methods.

Despite the widespread applicability and high capacitance values of activated carbon materials, other electrolyte and ion transport issues may arise and limit the performances of electrodes fabricated from these materials. Fabricating ordered electrode structures seems promising for overcoming this limitation. For example, CNTs have been extensively investigated due to their one-dimensional structures that result in porous electrode networks.

The primary challenge of CNT-based electrodes is due to the low overall surface area available for ion adsorption. Compositing CNTs with graphene is a recently reported method for overcoming this limitation by utilizing the expanded surface areas of graphene sheets and resulting capacitance capabilities [15]. Moreover, CNTs provide rigid conductive pathways in the active layer and prevent the aggregation of graphene sheets that causes electrode blockages and hinders electrolyte penetration. In fact, the aggregation of graphene sheets is an inherent issue when using pure graphene based electrodes. Features such as unique in-plane designs [16] are required to increase the utilization of the carbon surface area. CNTs and graphene can also be composited with psuedocapacitive materials, as discussed in Chapter 3.

Combining psuedocapacitive materials with CNTs and graphene can help overcome the inherent challenges associated with CNTs and graphene. The low active surfaces of CNTs available for double-layer formation can be counteracted by the incorporation of high performance psuedocapacitive materials, while the CNTs can maintain an electrode structure conducive to electrolyte access. With graphene, the psuedocapacitive material can act as a spacer to prevent agglomeration and also contribute to overall capacitance. Investigation of electric double-layer electrode materials is not limited to CNTs and graphene. Unique template mesoporous carbons [17,18], carbon nanofibers [19] and carbon onion [20] nanostructures are other examples of carbonaceous materials that have demonstrated applicability in ES technologies.

Further investigations are required to develop active carbon-based materials with higher capacitances and electrode arrangements favorable to mass transport. Moreover, inexpensive, upscalable fabrication techniques for these materials are also required to render them commercially viable.

9.3.3 Pseudocapacitor Electrode Materials

Energy storage pseudocapacitive materials that have traditionally been investigated involve transition metal oxides and conductive polymers. Despite promising results realized from both approaches, several technical and cost challenges remain to be overcome.

9.3.3.1 Transition Metal Oxides

Ruthenium oxide-based materials have been extensively investigated as psuedocapacitive materials based on their good operational stability and extremely high theoretical capacitance values [21]. The long term applicability of these materials is limited due to the high cost and limited availability of ruthenium. Therefore, it is of prime interest to investigate non-precious metal oxides to capitalize on their distinct cost advantages.

Several metal oxides with varying phase structures and nanostructures have been developed, including manganese oxide, iron oxide, molybdenum oxide, tin oxide, and titanium oxide. The most prominent issues surrounding transition metal oxide development efforts are (1) limited electronic conductivity and low theoretical capacitance in comparison to ruthenium oxide and (2) poor cyclability due to their redox nature and electrochemical instabilities. The varying electrochemically stable potential windows for some species of metal oxides completely eliminate their potential applications in ES devices.

Several researchers investigated unique methods to overcome the poor metal oxide electronic conductivity and energy storage limitations. Deliberate nanostructure control is a common technique utilized to circumvent these challenges. Particularly with manganese oxide, the specific crystal structures have been demonstrated to exert significant impacts on specific capacitance

[22,23] with the α crystallographic structure providing the highest energy storage capabilities [22]. Moreover, the impact of morphology has also been demonstrated [23,24] and deemed capable of tailoring the specific capacitances of these materials. To this end, various manganese oxide nanostructures have been investigated. Nanorods [25], nanoplates [26], and needle-like morphologies [24] demonstrated promising performance improvements.

Another common technique for improving the performances of transition metal oxides is using systems containing two or more metal species. These complex transition metal oxide systems exhibit beneficial electronic properties and show significant promise as psuedocapacitive materials. These include high surface area microporous $NiCo_2O_4$ spinel material [27,28], $FeTiO_3$ [29], and $MnFe_2O_4$ [30]. The complementary redox and electronic properties of these materials render their application in ES devices very promising.

Introducing nitride or sulfide species also has the potential for marked performance enhancement [31–34]. Further work is required to clearly understand the performances of these materials and develop different formulations and structures. Operational stability must also be considered. Issues such as volumetric changes during cycling and electrochemical instabilities will compromise the overall performance of a device.

9.3.3.2 Conductive Polymers

Conductive polymers can provide excellent capacitive storage capabilities and good electronic conductivity. They also are capable of storing energy throughout the bulk of a material, and thus nanostructure control is a required strategy to provide mass transport and ion access to improve the utilization of internal redox centers. Steady progress has been made on this front, for example, by developing aligned polyaniline nanowires [35] to improve reactant access.

However, the practical applicability of conductive polymers as psuedocapacitive materials is limited primarily by inherent instabilities caused by their specific redox charge storage mechanisms. Volumetric changes and irreversibilities induced during cycling will also limit the long term operational stability of conductive polymer-based electrodes. The literature notes that the impacts of these inherent challenges may be mitigated somewhat by methods such as incorporation of substituent groups [36–38] and ultrasonic irradiation during polymer processing [39,40].

Despite these advancements, the use of stand-alone conductive polymers will remain in the research and development phase unless significant improvements can be made. It is possible to composite conductive polymers with electric double-layer carbon-based materials. This is a viable approach to overcome the inherent disadvantages of conductive polymer materials while capitalizing on their specific energy storage capabilities. This promising approach will be discussed in the following section.

9.3.4 Composite Electrode Materials

The technical challenges associated with electric double-layer carbon and psuedocapacitive material development have been addressed innovatively by several techniques as discussed in the previous section. However, while advances in the energy storage capabilities of these materials have been marginal over recent years, breakthroughs in energy storage techniques are necessities. Compositing carbons with psuedocapacitive materials emerged as promising approaches to realize significantly higher energy storage capabilities and successfully bridge the gap between batteries and traditional capacitor materials. This method capitalizes on specific charge storage mechanisms that can complement each other in new ES designs.

As previously discussed, the poor cyclability of conductive polymer materials may hinder their application as stand-alone electrode materials in ESs. Compositing conductive polymers with carbon-based active materials may overcome these challenges while also increasing the electronic conductivity of the active layer. For example, combining psuedocapacitive polymer materials with CNTs [41–43] and graphene [44–46] are currently the most extensively investigated composite formulations for ES applications.

While CNTs possess limited surface areas compared to other microporous carbon blacks, they maintain rigid mechanically robust structures that can result in favorable three-dimensional electrode architectural configurations. The network arrangement can provide a porous structure that can readily facilitate the transport of electrolyte species and provide highly electronically conductive pathways to the redox centers of the active materials. The structure, porosity, and pore size distribution of this scaffold-like architecture can also be tailored by using CNTs with varying diameters, surface properties, and wall thicknesses or by modifying preparation techniques.

These unique electrode structures can effectively serve to increase the utilization of the redox centers present throughout the bulk of conductive polymer materials. Moreover, a flexible CNT network structure can cater to the volumetric changes resulting from charging and discharging processes. This structure can maintain the mechanical integrity of the active material layer, improving the cyclabilities of the materials [47]. The surfaces of CNTs can also interact favorably with many conjugated conductive polymer structures and these composites can readily form charge transfer complexes [48]. The exact nature of these interactions and the polymer–CNT interface are still unknown and should be investigated further.

Overcoming the low electronic conductivity and irreversibility of metal oxide species can also be achieved by developing carbon–metal oxide composites including manganese oxide–carbon nanotube thin films [49], graphene nanoplatelet-supported manganese oxide nanoparticles [50], manganese oxide coated carbon nanofoams [51], manganese oxide nanosheets dispersed on functionalized graphene [52], and nickel oxide nanoparticles supported on graphene oxide [53]. Incorporation of metal oxides into CNT-based electrodes is a commonly investigated approach.

Direct deposition of metal oxide species directly onto the surfaces of CNTs has been reported by techniques such as electrodeposition [54,55], microwave-assisted deposition [46], wet chemistry [56], and hydrothermal treatment [50]. Direct deposition creates a good interface between the two materials and allows the CNTs to act as electronic charge carriers with minimal interfacial resistance. While maintaining the inherent advantages of using CNTs as backbone structures, as discussed previously, these one-dimensional nanostructures can also be fabricated into aligned electrode morphologies exhibiting enhanced electrolyte transport and charge storage capabilities. The CNT arrangements can serve to suppress volumetric changes occurring during charge and discharge cycles and dramatically improve the cyclabilities of metal oxide materials.

To offset the drawback of low surface areas of CNTs, graphene-based composite materials can be formed to overcome this limitation by providing satisfactory surface areas and mechanical and electronic properties. Metal oxide species can be deposited directly on graphene surfaces to form complementary composite materials [24,50]. For example, manganese oxides with varying morphologies were deposited onto graphene by an aqueous precipitation method [24]. Specific capacitances for these composite materials exceeded those for the individual constituents and demonstrated the impacts of synthesis conditions and composite morphologies.

With this class of composite materials, graphene can provide electronically conductive pathways and act as a buffer in an electrode structure to mitigate the effects of volumetric changes during operation. Metal oxide species may serve as spacers to suppress agglomerations of graphene. When fabricated into electrode structures, these composite materials exhibit porous structures that are conducive to electrolyte access. The expected drop in prices of CNTs and graphene, along with the development of improved manufacturing technologies, may make these composites commercially viable alternative materials for ES electrodes.

The potential for application of composite materials in ES devices is remarkable. Coupling the distinct charge storage mechanisms of carbons and psuedocapacitive materials can allow researchers to achieve breakthroughs in energy storage capacitances while overcoming the challenges of poor cyclability. To do so, scientists must further their fundamental understanding of the behaviors of this broad class of composite materials during design and fabrication. The investigation of novel composite arrangements with unique morphologies and electrode architectures provides promise for future ES research and development. Furthermore, the physical properties of the electrode materials utilized should be tailored to the specific type of electrolyte utilized. Electrode materials exhibit very different performances in aqueous and organic electrolytes with different ions.

9.4 Electrolyte Innovations

Increasing the operating voltages of ES devices is an extremely practical way to improve energy storage density. Although aqueous electrolytes are preferred from a material handling and processing perspective, they are stable only up to a potential of ~1.0 V, above which decomposition in the form of gas evolution will occur. Organic and nonaqueous electrolytes can be used and will increase the stable potential window up to ~2.5 V. The most common organic electrolytes are acetonitrile and propylene carbonate, with the former more commonly used due to its higher ionic conductivity.

Advances in the understanding of ion desolvation and transport mechanisms have furthered the utility of these materials. However, environmental toxicity and safety issues associated with organic electrolytes coupled with their still limited operational potential windows shifted the focus to the development of ionic liquids as electrolytes for new ESs. With ionic liquid electrolytes, operating voltages can be increased to ~3.5 V or more without instability issues arising. Moreover, ionic liquids have well defined ion sizes and do not have ion salvation and desolvation mechanisms that plague aqueous and organic electrolyte systems.

The primary technical challenge facing their application is the limited ionic conductivity arising from the large sizes and structures of the constituent ion species. The initial belief was that ionic liquids were not effective for operations at low temperature because of these factors. Recent developments, however, demonstrated that with appropriate design and material selection, ionic liquid-based ESs can perform reasonably over a broad range of operating temperatures. Using carefully selected mixtures of anionic and cationic species, for example, the temperature-dependent behaviors and physical properties of ionic liquids can be modified significantly [57].

The development of ionic liquid electrolyte-based ESs is a promising approach but it is still in an infancy stage. With further research on carefully selected ionic liquid combinations and compatibility investigations with various electrode active materials, marked progress is expected for the ES industry.

9.5 Development of Computational Tools

On a fundamental level, we still lack a clear understanding of the specific mechanisms of energy storage in ES devices. Detailed computational tools including molecular scale modeling of the physical and chemical processes of charging, storage, and discharge are important, especially for new component materials such as composite active electrodes (carbons and psuedocapacitive materials) and electrolytes (organic and ionic liquids). Most recent computational studies focused on the effects of pore sizes on double-layer capacitance.

Empirical evidence suggests that pores smaller than the sizes of solvated ion species can contribute to capacitance. This indicates a clear lack of understanding of ion desolvation, transport, and adsorption in micropores. Density functional theory has been applied to these systems to seek insight into the specific mechanisms. Drastically different charge storage mechanisms were determined for pore sizes in different size regimes (microporous, mesoporous, and macroporous).

Studies of this nature can reveal fundamental information that will be invaluable for designing new electrode materials. Moreover, the type of electrolyte and ion species utilized, along with varying compatibilities with different electrode materials, will have significant effects on ES performance. Increased understanding of the electrolyte dynamics of a system will allow the design of novel electrolytes with enhanced ion transport properties and molecular interactions at the electrode interfaces, resulting in significantly improved performance.

9.6 Future Perspectives and Research Directions

Continuous progress in the field of ES devices has been realized in recent years and must continue if we are to bridge the gap between traditional capacitor devices and conventional batteries. The development of unique nanostructured electrode materials such as porous carbon, pseudocapacitive, and composite electrodes is a big step toward continuous improvement. Investigations of the utilization of organic and ionic liquid electrolytes should continue so that we can further increase the operating voltage stabilities of ES devices and improve energy storage capabilities.

These approaches will provide the opportunities to investigate new system configurations of various combinations of electrode materials, electrolytes, and their constituents. Investigations should be supplemented by detailed computational approaches to obtain further insights into the molecular scale behaviors of these systems and provide fundamental knowledge to aid in the designs of novel component materials. Steadily increasing the energy storage and cyclability of these devices while reducing costs will significantly increase their commercial viability. The following future research directions are suggested:

1. Development and investigation of active carbon electrode materials with tunable morphologies and pore structures with optimized electrolyte compatibilities

2. Development and investigation of various formulations and nanostructures of metal oxide pseudocapacitor materials to improve energy storage capabilities
3. Development and investigation of novel conductive polymer compositions and nanostructures to improve redox center utilization and energy storage capabilities
4. Development and investigation of composite electrode materials, integrating their combined benefits and overcoming associated challenges
5. Investigation of unique ionic liquid combinations and compositions to optimize physical properties, temperature dependence, and electrode compatibility
6. Further the fundamental understanding of computational tools examining specific physical and chemical phenomena occurring during ES charging, storage, and discharge of the aforementioned materials

References

1. Zhao, X. et al. 2011. The role of nanomaterials in redox-based supercapacitors for next generation energy storage devices. *Nanoscale*, 3, 839–855.
2. Azaïs, P. et al. 2007. Causes of supercapacitor aging in organic electrolytes. *Journal of Power Sources*, 171, 1046–1053.
3. Miller, J. R. and A. F. Burke. 2008. Electrochemical capacitors, challenges, and opportunities for real world applications. *Interface*, 17, 53–57.
4. Dangler, C. et al. 2011. Role of conducting carbon in electrodes for electric double-layer capacitors. *Materials Letters*, 65, 300–303.
5. Portet, C. et al. 2004. Modification of Al current collector surface by sol–gel deposit for carbon–carbon supercapacitor applications. *Electrochimica Acta*, 49, 905–912.
6. Wu, H. C. et al. 2009. High performance carbon-based supercapacitors using Al current collector with conformal carbon coating. *Materials Chemistry and Physics*, 117, 294–300.
7. Portet, C. et al. 2006. Modification of Al current collector–active material interface for power improvement of electrochemical capacitor electrodes. *Journal of the Electrochemical Society*, 153, A649–A653.
8. Zhou, R. et al. 2010. High performance supercapacitors using a nanoporous current collector made from super-aligned carbon nanotubes. *Nanotechnology*, 21, 345701.
9. Jang, J. H. et al. 2006. Electrophoretic deposition (EPD) of hydrous ruthenium oxides with PTFE and their supercapacitor performances. *Electrochimica Acta*, 52, 1733–1741.
10. Ma, S. B. et al. 2007. Synthesis and characterization of manganese dioxide spontaneously coated on carbon nanotubes. *Carbon*, 45, 375–382.

11. Wu, M. et al. 2004. Redox deposition of manganese oxide on graphite for supercapacitors. *Electrochemistry Ccommunications*, 6, 499–504.
12. Peng, C. J. Jin, and G. Z. Chen. 2007. A comparative study on electrochemical co-deposition and capacitance of composite films of conducting polymers and carbon nanotubes. *Electrochimica Acta*, 53, 525–537.
13. Sivakkumar, S. et al. 2007. Performance evaluation of CNT–polypyrrole–MnO_2 composite electrodes for electrochemical capacitors. *Electrochimica Acta*, 52, 7377–7385.
14. Chmiola, J. et al. 2006. Anomalous increase in carbon capacitance at pore sizes less than one nanometer. *Science*, 313, 1760–1763.
1. 5. Yu, A. et al. 2010. Ultrathin, transparent, and flexible graphene films for supercapacitor application. *Applied Physics Letters*, 96, 253105.
16. Yoo, J. J. et al. 2001. Ultrathin planar graphene supercapacitors. *Nanoletters*, 11, 1423–1427.
17. Kim, W. et al. 2009. Preparation of nitrogen-doped mesoporous carbon nanopipes for the electrochemical double-layer capacitor. *Carbon*, 47, 407–1411.
18. Monk, J. R. Singh, and F. R. Hung. 2011. Effects of pore size and pore loading on the properties of ionic liquids confined inside nanoporous CMK-3 carbon materials. *Journal of Physical Chemistry C*, 115, 3034–3042.
19. Li, Q. et al. 2012. In situ construction of potato starch based carbon nanofiber–activated carbon hybrid structure for high performance electrical double-layer capacitor. *Journal of Power Sources*, 207, 199–204.
20. Pech, D. et al. 2010. Ultrahigh power micrometre sized supercapacitors based on onion-like carbon. *Nature: Nanotechnology*, 5, 651–654.
21. Hu, C., C. W. C. Chen, and K. H. Chang. 2004. How to achieve maximum utilization of hydrous ruthenium oxide for supercapacitors. *Journal of the Electrochemical Society*, 151, A281–A290.
22. Devaraj, S. and N. Munichandraiah. 2008. Effect of crystallographic structure of MnO_2 on its electrochemical capacitance properties. *Journal of Physical Chemistry C*, 112, 4406–4417.
23. Yang, Y. and C. Huang. 2010. Effect of synthetical conditions, morphology, and crystallographic structure of MnO_2 on its electrochemical behavior. *Journal of Solid State Electrochemistry*, 14, 1293–1301.
24. Mao, L. et al. 2012 Nanostructured MnO_2–graphene composites for supercapacitor electrodes : effects of morphology, crystallinity, and composition. *Journal of Materials Chemistry*, 22, 1845–1851.
25. Li, Y. et al. 2011. Preparation and electrochemical performances of ⊠-MnO_2 nanorod for supercapacitor. *Materials Letters*, 65, 403–405.
26. Ai, Z. et al. 2008. Microwave-assisted green synthesis of MnO_2 nanoplates with environmental catalytic activity. *Materials Chemistry and Physics*, 111, 162–167.
27. Wei, T. Y. et al. 2010. Cost-effective supercapacitor material of ultrahigh specific capacitances : spinel nickel–cobaltite aerogels from an epoxide-driven sol–gel process. *Advanced Materials*, 22, 347–351.
28. Xiao, J. and S. Yang. 2011. Sequential crystallization of sea urchin-like bimetallic (Ni, Co) carbonate hydroxide and its morphology conserved conversion to porous $NiCo_2O_4$ spinel for pseudocapacitors. *RSC Advances*, 1, 588–595.
29. Tao, T. et al. 2011. Ilmenite $FeTiO_3$ nanoflowers and their pseudocapacitance. *Journal of Physical Chemistry C*, 115, 17297–17302.

30. Kuo, S. L. and N. L. Wu. 2007. Electrochemical capacitor of MnFeO with organic Li–ion electrolyte. *Electrochemical and Solid State Letters,* 10, A171–A175.
31. Bao, S. J. et al. 2008. Biomolecule-assisted synthesis of cobalt sulfide nanowires for application in supercapacitors. *Journal of Power Sources,* 180, 676–681.
32. Choi, D. G. E. Blomgren, and P. N. Kumta. 2006. Fast and reversible surface redox reaction in nanocrystalline vanadium nitride supercapacitors. *Advanced Materials,* 18, 1178–1182.
33. Glushenkov, A. M. et al. 2011. Structure and capacitive properties of porous nanocrystalline VN produced by NH_3 reduction of V_2O_5. *Chemistry of Materials,* 22, 914–921.
34. Tao, F. et al. 2007. Electrochemical characterization on cobalt sulfide for electrochemical supercapacitors. *Electrochemistry Communications,* 9, 1282–1287.
35. Wang, K. J. Huang, and Z. Wei. 2010. Conducting polyaniline nanowire arrays for high performance supercapacitors. *Journal of Physical Chemistry C,* 114, 8062–8067.
36. Arbizzani, C. et al. 1995. N- and p-doped polydithieno [3,4-B,3′,4′-D] thiophene, a narrow band gap polymer for redox supercapacitors. *Electrochimica Acta,* 40, 1871–1876.
37. Sivakkumar, S. and R. Saraswathi. 2004. Performance evaluation of poly (N-methylaniline) and polyisothionaphthene in charge storage devices. *Journal of Power Sources,* 137, 322–328.
38. Sivakumar, R. and R. Saraswathi. 2003. Redox properties of poly (N–methylaniline). *Synthetic Metals,* 138, 381–390.
39. Hussain, A. et al. 2006. Effects of 160 MeV Ni^{12+} ion irradiation on HCl doped polyaniline electrode. *Journal of Physics D,* 39, 750–755.
40. Li, W. et al. 2005. Application of ultrasonic irradiation in preparing conducting polymer as active materials for supercapacitor. *Materials Letters,* 59, 800–803.
41. Lee, H. et al. 2011. Fabrication of polypyrrole (PPy)–carbon nanotube (CNT) composite electrode on ceramic fabric for supercapacitor applications. *Electrochimica Acta,* 56, 7460–7466.
42. Xiao, Q. and X. Zhou. 2003 The study of multiwalled carbon nanotube deposited with conducting polymer for supercapacitor. *Electrochimica Acta,* 48, 575–580.
43. Zhou, Y. et al. 2004. Electrochemical capacitance of well-coated single-walled carbon nanotube with polyaniline composites. *Electrochimica Acta,* 49, 257–262.
44. Davies, A. et al. 2011. Graphene-based flexible supercapacitors: pulse electropolymerization of polypyrrole on free-standing graphene films. *Journal of Physical Chemistry C,* 115, 17612–17620.
45. Wang, D. W. et al. 2009. Fabrication of graphene/polyaniline composite paper via in situ anodic electropolymerization for high performance flexible electrode. *ACS Nano,* 3, 1745–1752.
46. Yan, J. et al. 2010. Preparation of a graphene nanosheet/polyaniline composite with high specific capacitance. *Carbon,* 48, 487–493.
47. Meng, C. C. Liu, and S. Fan. 2009. Flexible carbon nanotube–polyaniline paper-like films and their enhanced electrochemical properties. *Electrochemistry Communications,* 11, 186–189.
48. Feng, W. et al. 2003. Well-aligned polyaniline–carbon nanotube composite films grown by in situ aniline polymerization. *Carbon,* 41, 1551–1557.
49. Lee, S. W. et al. 2010. Carbon nanotube–manganese oxide ultrathin film electrodes for electrochemical capacitors. *ACS Nano,* 4, 3889–3896.

50. Yu, A. A. Sy, and A. Davies. 2011. Graphene nanoplatelet-supported MnO_2 nanoparticles for electrochemical supercapacitor. *Synthetic Metals,* 161, 2049–2054.
51. Long, J. W. et al. 2009. Multifunctional MnO_2: carbon nanoarchitectures exhibit battery and capacitor characteristics in alkaline electrolytes. *Journal of Physical Chemistry C,* 113, 17595–17598.
52. Zhang, J. J. Jiang, and X. Zhao. 2011. Synthesis and capacitive properties of manganese oxide nanosheets dispersed on functionalized graphene sheets. *Journal of Physical Chemistry C,* 115, 6448–6454.
53. Wu, M. S. et al. 2012. Formation of nano-scaled crevices and spacers in NiO-attached graphene oxide nanosheets for supercapacitors. *Journal of Materials Chemistry,* 22, 2442–2448.
54. Chou, S. L. et al. 2008. Electrodeposition of MnO_2 nanowires on carbon nanotube paper as free-standing flexible electrode for supercapacitors. *Electrochemistry Communications,* 10, 1724–1727.
55. Zhang, H. et al. 2008. Growth of manganese oxide nanoflowers on vertically aligned carbon nanotube arrays for high rate electrochemical capacitive energy storage. *Nanoletters,* 8, 2664–2668.
56. Chen, S. et al. 2010. Graphene oxide–MnO_2 nanocomposites for supercapacitors. *ACS Nano,* 4, 2822–2830.
57. Lin, R. et al. 2011. Capacitive energy storage from –50 C to 100 C using an ionic liquid electrolyte. *Journal of Physical Chemistry Letters,* 2, 2396–2401.

Index

For Product Safety Concerns and Information please contact our EU
representative GPSR@taylorandfrancis.com
Taylor & Francis Verlag GmbH, Kaufingerstraße 24, 80331 München, Germany

www.ingramcontent.com/pod-product-compliance
Ingram Content Group UK Ltd.
Pitfield, Milton Keynes, MK11 3LW, UK
UKHW020932180425
457613UK00012B/324